Fabless Semiconductor Manufacturing

Fabless Semiconductor Manufacturing

In the Era of Internet of Things

edited by
Chinmay K. Maiti

JENNY STANFORD
PUBLISHING

Published by

Jenny Stanford Publishing Pte. Ltd.
101 Thomson Road
#06-01, United Square
Singapore 307591

Email: editorial@jennystanford.com
Web: www.jennystanford.com

British Library Cataloguing-in-Publication Data
A catalogue record for this book is available from the British Library.

Fabless Semiconductor Manufacturing: In the Era of Internet of Things

ISBN 978-981-4968-27-0 (Hardcover)
ISBN 978-1-003-31497-4 (eBook)

Dedicated to

Professor N. B. Chakrabarti

Contents

Preface

The road to improving the performance of integrated circuits in the internet-of-things era entails fundamental device improvements such as novel architectures, scaling the size, and transistor features that increase speed and minimize power consumption. Intensive research efforts are being invested to co-integrate multiple functionalities, anticipating a limit to this continual shrinking via the More-Moore and the More-than-Moore eras. Flexible electronics have numerous possibilities in this context. The use of stress/strain tuning and high-mobility materials in combination with innovative device topologies has led to nanoelectronic devices and pushed the limits of CMOS. The book starts with an outlook on microelectronics' potential growth for the "Internet of Things". It focuses on computer-aided design and simulation of nanoscale transistors for forthcoming technology nodes (7 nm and below). The monograph focuses on advanced nanodevices design with technical depth and conceptual clarity, and it includes cutting-edge device design solutions for 7 nm technology nodes and below. Industry 4.0 intends to address a fast-changing and challenging manufacturing environment with diverse demands, short order lead-time and product life cycle, limited capacities, and highly complex process technologies. One of the advantages of employing artificial intelligence in the manufacturing industry is that it opens a possibility of fast decision-making without human intervention, which in turn, can significantly shorten the development cycle. Neuromorphism is one of the new "More-than-Moore" paradigms which the semiconductor manufacturers can use to continue technological progress while Moore's law reaches its limits.

This book compiles an overview of advanced nanodevices design employing advanced simulation tools into a single resource. Process simulation is analogous to virtual semiconductor device manufacturing. According to IRDS, 3D integration is one of the most important directions to include in device performance roadmaps. The readers are introduced to device design and simulation after a brief overview of basic device physics. As transistor sizes shrink, nanowires and nanosheets offer the advantage of having a smaller channel and a higher surface-to-volume ratio. The design, simulation, and analysis of Si, heterostructure silicon-germanium (SiGe), and III-N compound semiconductor devices are covered in this monograph. As we enter an era of atomic-scale device dimensions, it has become imperative to utilize advanced simulation tools. The monograph is divided into seven chapters: Fabless Intelligent Manufacturing, Simulation Environment, Nanowire Transistors, Nanosheet Transistors, III-Nitride Flexible Electronic Devices, FDSOI RF Flexible Electronics, and Simulation at Atomic Scale. The monograph bridges the key challenges in fabless manufacturing and designing in microelectronics for IoT and corresponding best-practice solutions using TCAD.

The goal of this book is to provide novice device engineers with a short introduction to device and technology design utilizing TCAD technologies. This book integrates a wide-ranging explanation of TCAD activities in the path to device design in Si, SiGe, and III-N materials, technology, and applications to fill a gap in the literature in a fast-expanding sector. It is aimed primarily at senior undergraduate and graduate students, as well as professors, who are looking for a technological CAD teaching reference book and others who want to learn about semiconductor device design through simulation. Engineers working on the advanced device and process design might use it as a reference.

I am especially grateful to all my research students whose association has been the source of learning for many new topics covered in this monograph. Special thanks go to Drs. A. R. Saha, S. Dey and Ms. E. Mohapatra, who have kindly contributed four chapters (Chapters 2, 4, 5, and 8). Finally, I would like to thank my family members (my wife Bhaswati, sons Ananda and Anindya) for their support, patience, and understanding during the preparation of the manuscript. The book is being dedicated to our beloved teacher, Professor N. B. Chakrabarti.

C. K. Maiti
Kolkata
May 2022

List of Abbreviations

ALD	atomic layer deposition
BEOL	back end of line
BOX	buried oxide
BTBT	band-to-band tunneling
CGP	contacted gate pitch
CM	compact model
CMP	chemical mechanical polishing
CVD	chemical vapor deposition
DUV	deep ultra violet
DIBL	drain-induced barrier lowering
DOS	density of states
DTCO	design-technology co-optimization
EDA	electronic design automation
E-beam	electronic beam
EOT	equivalent oxide thickness
EUV	extreme ultraviolet
FDSOI	fully depleted SOI
FEOL	front end of line
FET	field effect transistor
GAA	gate-all-around
GIDL	gate-induced drain leakage
GPU	graphics processing unit
HDD	high doped drain extensions
IC	integrated circuit
IoT	Internet of Things
LDD	low doped drain extensions
LOP	low operating power
LSTP	low standby power
MOS	metal oxide semiconductor
MOSFET	metal-oxide-semiconductor field-effect transistor
MP	metal pitch
NM	noise margin
NS	nanosheet
NW	nanowire

NWFET	nanowire FET
PDSOI	partially depleted SOI
PPAC	power-performance-area-cost
PSG	phosphorus silicate glass
RIE	reactive ion etching
RO	ring oscillator
SADP	self-aligned double patterning
SAQP	self-aligned quadrupole patterning
SCE	short channel effect
SEG	selective epitaxial growth
SoC	system on chip
SOI	silicon on insulator
SON	silicon on nothing
SOS	silicon on sapphire
SPICE	simulation program with integrated circuit emphasis
SRAM	static random access memory
SS	subthreshold slope
STI	shallow trench isolation
TCAD	technology computer aided design
TFET	tunnel field-effect transistor
TEM	transmission electron microscope
TLM	transmission line model
VDP	vertical device pitch
VFET	vertical MOSFET
VLSI	very large scale integration

Chapter 1

Introduction

Chinmay K. Maiti
SOA University, Bhubaneswar 751030, Odisha, India
ckmaiti@outlook.com

1.1 Introduction

The ever-increasing demand for faster, smaller, and more advanced products necessitates constant innovation from the microelectronics to nanoelectronics industries. At the very core of these products lie semiconductor-based integrated circuits (ICs), the building blocks of logic processors (or central processing units [CPUs]) and memory chips found in computing systems ranging from hand-held smartphones to communications satellites. ICs are serving as the backbone of the internet, social media, and the internet of things (IoT). This introductory chapter starts with an overview of the current trends of fabless smart manufacturing in the era of the IoT. Semiconductor manufacturers play an important role in economic development. Manufacturing of "smart" electronic products is required to create smart networks or "smart things" to enable smart

Fabless Semiconductor Manufacturing: In the Era of Internet of Things
Edited by Chinmay K. Maiti

ISBN 978-981-4968-27-0 (Hardcover), 978-1-003-31497-4 (eBook)
www.jennystanford.com

environments and integrated ecosystems. This leads to the need for novel device design and manufacturing technologies consisting of nanoscale features and three-dimensional (3D) architectures.

The world is currently on course for undergoing a transition to the so-called data economy [1, 2]. A good proportion of the economy today depends on data and this dependence will increase with the advent of 5G network infrastructure and the IoT. The IoT had been predicted to connect over 200 billion smart data-generating objects to the internet in 2020, which is expected to contribute to the creation of a massive 175 zettabytes of data by 2025 [3]. It was estimated that about 33 zettabytes (i.e., 33×10^{21} bytes) of data generated globally in 2018 (see Figure 1.1); or about 12 gigabytes per day for every single person on the planet [4]. The various sources of data creation include standalone desktop computers, portable smartphone devices, social media activities, augmented reality/virtual reality (AR/VR) devices, autonomous cars, IoT-based systems, and intelligent robots [5].

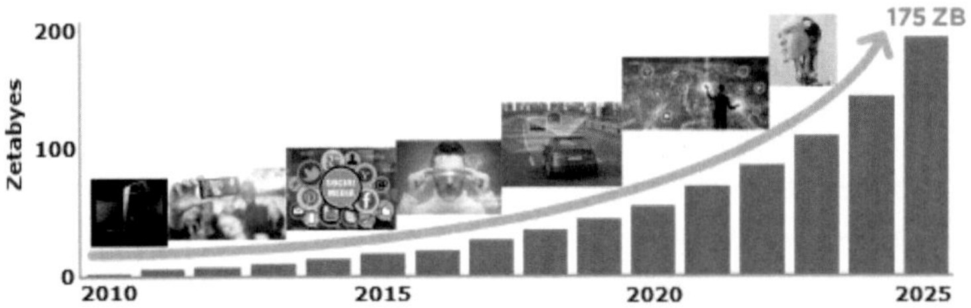

Figure 1.1 Measured and estimated quantities of global data generation on an annual basis (in 10^{21} bytes) from the year 2010 to 2025 [3].

These data show an increased integration of electronics in our daily lives. Indeed, this technology, developed primarily for military and aerospace applications, is intended to manufacture objects with high added values, in particular processors and electronic components to develop business sectors as diverse as telecommunications, transport, or even health. Thus with the rise of information and communication technologies, electronics is ubiquitous in everyday objects. However, integrating new electronic functions into each object remains a considerable challenge. Indeed, the incorporation of an electronic circuit in a cheap and little object represents a major constraint in terms of cost and integration of relevant technologies.

In April 2014, the redesign of the themes of the International Technology Roadmap for Semiconductors (ITRS), a consortium of microelectronics experts, clearly updated this paradigm shift [6]. The reorganization allows classifying all research topics targeted in seven themes among which we see a theme "More Moore" for continued miniaturization, performance gains, and two oriented themes: diversification with "heterogeneous components" and "beyond CMOS". Miniaturization is symbolized by Moore's law, empirically stating that the number of transistors in a processor doubles about every two years, while the price per component will be halved. This law, first proposed in 1965 and then refined in 1975 [7] by Gordon Moore, has been verified in practice until the end of the last century. Since the early 2000s, a simple reduction of device dimension is no longer sufficient to verify Moore's law [8]. New processes, architectures, and materials have emerged to continue the quest for doubling computing power every two years. It is about the "More Moore" era. Previously managed by the ITRS, the specification roadmaps logic circuits have been fixed since 2016 by the International Roadmap for Devices and Semiconductors (IRDS).

The two research orientations "More Moore" and "More than Moore" should not be put in the competition. From a purely scientific point of view, there are applications and interests in both types of research according to their need for miniaturization or diversification. Work on the miniaturization is to be done for new generation computers needing a power of calculation and an increasingly important speed, which requires finer nodes. On the other hand, MEMS works very well with advanced technology nodes but requires the use of various functions within a single chip. So it is needed that we work on the diversification of functions and dimensions of technologies. The use of 3D structures could make it possible to solve both the challenges in terms of diversification and miniaturization.

Nanoelectronics roadmap for Europe: identification and dissemination (NEREID) was launched in December 2015 [9]. This project aims to develop a new nanoelectronics roadmap that is centered on the needs of the European semiconductor and semiconductor-based application industries, as well as major societal challenges. This roadmap aims to cover advanced concepts developed by various research institutes and universities to identify promising novel technologies early. Energy, automotive, medical/life science, security, IoT, mobile convergence, and digital manufacturing are some of the industrial applications. Advanced logic and connectivity, functional diversification (smart sensors, smart energy, and energy for autonomous systems), beyond CMOS, heterogeneous integration, and system design, equipment, materials, and manufacturing science are all covered by the NEREID roadmap. In the fields of More Moore, beyond CMOS, and computing systems, the NEREID roadmap is being developed in conjunction with the recent international IRDS roadmap, and it is also complementary to IRDS in the More-than-Moore domain (e.g., smart sensors, smart energy, and energy harvesting), where the European industry has the leading-edge capability [10].

In recent years, novel techniques that rely on IC manufacturing technology have made rapid development, e.g., the IoT, artificial intelligence (AI) [11], virtual reality (VR), and cloud computing, and so on (see Figure 1.2). AI has emerged as a potential technology for addressing the onset of the data economy [12]. AI involves prediction, inference, and intuition based on accessing and processing large amounts of data. However, conventional computing systems built using the classic von Neumann architecture are unfortunately not suited for such tasks [13]. This is because systems designed using this architecture consist of a microprocessor (CPU) for performing logic operations that are physically separated from memory units for data storage on a short-term (dynamic random access memory [DRAM]) and long-term (hard disk/solid-state drive, HDD/SSD) basis [14]. Processing a set of data stored in the memory involves a sequence of steps in which the data is first transferred from the memory to the CPU, then processed by the CPU, and finally sent back to the memory for storage [15]. This leads to a significant quantity of data moving back and forth between the physically separate logic processor and memory units over bandwidth-limited and energy-inefficient electrical connections (bus) [13]. It has been reported that up to 40% of the power consumed by modern computing systems is wasted simply in transferring data back and forth instead of doing anything with it [16].

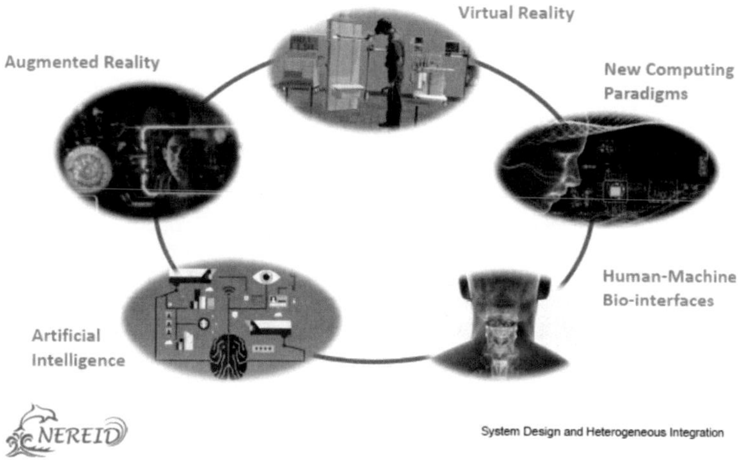

Figure 1.2 The emerging technologies, such as IoT, AI, VR, and cloud computing, are finding applications in heterogeneous integration and system design [9].

In pace with the maturity of novel application techniques, a substantial increase in demand for semiconductor products emerges. This emerging technological age, now known as the IoT, unveils new requirements for our daily life [17]. From AR/VR headsets to autonomous cars, agricultural drones, and intelligent robots, a wide variety of applications are expected to become new sources of an ever-increasing real-time flow of data [18]. Many of the wearable and mobile applications will have limited resources in terms of available device area and power source (e.g., battery capacity). Historically, advances in computational capabilities have been attained in terms of faster data processing rates enabled by the miniaturization of transistors. When simple two-dimensional (2D) areal scaling of transistors no longer increased processing speed, the use of innovative techniques enabled continued improvements in terms of higher power efficiencies and a lower cost per device [19]. Examples include the implementation of new material stacks and the shift to 3D device geometries. Since AI operates on a large amount of data, there is a need not only to process the data faster but also for accessing a large amount of data in a quick and energy-efficient manner. A broad range of new devices and novel device architectures (see Figure 1.3) are needed to deliver the performance and functionality gains for enabling next-generation computing in AI and IoT era.

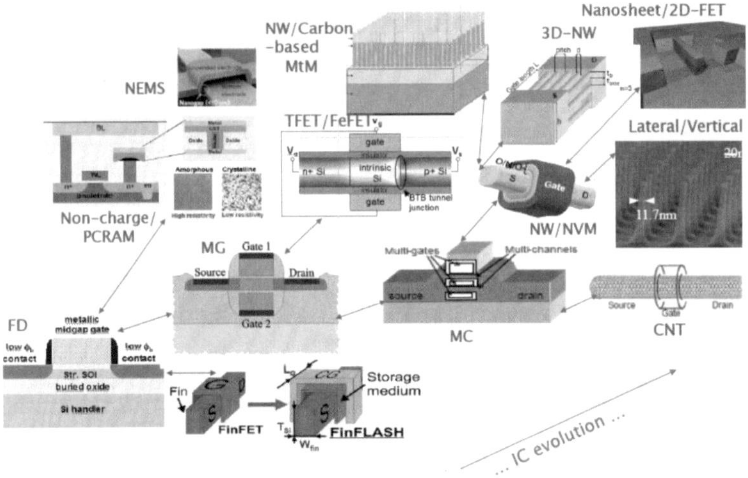

Figure 1.3 IC manufacturing technology predictions. A broad range of new devices and novel device architectures are shown [9].

The emerging and novel device architectures needed for addressing the data explosion in the IoT era are based on multi-material stacks having atomic-scale dimensions and 3D features. As a result, the fabrication of such devices calls for processing techniques that retain precise control over the thickness and properties of functional materials during either deposition or etching on both planar and 3D substrates. Furthermore, the increase in the quantity and type of materials involved together with the presence of 3D features during device fabrication and in final device architectures have simply raised the complexity of making such devices. Besides increasing the density on a chip of constant area, scaling the dimensions of transistors have other incentives that can improve their functionality.

As the lateral transistors rely primarily on the scaling of contacted gate pitch (CGP), which forces the scaling of gate length, S/D spacers, and contacts, the scaling of lateral transistors will soon reach its limit [20, 21]. Reduced electrostatic control, increased parasite capacitance, and increased access resistance are all undesirable outcomes of reducing any of these dimensions. There are lateral devices, such as nanowire (NW)-based field effect transistors (FETs), which can delay but not solve the problem of CGP budgeting. The so-called ground rules are used to allow area comparisons of different technologies without having to do actual designs. The CGP, the bottom layer metal pitch (MP), and the number of metal tracks required to complete the logic cell (Figure 1.4). However, very little research has been done on device evaluation that takes into account the limitations posed by these advanced technologies and benchmarks vertical and lateral devices [22–25]. The vertical devices are now in the focus. Because these devices are oriented vertically, they are less limited in terms of gate length and spacer thickness, and thus they should have greater scalability than lateral transistors. It is, thus, important to quantify the benefits of vertical devices in terms of power, performance, and area (PPA) metrics, combining design approaches, and technology constraints that are likely to exist at the 5 nm technology.

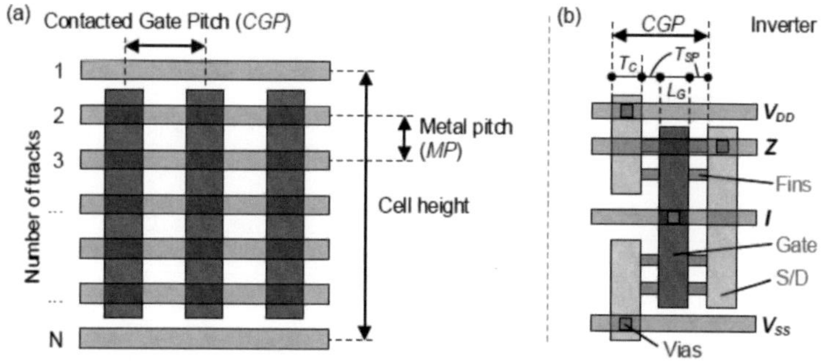

Figure 1.4 (a) Technology ground rules allow to quickly estimate a footprint of a standard cell. (b) Sketch of an inverter layout made with FinFETs. The definition of CGP is illustrated [26].

The semiconductor industry has so far relied on Moore's law, which states that every new generation of transistors outperforms the previous one in terms of power, performance, area, and cost (PPAC). However, this pattern is currently being challenged. The main problem is the sheer complexity of both technology and design, which drives up not only manufacturing but also research and development (R&D) costs significantly. In 1974, Dennard *et al.*

[27] published a seminal paper in which they demonstrated the benefits of scaling. The authors concluded that a small transistor can be switched between ON and OFF states with less power and at a higher frequency (or clock speed) than a larger one. This held for some time until the late 1990s after which scaling transistors no longer made them more energy-efficient. Since then, device scaling has continued through the implementation of alternative techniques that can deliver improved performance, within reasonable power limits, at a manufacturable chip area and with a reduced cost per device [28]. This combined approach to PPAC scaling is the current norm within the semiconductor industry that has yielded state-of-the-art logic processors comprising billions of transistors on a single die [5, 19]. Figure 1.5 outlines the scaling landscape at the recent ground rules toward scaling trends. Intel first introduced a 22 nm technology node featuring FinFETs. The transition from one node to another comes with roughly 0.5× scaling of MP and CGP product [29, 30]. Figure 1.4 indicates that various combinations of MP and CGP may be chosen for a certain technology [31, 32].

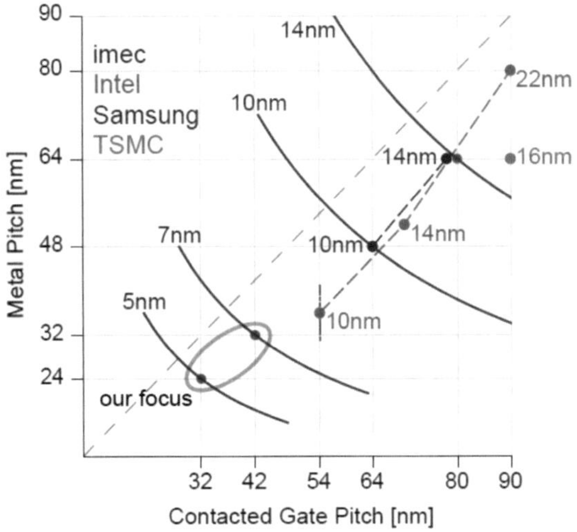

Figure 1.5 Technology nodes having the same name do not necessarily have the same ground rules [26].

One of the early strategies involved strain engineering whereby the silicon in the source and drain regions was mixed with germanium

[33]. This physically squeezed (or compressed) the silicon in the channel region that altered its crystalline and hence, electrical properties. This improved hole mobility in the p-MOSFET channels, which lowered channel resistance and waste heat generation (i.e., improved energy efficiency). Conversely, subjecting n-MOSFET channels to tensile stress improved electron mobility. A subsequent technique involved the adoption of new functional materials for the gate stack - a high-k gate dielectric (e.g., HfO_2) and metal gates with optimum work function (e.g., TiN gate for p-MOSFET, TiAlN gate for n-MOSFET) [34, 35]. This lowered leakage current from the tunneling of electrons between the gate and the channel through the previous gate oxide layer (SiO_2) that was a comparatively lower dielectric constant [34]. Although these new materials helped staunch the main source of leakage current, continued device scaling to a gate length below 30 nm caused the leakage between the source and the drain to become a significant issue when the transistor is in an OFF state.

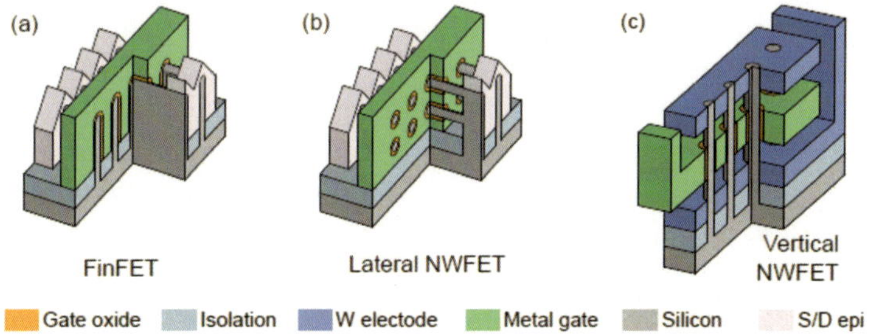

Figure 1.6 3D sketches of different devices: (a) FinFET, (b) lateral NWFET, and (c) vertical NWFET [26].

A way to regain control over channel current flow was to incorporate a new design architecture by raising the channel above the plane of the silicon substrate. This led to the formation of a 3D "fin" structure that is characteristic of the 3D FinFET design, which is shown in Figure 1.6a [36–39]. The gate wraps around the fin on three sides of the raised 3D channel, instead of only across one side of a planar 2D channel. This allows the gate to have better electrostatic control of the electric field in the channel, which reduces leakage current between source and drain. An additional benefit is a need for a lower gate voltage to switch the transistor, which lowers power consumption. State-of-the-art logic processors are manufactured using the 3D FinFET design.

An evolution of this 3D design architecture is the 3D gate-all-around FET (GAAFET), which is shown in Figure 1.6b [40–42]. In this device, the gate electrode is wrapped around all of the sides of the channel region. This enhances the electrostatic control of the electric field in the channel provided by the aforementioned 3D FinFET design, which can further improve transistor energy efficiency and reduce power consumption. The electrostatic control of the channel by the gate is so efficient in such gate-all-around (GAA) architectures that it is even possible to fabricate a MOSFET device without forming p–n junctions between the source, the channel region, and the drain. Such "junctionless" multi-gate transistors have the potential to greatly simplify the fabrication process of MOSFET devices [43]. Vertical devices are not new [44, 45]. A comprehensive review of the behavior of vertical devices in scaled technologies may be found in reference [46]. Because they are oriented vertically, vertical MOSFETs (VFETs) (Figure 1.6c) are less limited in terms of gate length and spacer thickness, and thus should have greater scalability than lateral devices.

The challenge is to find innovative solutions to add intelligence to communicating systems. The IoT [47–50] is a trend in the present-day industry that renders a variety of physical devices able to collect and process data and even to act in real-time in its environment. Thereby, designs with low-power consumption and high reliability are crucial for the proper development of all the benefits that IoT can provide. Microelectronics for the IoT [51] provides a systematic evaluation and analysis of microelectronics tailored for the IoT [52].

Neuromorphic computing is an advanced area of research that aims to find new algorithm solutions and applications in everyday life [53]. A new design of spiking neuron (SN) in 28 nm node FDSOI technology has been reported [54, 55]. The microelectronic industry is searching for a new paradigm to follow technical advances that have dominated our information society, as Moore's law is nearing its end [56]. Neuromorphic engineering, along with quantum computing, is one of the most exciting research fields. Very-large-scale integrated neural processors [57] mimic biological nervous systems at the IC level, making neuromorphic computation much more energy-efficient than conventional architecture based on von Neumann finite state machines. To create an artificial neural network (ANN) in such systems, a large number of neurons must be integrated and interconnected on a silicon die (ANN). One of the system's fundamental building blocks is the neuron circuit architecture [58].

Fabricating state-of-the-art and next-generation logic and memory devices entails overcoming numerous challenges. The scaling of such devices involves making use of self-aligned multiple patterning techniques. These require deposition processes that yield nanoscale and conformal films (e.g., SiO_x, SiN_x, and TiOx) on 3D substrates with excellent growth control. Functional components of logic devices, e.g., present generation FinFETs and next-generation GAAFETs, consist of multi-material gate stacks comprised 3D architectures. These stacks need to be encapsulated with conformal, etch-resistant, and nanoscale dielectric spacers (SiO_2, SiN_x). The metal contacts at the source/drain regions of these FETs require nanoscale, conductive liners (e.g., TiN_x). Emerging memory technologies such as PCRAM, MRAM, and RRAM consist of sensitive memory cell stacks that need to be isolated from each other by dielectric gap fills (e.g., SiO_2) in dense crossbar array configurations. They also need to be protected from subsequent processing environments by conformally encapsulating the memory cell stacks with dielectric films (e.g., SiN_x, SiO_2). Thermal and plasma ALD are atomic-scale processing techniques that can address these aforementioned challenges from the perspective of materials deposition.

1.2 Technology Computer-Aided Design (TCAD)

The electronic design automation that models IC fabrication and device operation is known as TCAD. It is the art of using critical analysis and detailed understanding to abstract the IC behavior of the process, device, and circuit simulation data. TCAD is one of the enabling methodologies that can support Moore's law-like advances in technology by minimizing development cycle times and costs in the semiconductor industry. Several problems labeled as challenging by the IRDS can be viewed as different manifestations of the same general trend, namely, rising problem dimensionality. In fact, as technology advances, it emphasizes the complexity and non-ideality of semiconductor devices' electrical behavior, piquing interest in alternatives to the traditional planar MOSFET architecture. To accurately characterize such devices, a 3D representation is required; as a result, 3D simulations have become a critical requirement for daily tasks. The scenario illustrates the need for meshing tools that can accurately and efficiently represent complex 3D geometries, resolving all essential features of the device structure while avoiding unacceptable grid size disadvantages. TCAD's predictive capabilities are also dependent on its contribution to the evaluation and mitigation of process variations, which become increasingly important as devices shrink into the nanometer range. Device parameter distributions are widened by phenomena like line-edge roughness (LER) and random dopant fluctuations (RDF), necessitating statistical treatment. This leads to computationally difficult 4D problems, with the extra dimension being the size of the ensemble under consideration.

As it takes over 10 years for a technology to go from the research laboratory to volume production, it is essential to use predictive TCAD tools for performance evaluation at the beginning. TCAD includes lithography modeling, which simulates lithography equipment imaging of the mask, photoresist characteristics, and processing; front end process modeling, which simulates the physical effects of manufacturing steps used to build transistors up to metallization; and device modeling, which uses a hierarchy of physically-based models to describe the operational characteristics of active devices. Grid generators, surface-advancement methods, solvers for systems of partial differential equations, and optimization routines are examples of numerical techniques [59]. Simulation data can be used to calculate the details of IC behavioral models at the transistor and circuit levels, as well as demonstrate physical constraints in processing and manufacturing [60, 61]. The extended TCAD tools can be used to improve IC fabrication technology from a product concept (i.e., from product specification to IC fabrication [62].

The numerical simulation of the physical effects of IC processing steps used to fabricate transistors up to metallization is referred to as process technology CAD. The process technology CAD is used to simulate semiconductor processing steps such as oxidation and diffusion, deposition and etching, ion implantation, and annealing, and generate input data files for the device simulator as realistically as possible. In the 1960s and 1970s, physical models for process technology were limited, and sophisticated process simulators with multidimensional models were not required for the large device geometry that was prevalent at the time. The physical understanding of IC processes has advanced considerably in recent years. Furthermore, as IC devices progress into the nanoscale regime, precise multidimensional process models are required.

The device CAD refers to numerical simulation of IC device operation. The device CAD, in general, consists of a set of physical models that describe carrier transport in materials. The energy balance, which solves a higher moment simplification of the Boltzmann transport equation (BTE), is a more complex and computationally difficult model than the simple drift-diffusion, which solves Poisson and continuity equations. Furthermore, the complex physics of today's nanoscale devices necessitates the use of Monte Carlo (MC) codes [63–65], which stochastically solve BTE, and Schrödinger solvers, which account for quantum mechanical (QM) effects in nanoscale devices. The best model to use is determined by the problem and the level of detail needed. Despite substantial advances in both numerical techniques and physics, the industry needs device exploration, scaling, and optimization that are becoming increasingly difficult to meet. As a result, device CAD's ability to accurately model today's device performance and predict tomorrow's device simulations is critical [66]. The terms process CAD and device CAD both refer to numerical simulations of various aspects of computational electronics, but they are related in a synergistic way. The ability to predict device and circuit performance requires a thorough understanding of the different IC processing steps.

The extended TCAD tools can also be used to determine the manufacturability of IC fabrication processes. Integrating TCAD in semiconductor foundry companies for efficient customization of IC manufacturing technology is also important [62, 67–73]. The innovations of CAD tools are usually aimed at achieving the best compromise between precision and calculation time. Despite being useful for providing high-quality solutions within limited computational budgets, these tools have not received broader attention in the design of IoT components. The possibility of using

TCAD to assist machine learning (ML) for semiconductor device failure troubleshooting and device reverse engineering is also emerging [11, 12, 74–77].

1.3 Physical Model Calibration

The underlying physics, numerical discretization problems, and proper characterization of the models implemented in the process and device CAD tools all affect the accuracy of simulation results. The capabilities of TCAD point tools have greatly improved over time. The simulation results, however, are not 100% predictable and reliable due to flaws in physical models. Even though the development and implementation of advanced physical models have followed a steady-state growth pattern, the characterization of existing models is vital for TCAD accuracy. As a result, proper calibration of the process and device models implemented in TCAD point tools are required to address the predictability problem. Calibration is a difficult and time-consuming process. The CAD tool's physical process models must be calibrated for the target technology during development. Designing short loop wafer fabrication experiments to calibrate all critical as-implanted as well as final doping profiles of all related impurities in the simulation structures is part of the calibration approach.

A set of suitable device characteristics are measured from the manufactured wafers of the target technology to calibrate the physical device models introduced in device CAD tools. To classify the physical device models in typical CMOS technology, this includes I_{DS} versus V_{GS}, I_{DS} versus V_{DS}, substrate current, I_{sub} versus V_{GS}, and so on [78, 79]. Two levels of calibration are used to improve the accuracy and predictability of simulation results: (i) local calibration and (ii) global calibration. IC designers should use the coupled process and device simulation to explore the impact of process requirements on electrical variations in devices and circuits. Process CAD's main goal is to capture the features that accurately represent a fabrication technology's performance constraints.

For predictive process and device simulation, the physical models introduced in process and device CAD tools can be identified with measured data from the target technology. A large amount of measured data is produced during the IC fabrication process development cycle. This target technology experimental database can be used to calibrate process and device models and correlate simulation data with target technology measured data. TCAD tools must be accurate and accurately predict new physical phenomena that may arise during the development phase to create a new IC fabrication technology. Unfortunately, efficient calibration methodologies must be developed for the successful application of TCAD in the design of new technology due to a lack of accurate physical models.

1.4 Database for Calibration

The measured doping profiles and device characteristics must be included in the calibration database used in the simulation. The doping profiles must be obtained under a wide range of processing conditions to account for the target technology's expected process variations. The device characteristics must be obtained under a variety of circuit operation biasing conditions. Any new physical phenomena observed in the fabrication facility must also be entered into the database. The database must also contain measured data of the anticipated doping materials under the expected processing conditions of the target technology to improve next-generation IC fabrication technology. As a result, the calibration database must include experimental data for all potential impacts of current and target next-generation technology.

1.5 Design Technology Co-optimization (DTCO)

Various architecture and technology dimensions should be considered for holistic device exploration, whether vertical or lateral. The design and technology should be co-optimized to find the next node system that is feasible to produce, has enough drive to satisfy application performance criteria, consumes minimal energy per switch, and is cost-effective. PPAC metrics are often used to describe these specifications. This work is part of a larger effort to optimize these metrics, to determine whether a transistor with vertically oriented channels might be such a future unit. As more and more applications are related to mobile, battery-limited situations, power is of particular interest. Since power is proportional to the square of supply voltage, V_{DD} scaling it for power reduction will be extremely efficient. A certain overdrive (V_{DD} – V_{TH}) should be maintained to scale V_{DD} and maintain fair system efficiency. Traditional device simulation is used in current DTCO workflows. As a result, they face the same issue as traditional device simulation: low accuracy and predictive power. To maintain the necessary precision, an entirely empirical, geometry-dependent mobility is typically used [80, 81]. Empirical mobility modeling is finding it increasingly difficult to keep up with shrinking device dimensions and new technical approaches for new technology nodes. To solve this problem, physical device modeling is being introduced into the DTCO framework [82].

The microelectronic industry is searching for a new paradigm to follow technical advances that have dominated our information society, as Moore's law is nearing its end. Neuromorphic engineering, along with quantum computing, are currently the most exciting fields [83]. Very-large-scale integrated neural processors mimic biological nervous systems at the IC level, making neuromorphic computation much more energy-efficient than conventional architecture based on von Neumann finite state machines [84]. To create an ANN in such systems, a large number of neurons must be integrated and interconnected on a silicon die, one of the system's fundamental building blocks in the neuron circuit architecture.

1.6 Industry 4.0

Industry 4.0 is a German government initiative that encourages the fourth generation of manufacturing by incorporating concepts like cyber-physical structures, virtual copies of actual equipment and processes, and decentralized decision-making to build a smart factory [85]. Industry 4.0 aims to resolve the dynamic and demanding manufacturing landscape characterized by a wide range of demands, short order lead times and product life cycles, restricted capacities, and increasingly complex process technologies. A smart manufacturing ecosystem built on Industry 4.0 technologies, like advanced automation, AI, ML, big data analytics, and IoT, can monitor and optimize manufacturing processes in real-time in a variety of ways, from high-level strategic resource and production planning to real-time equipment-level smart dispatching and predictive maintenance. Smart manufacturing and Industry 4.0 offer major cost-cutting, productivity-boosting, and quality-improving prospects in microelectronics manufacturing. However, it also poses fresh and special obstacles to the industry. Deep learning models with process conditions and doping profile/electrical characteristics as input and output are trained using AI algorithms that can characterize the semiconductor domain. Emerging technologies, such as the IoT [86, 87], wireless sensor networks [88, 89], big data [90], cloud computing [91, 92], embedded systems [93], and mobile internet [94], are being integrated into the semiconductor manufacturing, ushering in the fourth industrial revolution.

At the end of each chapter, a comprehensive list of references is presented to help readers understand the evolution and development of nanoscale device manufacturing technology for ICs.

1.7 Fabless Intelligent Manufacturing (Chapter 2)

Chips or ICs refer broadly to the two main types of semiconductor devices. Getting an IC designed and manufactured is becoming incredibly expensive. We shall discuss several phases of development in the semiconductor industry. Until the mid-1980s, all semiconductor manufacturing companies were integrated device manufacturers (IDMs), that is, they controlled both the design and manufacture of their chips. These companies, like Intel and Samsung, design and manufacture standard ICs that systems companies use in their products. A systems company makes a consumer product from chips that other companies have designed or fabricated. This scenario has changed gradually, and now there are only a few IDMs (Intel and Samsung notably). All other chip makers outsource the manufacturing of their designs to a foundry.

Because the wafer fabrication system is the most complicated and expensive part of the semiconductor manufacturing process, its scheduling has a huge impact on economic efficiency. The production management standard does not correspond to the industrial scale. Low production efficiency, low equipment utilization, and low product yield have often hampered the long-term growth of businesses as scheduling technology has lagged. To ensure the quality of finished products, many types of controls are needed in semiconductor manufacturing, which is used to monitor the process to reduce defects and increase yield. Since inspection operations have a direct effect on product cycle times, sampling techniques are used to minimize the number of inspected lots while maintaining quality. There are many sampling techniques available, each of which can be categorized based on their ability to deal with factory dynamics. Several algorithms have been proposed and validated to efficiently handle defect inspection queues by skipping lots that no longer provide sufficient details. An optimization model for defect inspection capability preparation has been proposed to help strategic and tactical decisions.

The multi-cluster tool is a new type of wafer fabrication system that is widely used in the fabrication of 300 mm wafers. Multi-cluster tools, unlike other manufacturing systems, have large-scale, complex wafer flow patterns, strict residency time limits, and intense resource conflict. Wafer production, wafer fabrication, wafer probe, assembly or packaging, and final test are the five major components of the IC product manufacturing process. For complex technology, capital-intensive, and high value-added features, wafer fabrication is at the heart of the IC product manufacturing process. The quality of an IC is solely determined by the wafer fabrication process.

The future growth potential of intelligent semiconductor manufacturing is enormous. To manage such a large amount of data, smart manufacturing research needs better computational resources as well as futuristic ever-growing data processing capabilities. Industry leaders are going to use AI and ML-enabled TCAD platforms to develop their AI-CHIPs and AI-ICs in the future [69].

Technology CAD simulators have been effective so far, particularly in terms of predicting transistor efficiency, even though proper calibration of model parameters is needed to match simulation results with measurements. However, TCAD's function has been limited to predicting transistor performance in the R&D process, not in mass production, due to several hours of simulation time. Adopting AI technologies to reduce TCAD simulation turn-around time has been explored as a solution to this issue [95–98]. One of the benefits of using AI in the manufacturing industry is that it allows for quick decision-making without the need for human interference, which can greatly reduce the development time [12, 75–77, 99–101]. ML advances have impacted many research areas, resulting in changes that would not have been possible using traditional optimization techniques.

Industrial manufacturing can expect substantial benefits from increased process automation, which is one of the industries where AI/ML unlocks a multitude of opportunities. At the same time, the fourth industrial revolution, also known as Industry 4.0, has come to an end with the launch of the industrial internet of things (IIoT), which provides increased wireless communication for real-time manufacturing data collection and processing.

Even though AI has been around as an academic and scientific discipline since the 1950s, AI technology has only recently gained traction in the semiconductor industry. The current explosion in AI science, investment, and real-world application is unparalleled. Semiconductor manufacturers may use AI systems to predict when or if usable equipment may malfunction, allowing maintenance and repair to be planned ahead of time. Manufacturers can increase performance while lowering the cost of machine failure thanks to AI-powered predictive maintenance. The ecosystem will assist manufacturers in shortening production and R&D cycles, increasing production capability, lowering production costs, ensuring product quality, and improving product yield by fully using real-time data and AI. It can benefit both high-tech industries like semiconductor wafer fabrication and traditional labor-intensive industries.

1.8 Simulation Environment (Chapter 3)

The monograph is planned for meeting the recommendations of the IRDS roadmap [10] concerning the efforts made by the industry and academic partners in TCAD modeling of processes and devices for future generations of transistors. To create a foundation for the simulation environment and to ensure reproducibility, the used material models and tools are introduced in this chapter. Technology CAD was originally pioneered at Stanford University [102]. Currently available advanced semiconductor device/process simulation tools are introduced first. There are several commercial TCAD software platforms for process/device simulation around the world. Most of them were first developed by the universities worldwide and later integrated into a suite by the TCAD vendors. Since the early 1990s, commercial vendors such as SILVACO and SYNOPSYS have developed their graphical user interfaces around the existing frameworks [103–108].

The theoretical modeling of 2D and 3D nanoscale transistors demands a proper treatment of quantum effects such as the energy level quantization caused by strong quantum confinement of electrons and band structure non-parabolicity. Electron-phonon scattering, surface roughness, alloy disorder, and tunneling should be included in the simulation. To address these issues, a multidimensional quantum transport solver based on the self-consistent solution of the Schrödinger and Poisson equations in the real space-effective mass approximation with a tight-binding extraction of the effective mass values is used to simulate III-V [109] and strained-Si devices [110]. In the nanodevices transport simulation framework, approaches that incorporate important quantum effects into semi-classical models have become very popular due to their lower computational cost in comparison to the purely quantum transport simulation techniques. An overview of currently available commercial TCAD tools, viz., GTS Framework, SILVACO, and SYNOPSYS has been presented. State-of-the-art device simulation tools combine: (i) accounting for quantum effects based on the rates of the relevant multi-subband scattering mechanisms [111] and (ii) using the semi-classical BTE in the relaxation time approximation by adopting the Kubo–Greenwood formalism [112, 113].

1.9 Nanowire Transistors (Chapter 4)

Nanotechnology has significantly improved and revolutionized nearly every technology and industrial sector in the last few decades, including electronics, information technology, medical and healthcare, energy, environmental remediation, transportation, and so on. As IoT applications are extremely diversified, there is a strong need for flexible solutions at the design, architectural, and technological levels. One compromised solution could be the electrically GAA NW architecture with much reduced parasitic capacitance and increased effective width for better short channel control and large drive current [114]. GAA transistors are based on NW and nanosheet, which have the advantage of having a smaller channel and a high surface-to-volume ratio as transistor sizes shrink. GAA architectures are thought to be the pinnacle of MOSFET refinement and are likely candidates for CMOS scaling beyond the 10 nm node. A GAA structure gives rise to better channel electrostatic control, allowing for a sub-threshold slope that is close to the theoretical minimum. The main concern, however, is achieving an on-current large enough for practical applications. This issue was addressed in a recent work [42], which demonstrated a GAAFET with two vertically stacked NWs as channels and a common gate, effectively doubling the on-current compared to a single wire. In this chapter, several proposed architectures are investigated using a comprehensive framework of advanced simulators to evaluate the fabrication process as well as the device characteristics, performance, variability, and reliability. Some design rules from the IRDS technology roadmap at a certain technology node are given in Table 1.1.

Table 1.1 IRDS technology roadmap for design guidelines for NWFETs

Technology Node	N7 (7 nm Node)	N5 (5 nm Node)
Gate Length (nm)	14	10
NW Diameter (nm)	7	5
NW Vertical Pitch (nm)	14	10
CGP (nm)	42	32

Source: IRDS, 2020.

With scaling down the device below 7 nm technology node, NWFETs are considered as the suitable candidate for the devices because of superior electrostatic integrity. Using GAAFET, IBM successfully demonstrated 5 nm process technology. A nanosheet-based GAAFET is an multi-bridge channel field effect transistor (MBCFET) [115]. As a result, the GAAFET is an excellent candidate for pushing the scalability flow to the sub-7 nm node regime, ushering in a new era of ultra-scaled transistors. Issues of scaling the approaches required for boosting the performance are shown in Figure 1.7. According to IRDS, lateral GAA devices are the ultimate structure at a 5 nm technology node [5]. The reduced dimension of the device leads to variation. Despite considerable efforts to control process variations, there exist intrinsic sources of dispersion in devices. Studies using computer-aided simulations provide an excellent way of understanding and evaluating the impact of these various sources on device performance.

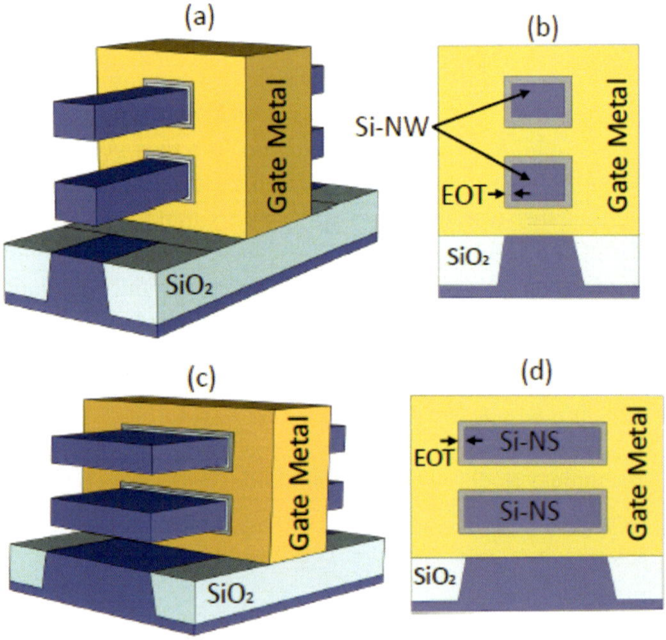

Figure 1.7 GAAFET (a) structure and (b) cross-sectional view with Si-NW channel. MBCFET (c) structure and (d) cross-sectional view with Si nanosheet channel [120].

In this chapter, the NW transistor design issues are presented [116]. The impact of the quantum effects has been taken into account via the use of the density gradient model in simulation, and the comparison of two models (i.e., drift-diffusion and density gradient models) is made. The impact of geometrical variation on the performance of NW transistors has also been discussed. Stress techniques have been used to further enhance NW transistor performance. SiGe stressor has been introduced in the source/drain and channel regions. The effects of stressors by changing the channel material have been presented.

ICs are affected by a wide variety of variations. These variations are different firstly by their origin. We can distinguish as (a) the parametric variations that result from the manufacturing process and affect the parameters like the length of the channel, the thickness of the oxide, the concentration of the dopants, or the dimensions of the interconnections, (b) variations in environmental conditions during circuit operation such as the temperature and the supply voltage, and (c) variations caused by aging phenomena, which affect the reliability of the components such as hot carrier injection, electromigration, or BTI degradation. It is important to investigate the performance, variability, and reliability limits of sub-10 nm technology based on stacked NWFETs using a detailed set of advanced physical models [111]. We shall identify problems that make current NWFET technology inferior to current FinFET technology, including I_{ON} current degradation due to gate stack impurities, (ii) increased time-zero variability, (iii) improved BTI-induced device degradation, and (iv) additional process-induced parasitic capacitances. In this chapter, the impacts of metal grain granularity (MGG) and random discrete dopant (RDD) in the channel, source, and drain extensions on device performance have been discussed. The modeling of MGG and RDD is presented [117, 118]. The impacts of MGG in a single NW transistor have been addressed and a comparison of MGG impacts in single and stacked NWs has been made. Finally, the combined effects of RDD and MGG will also be considered in this chapter [119].

1.10 Nanosheet Transistors (Chapter 5)

Due to the extreme downscaling, the transistor dimensions have now reached several physical limits. As a result, both the manufacturing process and transistor designs are struggling to keep up with the scaling while still delivering adequate performance. The introduction of FinFETs has emerged as a defending contender to overcome the short channel effect (SCE) along with further miniaturization. The channel electrostatics was improved by using tri-gate structures with scaling of gate length down to 7 nm technology node [121]. However, the performance of tri-gate structures has been found to degrade below 7 nm technology node [122]. The scaling of gate length can be continued in sub 10 nm node with improved electrostatics using GAA NW transistors [123]. The best electrostatic control of the gate over the channel can be achieved only when it can be controlled from all sides of the channel [124–128]. Thus, GAA NW/nanosheet transistors are now considered to be the most suitable and feasible alternative structures for ongoing rapid scaling even below 7 nm technology nodes [129].

GAA transistors are based on NWs and nanosheets [130]. A nanosheet-based GAAFET uses nanosheets instead of NWs as the channel. Today's emphasis on low-power and low-voltage operation in industry and academe requires an understanding of the nanodevices at ultimate scaling [131]. The technology generation of logic devices in the industry is expected to reach 5 nm and 3 nm in the years 2021 and 2024, respectively. Using GAAFET, IBM successfully demonstrated 5 nm process technology. A nanosheet-based GAAFET is a MBCFET. Except for the fact that MBCFET uses nanosheets instead of NWs as a channel, they are architecturally similar. MBCFETs have many advantages in terms of design and structure. First, the nanosheets can be stacked vertically to increase speed without taking up more space. Second, since they use the same manufacturing process, they are extremely compatible with current technology.

In contrast to planar FETs, channels in NW and nanosheet transistors (NSFETs) have a finite size effect. As a result, their behavior and characteristics must be thoroughly investigated before they can be used as transistor channels in current technology. Since the device is ultra-scaled in size, the role of interface state interference in assessing device operation and reliability becomes important. As a result, band structures and quantum of conductance, as well as density of states, must be investigated. In this chapter, using physics-based 3D TCAD simulation and device optimization beyond the 7 nm technology node, we can show possible improvements in critical design parameters for the vertically stacked NSFETs to surpass NWFET technology. We have shown that stacked NWs offer versatile design options for the performance and power management essential for More-than-Moore applications beyond the 7 nm technology node.

With the use of a high-k dielectric material (e.g., HfO_2), the ever-increasing gate leakage current can be decreased and the gate-to-channel capacitive coupling can be improved. In addition, the thinner gate oxide layer can provide better channel control capability, which can effectively suppress the SCE and maintain excellent compatibility with traditional process technology [132, 133]. MGG induced random work-function variation is another important source of random variability for the technology nodes that use a high-k/metal gate stack [134]. The effect of MGG on device performance is closely correlated with the metal grain size, concerning the overall gate size. Metal gate granularity is an issue mainly connected with the "gate first" process technology, whereby the gate metal is deposited before any high-temperature annealing procedure [135]. Depending on their crystal orientations, the work function of each metal grain varies randomly. Therefore, instead of a single work-function value, the transistor gate contains multiple grains with different work-function values that cause random fluctuation of device performance parameters [136, 137].

The random and discrete nature of dopant charges in the channel region changes the device characteristics in highly scaled multigate FETs. The impacts of RDD and the presence of discrete dopants in different regions of NSFETs have been studied in detail. The degradation of performance caused by reducing the gate length can be improved by enhancing the sheet width. The sheet width has been varied from 3 nm to 12 nm. In addition, the effects of stacking multiple nanosheets for possible performance improvement of lateral nanosheet transistors using predictive simulation have been performed. A comparison of discrete dopants in single, 2-stacked, and 3-stacked devices has been performed. In conventional MOSFETs, the use of a high-k/metal stack modifies the behavior of a transistor and the carrier mobility can get degraded by the presence of charges trapped in the dielectric [138, 139]. To meet performance needs, the use of the technology booster is common [140]. To enhance the device performance, the use of intentional mechanical stress within a device has been proposed [141]. Finally, the performance of NSFET has been analyzed by comparing it with NWFET.

1.11 III-Nitride Flexible Electronic Devices (Chapter 6)

The conventional semiconductor and printed electronics industries have both shown interest in the field of smart and communicating flexible systems. With the rise of information and communication technologies, electronics are ubiquitous in everyday objects. However, integrating new electronic functions into each object remains a challenge. Indeed, the incorporation of an electronic circuit in a cheap and little object represents a major constraint in terms of cost and integration of relevant technologies. Because of the shape of the objects, often complex and non-rigid circuits need to be integrated into minimum volume, the development of flexible, compact, low-cost electronics has become essential [142]. The exploration implies the review of the paradigm compared to rigid electronics. This involves the search for new materials to compose the transistors of this new type of electronics. At present, the state-of-the-art manufacturing and industrialization of this type of electronics are being developed and show that we are still far enough from flexible products manufactured in mass production [143, 144].

Flexible electronics will allow a wide range of devices and applications that are not possible today. Interest in devices on flexible substrates is flexibility, bendability, and conformability. These characteristics allow it to be used in numerous applications, for example, large-area flexible displays, RFID tags, and flexible sensors that can be applied in wearable electronics, smart labels, or electronic skin [145–147]. Over the next decade, flexible electronics is expected to expand into a multibillion-dollar industry that will change the way we think about electronics. Flexible electronics' unique properties, such as compliant structures, ultra-thin profiles, low weight, low-cost potential, and high reliability, have the potential to have a huge impact on consumer electronics, aviation and space electronics, life sciences, military applications, and telecommunications. Flexible electronics would allow for a wide range of devices and applications that are currently impossible to achieve [148]. The goal is to manufacture foldable, lightweight thin circuits, stretchy and with performances close to those currently achieved by conventional components and circuits.

Flexible devices based on organic materials and printed electronics have the advantage of offering a low cost and also a large area. However, these strategies are limited by poor carrier mobility and relatively low print resolution, which cannot satisfy high operating frequencies. The appearance of new material such as graphene opens the possibility of manufacturing direct components on flexible substrate allowing to reach frequencies much higher than for organic materials up to gigahertz. To exceed frequency restrictions, components from the GaN sector, for example, flexible AlGaN/GaN HEMTs have been explored by transferring them from the rigid substrate to a flexible ribbon [149].

In this chapter, to solve for new applications in the field of flexible electronics, GaN-on-silicon-substrate technology will be presented toward flexible GaN-based HEMTs fabricated on non-planar surfaces using layer transfer technology on adhesive tape, which is a significant benefit for the production of telecom and military applications. The proposed approach is expected to pave the way for the use of GaN-based flexible technology for electronics that require mechanical flexibility, high-frequency operation, and high power efficiency.

1.12 FDSOI RF Flexible Electronics (Chapter 7)

For flexible electronics, normally organic layer materials are used not just for the dielectric but also for the substrate and active layers due to interface and compatibility between the layers [150, 151]. Even though organic TFTs are receiving a lot of attention, they have some limitations mostly because of the organic semiconductor used, such as their low mobility and instability. These can be overcome by combining inorganic semiconductors, like mixed metal oxides such as In-Ga-Zn-O (IGZO) having high mobility, and organic insulators making a hybrid device [152].

Flexible electronics have received a lot of attention in recent days due to the several possible applications that are being envisaged. A promising technology for IoT applications is the FDSOI technology. As illustrated in reference [153], using a back biasing voltage in FDSOI allows for a compromise between very low leakage and high speed, as well as changing the operating conditions of the device. The critical and widespread deployment of IoT necessitates efficient energy flexibility. As a result, adaptive solutions that can operate at various voltage supplies, including very low voltages near-threshold levels, are required. This versatility is provided by the FDSOI technology, which performs well at low voltage levels as well as at normal voltage levels. Furthermore, when the node is idle, this technology has the potential to design systems with very low leakage.

In the last decades, silicon-on-insulator (SOI) technology, which is based on the introduction of a buried insulator layer, known as the buried oxide (BOX), in the Si substrate, has been recognized as the best alternative to conventional bulk technology [154, 155]. The main objective of SOI technology is the elimination of the drawbacks of conventional Si devices [156]. From the point of view of manufacturing, the compatibility between SOI and conventional technology resulted in a rapid consolidation of both structures, hence it is extensively used in the industry [157] as well as in the academy area [158]. Some of the indispensable steps in conventional MOSFETs guarantee the correct isolation between devices and reduce the parasite effects that are unnecessary in SOI thanks to the BOX and the lateral dielectric. For this reason, the chips based on SOI are compact and fabricated more easily. Nevertheless, the manufacturing cost of SOI is slightly higher than for conventional devices because wafers have to be pre-processed to get the substrate. Another problem of this technology is the influence of Si film thickness on electron mobility degradation due to phonon confinement, Coulomb scattering, and doping fluctuations [159]. However, these drawbacks can be tolerated instead of a large number of advantages, which are summarized as follows [39, 156]:

- This technology is fully compatible with existing manufacturing facilities.
- It can significantly reduce the number of steps in the device fabrication process due to the use of an insulator substrate.
- The density of integration is higher due to the easier circuit design.
- SOI technology is tolerant to ionizing radiation.
- A specific operation requires a lower bias, hence a higher operation velocity is required for a specific bias.

- Higher control over SCE.
- Parasitic capacitance is reduced.
- Both planar and 3D devices can be manufactured.

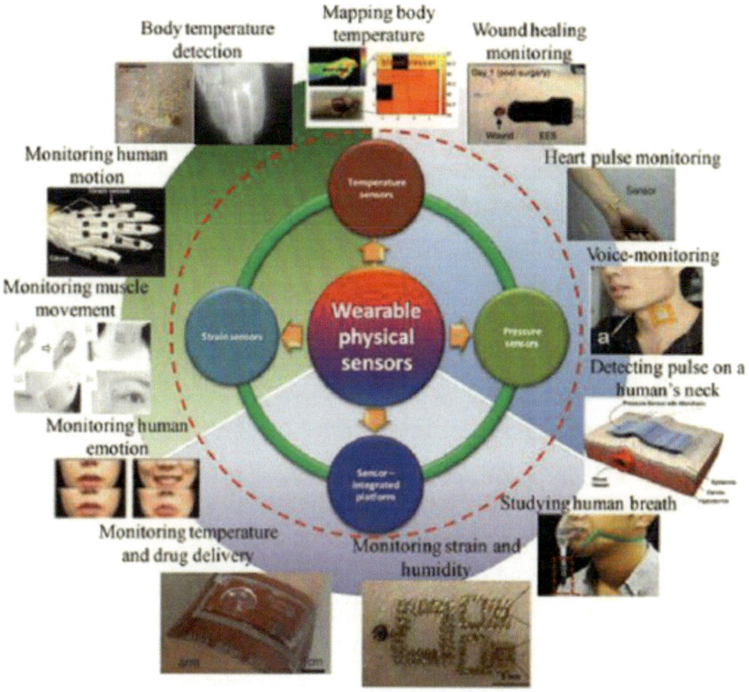

Figure 1.8 Application domains of the IoT [160].

The objective of this chapter is to combine the potentials of a mature and reliable rigid silicon technology with the promising opportunities of mechanical bendability. A well-established 65 nm node CMOS technology on SOI has been chosen for this work for (i) its high frequency, (ii) good understanding of its characterization and modeling, and finally (iii) the presence of a buried oxide that helps the transfer on a plastic film [160]. SOI technology refers to technology based on wafers featuring a top monocrystalline silicon layer above an insulating film, typically silicon dioxide. The handler or backside of the wafer usually consists of a thick silicon substrate.

Among the various SOI material fabrication strategies, three different methods have been widely used to produce SOI wafers during the last decades, namely SIMOX (Separation by IMplantation of OXygen) [161, 162], BESOI (Bonded and Etch-back SOI) [163, 164], and Smart Cut [165, 166] processes. The former relies on directly implanting oxygen to form a buried oxide layer upon thermal annealing through the top silicon layer, whereas the BESOI and Smart Cut processes rely on wafer bonding.

1.13 Simulation at Atomic Scale (Chapter 8)

Process/device simulation mainly consists of two stages: the modeling of the observed phenomena and the application of these models to the studied systems. In microelectronics, the development of new technology is very expensive. It is, therefore, judicious to resort to numerical simulations, to limit the number of experiments required to develop the device/process. In addition, simulation makes it possible to overcome certain uncontrolled experimental parameters and to have access to information that is not easy, if at all, measurable experimentally. Many methods have been developed to simulate these semiconductor manufacturing processes. Not all are based on the same principles, and their applications, therefore, differ according to their "capabilities". It is both convenient and relevant to classify the different simulation methods according to the scales they can handle. Spatial scale first: simulation of a few atoms or macroscopic systems? Time scale then, the time of the simulation experiment can vary from the order of the picoseconds to the order of the hours. Figure 1.9 summarizes this classification.

It is possible to simulate all manufacturing processes, from implantations to deposition and annealing through lithography. The calculation time required for the simulation depends directly on the desired degree of precision. So there is always a compromise to search between calculation time and precision of the result. Each method in Figure 1.9 represents a choice given this compromise. In this chapter, we are interested in the diffusion of dopants during the manufacturing process. We shall briefly present the different existing methods to simulate these phenomena, before looking into more detail and justifying this choice.

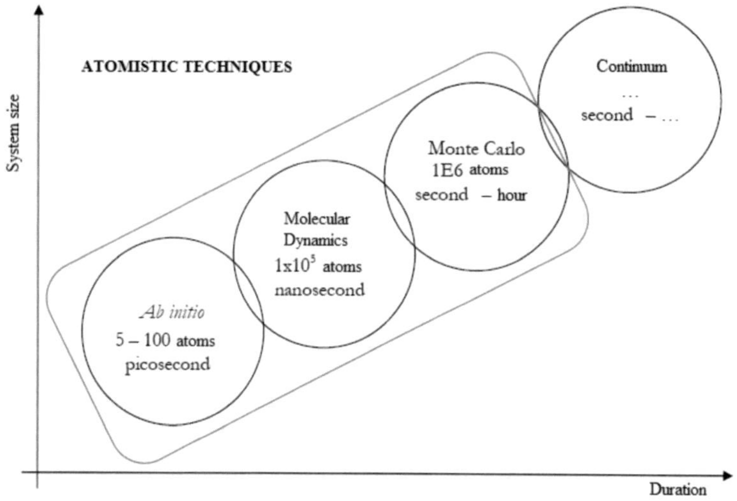

Figure 1.9 Summary of simulation classification.

The characteristic that distinguishes this type of simulation from others is that it does not require the use of input empirical parameters to be adjusted or calibrated. The ab initio simulations consist of processing the evolution of the electrons of the atoms of the system. They are based on the wave function of each particle, the square of which represents the probability density of presence at a point in space, and therefore provides complete information on the state of charge of a system of atoms. The problem lies in solving the Schrödinger equation and "it is not a small task". No sooner have we left the simpler cases of hydrogen and helium than approximations are essential. The method then takes on a more empirical character only initially, which does not prevent it from being effective. In addition, to simplify the necessary calculations, the theory of the density functional density (DFT) [167–169], as well as the pseudo-potentials [170], whose principle is to explicitly treat only the electrons in devices. For more details, one can refer to classic references [171–174].

1.14 Summary

This book offers a thorough examination and study of nanoelectronics, with a focus on the IoT. It provides a large, comprehensive, and practical view of design principles, as well as validated techniques for developing microelectronic products for IoT applications in the semiconductor industry. It bridges the main obstacles in developing microelectronics for IoT and corresponding best-practice solutions. It is almost indispensable to develop a new architecture to continue to follow Moore's law and reduce the dimensions of transistors. To meet the expectations of the ITRS/IRDS regarding aggressive gate dimensions (scaling), new architectures need to be developed. The book deals with an outlook on the future development of nanodevices for the era of the IoT. The monograph bridges the key challenges in intelligent fabless manufacturing and designing in microelectronics for IoT and corresponding best-practice solutions using TCAD. Several real-world examples are presented in detail to demonstrate how to design, prototype, and fabless manufacture microelectronic devices for various IoT applications across multiple industries. Several solutions have been proposed for improving the device performance at advanced technology nodes down to 3 nm.

References

1. "Fuel of the future: Data is giving rise to a new economy | Briefing | The Economist." https://www.economist.com/briefing/2017/05/06/data-is-giving-rise-to-a-new-economy (accessed Feb. 22, 2021).
2. "Report: The Digitization of the World - From Edge to Core." https://www.platinasystems.com/report-the-digitization-of-the-world-from-edge-to-core (accessed Feb. 22, 2021).
3. T. Faraz, "Expanding the toolbox of atomic scale processing," PhD Thesis, Technische Universiteit Eindhoven, 2021.
4. "Data Age 2025: The datasphere and data-readiness from edge to core." https://www.i-scoop.eu/big-data-action-value-context/data-age-2025-datasphere/ (accessed Feb. 23, 2021).
5. IEEE, International Roadmap for Devices and Systems (IRDS) 2020 Edition: Executive Summary.
6. ITRS, ITRS Roadmap 2015 Edition, More Moore, 2015.
7. G. E. Moore, "Cramming more components onto integrated circuits," *Electronics*, vol. 38, no. 8, p. 114, 1965. doi: 10.1109/N-SSC.2006.4785860.

8. D. Bergeron, "More than Moore," in *2008 IEEE Custom Integrated Circuits Conference*, 2008, pp. xxv–xxvi. doi: 10.1109/CICC.2008.4672003.
9. "NEREID Roadmap - Download | NEREID." https://www.nereid-h2020.eu/roadmap (accessed Apr. 19, 2021).
10. J. Ahopelto *et al.*, "Nanoelectronics roadmap for Europe: From nanodevices and innovative materials to system integration," *Solid. State. Electron.*, vol. 155, pp. 7–19, May 2019. doi: https://doi.org/10.1016/j.sse.2019.03.014.
11. S. Myung *et al.*, "Real-Time TCAD: A new paradigm for TCAD in the artificial intelligence era," in *2020 International Conference on Simulation of Semiconductor Processes and Devices (SISPAD)*, 2020, pp. 347–350. doi: 10.23919/SISPAD49475.2020.9241622.
12. C. Teo, K. L. Low, V. Narang, and A. V Thean, "TCAD-enabled machine learning defect prediction to accelerate advanced semiconductor device failure analysis," in *2019 International Conference on Simulation of Semiconductor Processes and Devices (SISPAD)*, 2019, pp. 1–4. doi: 10.1109/SISPAD.2019.8870440.
13. C. D. Wright, "Precise computing with imprecise devices," *Nature Electronics*, vol. 1, no. 4. Nature Publishing Group, pp. 212–213, Apr. 2018. doi: 10.1038/s41928-018-0061-9.
14. R. Nair, "Evolution of memory architecture," *Proc. IEEE*, vol. 103, no. 8, pp. 1331–1345, Aug. 2015. doi: 10.1109/JPROC.2015.2435018.
15. A. Sebastian *et al.*, "Temporal correlation detection using computational phase-change memory," *Nat. Commun.*, vol. 8, no. 1, pp. 1–10, Dec. 2017. doi: 10.1038/s41467-017-01481-9.
16. G. Kestor, R. Gioiosa, D. J. Kerbyson, and A. Hoisie, "Quantifying the energy cost of data movement in scientific applications," in *Proceedings - 2013 IEEE International Symposium on Workload Characterization, IISWC 2013*, 2013, pp. 56–65. doi: 10.1109/IISWC.2013.6704670.
17. D. Miorandi, S. Sicari, F. De Pellegrini, and I. Chlamtac, "Internet of things: Vision, applications and research challenges," *Ad Hoc Networks*, vol. 10, no. 7. Elsevier B.V., pp. 1497–1516, Sep. 2012. doi: 10.1016/j.adhoc.2012.02.016.
18. "Intel | Data Center Solutions, IoT, and PC Innovation." https://www.intel.com/content/www/us/en/homepage.html?wapkw=infographic+guide+to+internet+of+things (accessed Feb. 23, 2021).
19. R. Clark *et al.*, "Perspective: New process technologies required for future devices and scaling," *APL Mater.*, vol. 6, no. 5, May 2018. doi: 10.1063/1.5026805.

20. H. J. Cho *et al.*, "Si FinFET based 10 nm technology with multi Vt gate stack for low power and high performance applications," in *Digest of Technical Papers - Symposium on VLSI Technology*, Sep. 2016, vol. 2016-September. doi: 10.1109/VLSIT.2016.7573359.
21. S. Y. Wu *et al.*, "Demonstration of a sub-0.03 um2 high density 6-T SRAM with scaled bulk FinFETs for mobile SOC applications beyond 10nm node," in *Digest of Technical Papers - Symposium on VLSI Technology*, Sep. 2016, vol. 2016-September. doi: 10.1109/VLSIT.2016.7573390.
22. H. Liu, D. K. Mohata, A. Nidhi, V. Saripalli, V. Narayanan, and S. Datta, "Exploration of vertical MOSFET and tunnel FET device architecture for Sub 10nm node applications," in *Device Research Conference - Conference Digest, DRC*, 2012, pp. 233–234. doi: 10.1109/DRC.2012.6256990.
23. G. Larrieu and X. L. Han, "Vertical nanowire array-based field effect transistors for ultimate scaling," *Nanoscale*, vol. 5, no. 6, pp. 2437–2441, Mar. 2013. doi: 10.1039/c3nr33738c.
24. S. Maheshwaram, S. K. Manhas, G. Kaushal, B. Anand, and N. Singh, "Vertical nanowire CMOS parasitic modeling and its performance analysis," *IEEE Trans. Electron Devices*, vol. 60, no. 9, pp. 2943–2950, 2013. doi: 10.1109/TED.2013.2272651.
25. T. H. Bao *et al.*, "Circuit and process co-design with vertical gate-all-around nanowire FET technology to extend CMOS scaling for 5nm and beyond technologies," in *2014 44th European Solid State Device Research Conference (ESSDERC)*, Nov. 2014, pp. 102–105. doi: 10.1109/ESSDERC.2014.6948768.
26. D. Yakimets, "Vertical transistors: A slippery path towards the ultimate CMOS scaling," PhD Thesis, KU Leuven, 2017.
27. R. H. Dennard, F. H. Gaensslen, H. N. Yu, V. L. Rideout, E. Bassous, and A. R. Leblanc, "Design of ion-implanted MOSFET's with very small physical dimensions," *IEEE J. Solid-State Circuits*, vol. 9, no. 5, pp. 256–268, 1974. doi: 10.1109/JSSC.1974.1050511.
28. M. Badaroglu *et al.*, "PPAC scaling enablement for 5nm mobile SoC technology," in *2017 47th European Solid-State Device Research Conference (ESSDERC)*, 2017, pp. 240–243. doi: 10.1109/ESSDERC.2017.8066636.
29. "Intel Process Technology Update: 10nm Server Products in 1H 2020, Accelerated 7nm in 2021." https://www.anandtech.com/show/14311/intel-process-technology-update-10nm-server-products-in-1h-2020-accelerated-7nm-in-2021 (accessed Mar. 28, 2021).

30. A. Mallik, J. Ryckaert, A. Mercha, D. Verkest, K. Ronse, and A. Thean, "Maintaining Moore's law: Enabling cost-friendly dimensional scaling," in *Extreme Ultraviolet (EUV) Lithography VI*, Apr. 2015, vol. 9422, p. 94221N. doi: 10.1117/12.2086085.
31. J. Ryckaert *et al.*, "DTCO at N7 and beyond: Patterning and electrical compromises and opportunities," *Proc. SPIE 9427, Des. Co-optimization Manuf. IX*, vol. 9427, p. 94270C, Mar. 2015. doi: 10.1117/12.2178997.
32. L. Liebmann, J. Zeng, X. Zhu, L. Yuan, G. Bouche, and J. Kye, "Overcoming scaling barriers through design technology co-optimization," in *2016 IEEE Symposium on VLSI Technology*, 2016, pp. 1–2. doi: 10.1109/VLSIT.2016.7573398.
33. T. Ghani *et al.*, "A 90nm high volume manufacturing logic technology featuring novel 45nm gate length strained silicon CMOS," in *IEEE IEDM Tech. Dig.*, 2003, pp. 978–980. doi: 10.1109/iedm.2003.1269442.
34. M. T. Bohr, R. S. Chau, T. Ghani, and K. Mistry, "The high-k solution," *IEEE Spectr.*, vol. 44, no. 10, pp. 29–35, Oct. 2007. doi: 10.1109/MSPEC.2007.4337663.
35. D. James, "From strain to high-K/metal gate - The 65-45 nm transition," in *ASMC (Advanced Semiconductor Manufacturing Conference) Proceedings*, 2008, pp. 76–81. doi: 10.1109/ASMC.2008.4529013.
36. N. Collaert, "CMOS nanoelectronics: Innovative devices, architectures, and applications," CRC Press, Boca Raton, Florida, 2012.
37. M. L. Talley, S. Shannon, L. Chen, and J. P. Verboncoeur, "IEDF distortion and resolution considerations for RFEA operation at high voltages," *Plasma Sources Sci. Technol.*, vol. 26, no. 12, p. 125001, Nov. 2017. doi: 10.1088/1361-6595/aa9465.
38. D. Hisamoto *et al.*, "FinFET—A self-aligned double-gate MOSFET scalable to 20 nm," *IEEE Trans. Electron Devices*, vol. 47, no. 12, pp. 2320–2325, 2000. doi: 10.1109/16.887014.
39. J. P. Colinge, "Multiple-gate SOI MOSFETs," *Solid. State. Electron.*, vol. 48, no. 6, pp. 897–905, Jun. 2004. doi: 10.1016/j.sse.2003.12.020.
40. Loubet *et al.*, "Stacked nanosheet gate-all-around transistor to enable scaling beyond FinFET," in *2017 Symposium on VLSI Technology*, Jul. 2017, pp. T230–T231. doi: 10.23919/VLSIT.2017.7998183.
41. J. P. Colinge *et al.*, "Silicon-on-insulator 'gate-all-around device," in *IEEE IEDM Tech. Dig.*, Dec. 1990, pp. 595–598. doi: 10.1109/iedm.1990.237128.
42. H. Mertens *et al.*, "Gate-all-around MOSFETs based on vertically stacked horizontal Si nanowires in a replacement metal gate process

on bulk Si substrates," in *Digest of Technical Papers - Symposium on VLSI Technology*, Sep. 2016, vol. 2016-Sept. doi: 10.1109/VLSIT.2016.7573416.

43. J. P. Colinge *et al.*, "Nanowire transistors without junctions," *Nat. Nanotechnol.*, vol. 5, no. 3, pp. 225–229, 2010. doi: 10.1038/nnano.2010.15.

44. J. M. Hergenrother *et al.*, "50 nm vertical replacement-gate (VRG) nMOSFETs with ALD HfO_2 and Al_2O_3 gate dielectrics," in *International Electron Devices Meeting. Technical Digest (Cat. No.01CH37224)*, pp. 3.1.1–3.1.4. doi: 10.1109/IEDM.2001.979400.

45. S. H. Oh *et al.*, "50 nm vertical replacement-gate (VRG) pMOSFETs," *Tech. Dig. - Int. Electron Devices Meet.*, pp. 65–68, 2000. doi: 10.1109/IEDM.2000.904260.

46. D. Yakimets *et al.*, "Vertical GAAFETs for the ultimate CMOS scaling," *IEEE Trans. Electron Devices*, vol. 62, no. 5, pp. 1433–1439, May 2015. doi: 10.1109/TED.2015.2414924.

47. R. Possamai Bastos, L. A. Guimarães, F. Sill Torres, and L. Fesquet, "Architectures of bulk built-in current sensors for detection of transient faults in integrated circuits," *Microelectronics J.*, vol. 71, pp. 70–79, Jan. 2018. doi: 10.1016/j.mejo.2017.11.006.

48. L. A. Guimares, R. P. Bastos, and L. Fesquet, "Detection of layout-level trojans by monitoring substrate with preexisting built-in sensors," in *2017 IEEE Computer Society Annual Symposium on VLSI (ISVLSI)*, 2017, pp. 290–295. doi: 10.1109/ISVLSI.2017.58.

49. L. A. Guimares, R. P. Bastos, T. F. de Paiva Leite, and L. Fesquet, "Simple tri-state logic Trojans able to upset properties of ring oscillators," in *2016 International Conference on Design and Technology of Integrated Systems in Nanoscale Era (DTIS)*, 2016, pp. 1–6. doi: 10.1109/DTIS.2016.7483811.

50. H. Salmani, M. Tehranipoor, and J. Plusquellic, "New design strategy for improving hardware Trojan detection and reducing Trojan activation time," in *2009 IEEE International Workshop on Hardware-Oriented Security and Trust*, 2009, pp. 66–73. doi: 10.1109/HST.2009.5224968.

51. M. Liu, *Microelectronics for the Internet of Things*, 2014.

52. Y. Shacham-Diamand, "Internet of things: What's in it for the microelectronics industry." http://www.chipex.co.il/_Uploads/dbsAttachedFiles/Whatsinitforthemicroelectronicsindustry.pdf

53. T. Bedecarrats, "Etude et intégration d'un circuit analogique, basse consommation et à faible surface d'empreinte, de neurone

impulsionnel basé sur l'utilisation du BIMOS en technologie 28 nm FD-SOI." PhD Thesis, Université Grenoble Alpes, 2019.

54. T. Bédécarrats, C. Fenouillet-Béranger, S. Cristoloveanu, and P. Galy, "A BIMOS-based 2T1C analogue spiking neuron circuit integrated in 28 nm FD-SOI technology for neuromorphic application," *Solid. State. Electron.*, vol. 168, p. 107717, Jun. 2020. doi: 10.1016/j.sse.2019.107717.

55. T. Bédécarrats, P. Galy, C. Fenouillet-Béranger, and S. Cristoloveanu, "Investigation of built-in bipolar junction transistor in FD-SOI BIMOS," *Solid. State. Electron.*, vol. 159, pp. 177–183, Sep. 2019. doi: 10.1016/j.sse.2019.03.057.

56. M. M. Waldrop, "More Than Moore," *Nature*, vol. 530, no. 7589. Nature Publishing Group, pp. 144–147, Feb. 09, 2016. doi: 10.1038/530144a.

57. C. S. Thakur *et al.*, "Large-scale neuromorphic spiking array processors: A quest to mimic the brain," *Frontiers Research Foundation*, vol. 12, p. 891, 2007. doi: 10.3929/ethz-b-000310459.

58. P. Galy, P. Dehan, J. Jimenez, and B. Heitz, "Preliminary results on low power sigmoid neuron transistor response in 28 nm high-k metal gate fully depleted SOI technology," *Solid. State. Electron.*, vol. 89, pp. 17–21, Nov. 2013. doi: 10.1016/j.sse.2013.06.002.

59. S. K. Saha, "Managing technology CAD for competitive advantage: An efficient approach for integrated circuit fabrication technology development," *IEEE Trans. Eng. Manag.*, vol. 46, no. 2, pp. 221–229, 1999. doi: 10.1109/17.759152.

60. R.W. Dutton and Z. Wu, "Technology-oriented CAD," *Technology CAD — Computer Simulation of IC Processes and Devices*, Springer US, pp. 1–36, 1993. doi: 10.1007/978-1-4615-3208-8_1.

61. R. W. Dutton and A. J. Strojwas, "Perspectives on technology and technology-driven CAD," *IEEE Trans. Comput. Des. Integr. Circuits Syst.*, vol. 19, no. 12, pp. 1544–1560, 2000. doi: 10.1109/43.898831.

62. S. Saha and S. K. Saha, "Modelling the effectiveness of computer-aided development projects in the semiconductor industry," *Int. J. Eng. Manag. Econ.*, vol. 1(2/3), no. 2/3, pp. 162–178, 2010. doi: 10.1504/ijeme.2010.036948.

63. M. Karner, Z. Stanojevic, F. Mitterbauer, C. Kernstock, and H. Demel, "Bringing physics to device design: A fast and predictive device simulation framework," in *2015 Silicon Nanoelectronics Workshop (SNW)*, 2015, pp. 1–2.

64. S. Bandyopadhyay *et al.*, "A rigorous technique to couple Monte Carlo and drift-diffusion models for computationally efficient device simulation," *IEEE Trans. Electron Dev.*, vol. ED-34, pp. 392–399, 1987. doi: 10.1109/T-ED.1987.22935.
65. H. Kosina and S. Selberherr, "Coupling of monte carlo and drift diffusion method with applications to metal oxide semiconductor field effect transistors," *Jpn. J. Appl. Phys.*, vol. 29, no. 12, pp. 2283–2289, Dec. 1990. doi: 10.1143/JJAP.29.L2283.
66. G. A. Armstrong and C. K. Maiti, *TCAD for Si, SiGe and GaAs Integrated Circuits*. The Institution of Engineering and Technology (IET), UK, 2008.
67. S. K. Saha, "The role of semiconductor foundries in advanced integrated circuit product development," in *2012 IEEE International Technology Management Conference, ITMC 2012*, 2012, pp. 32–35. doi: 10.1109/ITMC.2012.6306393.
68. S. K. Saha, "Integrating TCAD in semiconductor foundry companies for an efficient customization of IC manufacturing technology," in *Proceedings of the 2000 IEEE Engineering Management Society, EMS 2000*, 2000, pp. 179–183. doi: 10.1109/EMS.2000.872497.
69. S. K. Saha, "Modeling process variability in scaled CMOS technology," *IEEE Des. Test Comput.*, vol. 27, no. 2, pp. 8–16, Mar. 2010. doi: 10.1109/MDT.2010.50.
70. F. C. Tseng, "Foundry technologies," in *Technical Digest - International Electron Devices Meeting*, 1996, pp. 19–24. doi: 10.1109/iedm.1996.553030.
71. W. B. Han, G. J. Ko, J. W. Shin, and S. W. Hwang, "Advanced manufacturing for transient electronics," *MRS Bull.*, vol. 45, no. 2, pp. 113–120, Feb. 2020. doi: 10.1557/mrs.2020.22.
72. H. Iwai, "Future semiconductor manufacturing: Challenges and opportunities," in *Technical Digest - International Electron Devices Meeting, IEDM*, 2004, pp. 11–16. doi: 10.1109/iedm.2004.1419051.
73. L. Mönch, J. W. Fowler, S. Dauzère-Pérès, S. J. Mason, and O. Rose, "A survey of problems, solution techniques, and future challenges in scheduling semiconductor manufacturing operations," *J. Sched.*, vol. 14, no. 6, pp. 583–599, Dec. 2011. doi: 10.1007/s10951-010-0222-9.
74. Y. S. Bankapalli and H. Y. Wong, "TCAD augmented machine learning for semiconductor device failure troubleshooting and reverse engineering," in *2019 International Conference on Simulation of Semiconductor Processes and Devices (SISPAD)*, Sep. 2019, vol. 2019-Sept., pp. 1–4. doi: 10.1109/SISPAD.2019.8870467.

75. M. Shi, P. Mo, and J. Liu, "Deep neural network for accurate and efficient atomistic modeling of phase change memory," *IEEE Electron Device Lett.*, vol. 41, no. 3, pp. 365–368, Mar. 2020. doi: 10.1109/LED.2020.2964779.

76. J. Chen *et al.*, "Powernet: SOI lateral power device breakdown prediction with deep neural networks," *IEEE Access*, vol. 8, pp. 25372–25382, 2020.

77. H. Carrillo-Nuez, N. Dimitrova, A. Asenov, and V. Georgiev, "Machine learning approach for predicting the effect of statistical variability in si junctionless nanowire transistors," *IEEE Electron Device Lett.*, vol. 40, no. 9, pp. 1366–1369, 2019. doi: 10.1109/LED.2019.2931839.

78. S. Saha, "Extraction of substrate current model parameters from device simulation," *Solid State Electron.*, vol. 37, no. 10, pp. 1786–1788, 1994. doi: 10.1016/0038-1101(94)90232-1.

79. S. Saha *et al.*, "Impact ionization rate of electrons for accurate simulation of substrate current in submicron devices," *Solid-State Electron.*, vol. 36, no. 10, pp. 1429–1432, 1993. doi: 10.1016/0038-1101(93)90051-Q.

80. M. Choi, V. Moroz, L. Smith, and J. Huang, "Extending drift-diffusion paradigm into the era of FinFETs and nanowires," in *2015 International Conference on Simulation of Semiconductor Processes and Devices (SISPAD)*, Oct. 2015, vol. 2015-Octob, pp. 242–245. doi: 10.1109/SISPAD.2015.7292304.

81. A. Erlebach, K. H. Lee, F. M. Bufler, H. Lee, and F. M. Bufler, "Empirical ballistic mobility model for drift-diffusion simulation," in *2016 46th European Solid-State Device Research Conference (ESSDERC)*, 2016, vol. 46, pp. 420–423. doi: 10.1109/ESSDERC.2016.7599675.

82. Z. Stanojevic *et al.*, "Physical modeling: A new paradigm in device simulation," in *2015 IEEE International Electron Devices Meeting (IEDM)*, 2015, pp. 5.1.1–5.1.4. doi: 10.1109/IEDM.2015.7409631.

83. C. Mead, "Analog VLSI and neural systems," Addison-Wesley, 1989.

84. J. Von Neumann and M. D. Godfrey, "First draft of a report on the EDVAC," *IEEE Ann. Hist. Comput.*, vol. 15, no. 4, pp. 27–75, 1993. doi: 10.1109/85.238389.

85. L. Canetta, A. Barni, and E. Montini, "Development of a digitalization maturity model for the manufacturing sector," Aug. 2018. doi: 10.1109/ICE.2018.8436292.

86. F. Tao, Y. Zuo, L. Da Xu, and L. Zhang, "IoT-based intelligent perception and access of manufacturing resource toward cloud manufacturing,"

IEEE Trans. Ind. Informatics, vol. 10, no. 2, pp. 1547–1557, 2014. doi: 10.1109/TII.2014.2306397.

87. S. Wang, J. Wan, D. Li, and C. Zhang, "Implementing smart factory of industry 4.0: An outlook," *Int. J. Distrib. Sens. Networks*, vol. 12, no. 1, p. 3159805, Jan. 2016. doi: 10.1155/2016/3159805.
88. M. Qiu, C. Xue, Z. Shao, Q. Zhuge, M. Liu, and E. H. M. Sha, "Efficent algorithm of energy minimization for heterogeneous wireless sensor network," in *Lecture Notes in Computer Science (including subseries Lecture Notes in Artificial Intelligence and Lecture Notes in Bioinformatics)*, 2006, vol. 4096 LNCS, pp. 25–34. doi: 10.1007/11802167_5.
89. M. Qiu and E. H. M. Sha, "Energy-aware online algorithm to satisfy sampling rates with guaranteed probability for sensor applications," in *Lecture Notes in Computer Science (including subseries Lecture Notes in Artificial Intelligence and Lecture Notes in Bioinformatics)*, 2007, vol. 4782 LNCS, pp. 156–167. doi: 10.1007/978-3-540-75444-2_20.
90. M. Chen, S. Mao, and Y. Liu, "Big data: A survey," in *Mobile Networks and Applications*, Apr. 2014, vol. 19, no. 2, pp. 171–209. doi: 10.1007/s11036-013-0489-0.
91. X. Xu, "From cloud computing to cloud manufacturing," *Robot. Comput. Integr. Manuf.*, vol. 28, no. 1, pp. 75–86, Feb. 2012. doi: 10.1016/j.rcim.2011.07.002.
92. "Cloud manufacturing service system for industrial-cluster-oriented application", *Liu | Journal of Internet Technology*, May 03, 2021. https://jit.ndhu.edu.tw/article/view/993.
93. J. Wan, D. Li, H. Yan, and P. Zhang, "Fuzzy feedback scheduling algorithm based on central processing unit utilization for a software-based computer numerical control system," *Proc. Inst. Mech. Eng. Part B J. Eng. Manuf.*, vol. 224, no. 7, pp. 1133–1143, Jul. 2010. doi: 10.1243/09544054JEM1690.
94. F. Soliman and M. A. Youssef, "Internet-based e-commerce and its impact on manufacturing and business operations," *Ind. Manag. Data Syst.*, vol. 103, no. 8–9, pp. 546–552, 2003. doi: 10.1108/02635570310497594.
95. L. A. Guimares, T. F. de Paiva Leite, R. P. Bastos, and L. Fesquet, "Non-intrusive testing technique for detection of Trojans in asynchronous circuits," in *2018 Design, Automation Test in Europe Conference Exhibition (DATE)*, 2018, pp. 1516–1519. doi: 10.23919/DATE.2018.8342255.
96. Rakesh Kumar, *Fabless Semiconductor Implementation*, McGraw-Hill Professional, 2008.
97. D. Nenni and P. McLellan, *Fabless: The Transformation of The Semiconductor Industry,* CreateSpace, 2014.

98. S. Myung *et al.*, "Real-time TCAD: A new paradigm for TCAD in the artificial intelligence era," in *2020 International Conference on Simulation of Semiconductor Processes and Devices (SISPAD)*, Sep. 2020, vol. 2020-Sept., pp. 347–350. doi: 10.23919/SISPAD49475.2020.9241622.
99. L. Zhang and M. Chan, "Artificial neural network design for compact modeling of generic transistors," *J. Comput. Electron.*, vol. 16, no. 3, pp. 825–832, Sep. 2017. doi: 10.1007/s10825-017-0984-9.
100. S. S. Raju, B. Wang, K. Mehta, M. Xiao, Y. Zhang, and H.-Y. Wong, "Application of noise to avoid overfitting in TCAD augmented machine learning," in *2020 International Conference on Simulation of Semiconductor Processes and Devices (SISPAD)*, 2020, pp. 351–354. doi: 10.23919/SISPAD49475.2020.9241654.
101. H. Y. Wong *et al.*, "TCAD-machine learning framework for device variation and operating temperature analysis with experimental demonstration," *IEEE J. Electron Devices Soc.*, vol. 8, pp. 992–1000, 2020. doi: 10.1109/JEDS.2020.3024669.
102. R. W. Dutton and Z. Yu, "Technology CAD — Computer simulation of IC processes and devices," *Springer*, Springer US, 1993. doi: 10.1007/978-1-4615-3208-8.
103. P. J. Hopper and P. A. Blakely, "The MASTER framework," *Microelectronics J.*, vol. 26, no. 2–3, pp. 177–190, 1995. doi: 10.1016/0026-2692(95)98920-m.
104. R. W. Dutton and R. J. G. Goossens, "Technology CAD at Stanford University: Physics, algorithms, software and applications," *Microelectronics J.*, vol. 26, no. 2–3, pp. 99–111, 1995. doi: 10.1016/0026-2692(95)98916-F.
105. J. Ueda, S. Kuroda, K. Fukuda, K. Kai, and K. Nishi, "Technology CAD at OKI," *Microelectronics J.*, vol. 26, no. 2–3, pp. 159–175, 1995. doi: 10.1016/0026-2692(95)98919-I.
106. P. Lloyd *et al.*, "Technology CAD at AT&T," *Microelectronics J.*, vol. 26, no. 2–3, pp. 79–97, 1995. doi: 10.1016/0026-2692(95)98915-E.
107. H. Stippel *et al.*, "Process simulation for the 1990s," *Microelectron. J.*, vol. 26, no. 2, pp. 203–215, 1995. doi: https://doi.org/10.1016/0026-2692(95)98922-E.
108. J. Meindl, "Low power microelectronics: Retrospect and prospect," *Proc. IEEE*, vol. 83, pp. 619–635, 1995.
109. M. D. Ganeriwala, C. Yadav, N. R. Mohapatra, S. Khandelwal, C. Hu, and Y. S. Chauhan, "Modeling of charge and quantum capacitance in low effective mass III-V FinFETs," *IEEE J. Electron Devices Soc.*, vol. 4, no. 6, pp. 396–401, 2016. doi: 10.1109/JEDS.2016.2586116.

110. G. Curatola and G. Iannaccone, "Ballistic transport in SiGe and strained-Si MOSFETs," *J. Comput. Electron.*, vol. 2, no. 2–4, pp. 309–312, 2003. doi: 10.1023/B:JCEL.0000011443.60531.10.

111. J. Lee *et al.*, "Nanowire FETs," in *2018 IEEE 13th Nanotechnology Materials and Devices Conference (NMDC)*, 2018, pp. 1–4. doi: 10.1109/NMDC.2018.8605884.

112. C. Medina-Bailon *et al.*, "Study of the 1D scattering mechanisms' impact on the mobility in Si nanowire transistors," in *2018 Joint International EUROSOI Workshop and International Conference on Ultimate Integration on Silicon (EUROSOI-ULIS)*, May 2018, vol. 2018-Jan, pp. 1–4. doi: 10.1109/ULIS.2018.8354723.

113. O. Badami *et al.*, "Comprehensive study of cross-section dependent effective masses for silicon based gate-all-around transistors," *Appl. Sci.*, vol. 9, no. 9, 2019. doi: 10.3390/app9091895.

114. S. Y. Wu *et al.*, "A 7nm CMOS platform technology featuring 4th generation FinFET transistors with a 0.027um2 high density 6-T SRAM cell for mobile SoC applications," in *2016 IEEE International Electron Devices Meeting (IEDM)*, Jan. 2016, pp. 2.6.1–2.6.4. doi: 10.1109/IEDM.2016.7838333.

115. G. Bae *et al.*, "3nm GAA technology featuring multi-bridge-channel FET for low power and high performance applications," in *2018 IEEE International Electron Devices Meeting (IEDM)*, Jan. 2018, vol. 2018-Dec., pp. 28.7.1–28.7.4. doi: 10.1109/IEDM.2018.8614629.

116. S. Dey, J. Jena, E. Mohapatra, T. P. Dash, S. Das, and C. K. Maiti, "Design and simulation of vertically-stacked nanowire transistors at 3nm technology nodes," *Phys. Scr.*, vol. 95, no. 1, p. 14001, Dec. 2020. doi: 10.1088/1402-4896/ab4621.

117. T. P. Dash, E. Mohapatra, J. Jena, S. Das, S. Dey, and C. K. Maiti, "Nanosized metal-grain-granularity induced characteristics fluctuation in gate-all-around Si-nanowire transistors at 1nm technology node," *Int. J. Nanosci.*, Jul. 2019. doi: 10.1142/s0219581x20400050.

118. S. M. Nawaz, S. Dutta, A. Chattopadhyay, and A. Mallik, "Comparison of random dopant and gate-metal workfunction variability between junctionless and conventional FinFETs," *IEEE Electron Device Lett.*, vol. 35, no. 6, pp. 663–665, 2014. doi: 10.1109/LED.2014.2313916.

119. T. P. Dash, S. Dey, J. Jena, S. Das, E. Mohapatra, and C. K. Maiti, "Metal grain granularity induced variability in gate-all-around Si-nanowire transistors at 1nm technology node," in *2019 Devices for Integrated Circuit (DevIC)*, 2019, pp. 286–290. doi: 10.1109/DEVIC.2019.8783717.

120. O. Sikder, "Influence of size and interface effects of silicon nanowire and nanosheet for ultra-scaled next generation transistors," PhD Thesis, Purdue University, 2020.

121. C. Auth *et al.*, "A 22nm high performance and low-power CMOS technology featuring fully-depleted tri-gate transistors, self-aligned contacts and high density MIM capacitors," in *Digest of Technical Papers - Symposium on VLSI Technology*, 2012, pp. 131–132. doi: 10.1109/VLSIT.2012.6242496.

122. A. Razavieh, P. Zeitzoff, and E. J. Nowak, "Challenges and limitations of CMOS scaling for FinFET and beyond architectures," *IEEE Trans. Nanotechnol.*, vol. 18, pp. 999–1004, 2019. doi: 10.1109/TNANO.2019.2942456.

123. J. Appenzeller, J. Knoch, M. T. Björk, H. Riel, H. Schmid, and W. Riess, "Toward nanowire electronics," *IEEE Trans. Electron Devices*, vol. 55, no. 11, p. 2827, 2008. doi: 10.1109/TED.2008.2008011.

124. W. C. Ma, Y. Huang, P. Chen, J. Jhu, Y. Chang, and T. Chang, "Impacts of vertically stacked monolithic 3D-IC process on characteristics of underlying thin-film transistor," *IEEE J. Electron Devices Soc.*, vol. 8, pp. 724–730, 2020. doi: 10.1109/JEDS.2020.3009350.

125. X. Yin *et al.*, "Vertical sandwich gate-all-around field-effect transistors with self-aligned high-k metal gates and small effective-gate-length variation," *IEEE Electron Device Lett.*, vol. 41, no. 1, pp. 8–11, Jan. 2020. doi: 10.1109/LED.2019.2954537.

126. D. Jang *et al.*, "Device exploration of nanosheet transistors for sub-7-nm technology node," *IEEE Trans. Electron Devices*, vol. 64, no. 6, pp. 2707–2713, Jun. 2017. doi: 10.1109/TED.2017.2695455.

127. A. Veloso *et al.*, "Nanowire & nanosheet FETs for ultra-scaled, high-density logic and memory applications," *Solid. State. Electron.*, vol. 168, p. 107736, Jun. 2020. doi: 10.1016/j.sse.2019.107736.

128. P. Weckx *et al.*, "Stacked nanosheet fork architecture for SRAM design and device co-optimization torward 3nm," in *IEEE International Electron Devices Meeting*, 2017, pp. 505–508. doi: 10.1109/IEDM.2017.8268430.

129. A. Asenov *et al.*, "Nanowire transistor solutions for 5nm and beyond," in *2016 17th International Symposium on Quality Electronic Design (ISQED)*, 2016, vol. 2016-May, pp. 269–274. doi: 10.1109/ISQED.2016.7479212.

130. V. Vashishtha and L. T. Clark, "Comparing bulk-Si FinFET and gate-all-around FETs for the 5 nm technology node," *Microelectronics J.*, vol. 107, p. 104942, Jan. 2021. doi: 10.1016/j.mejo.2020.104942.

131. D. Nagy, G. Indalecio, A. J. Garc, A-Loureiro, M. A. Elmessary, K. Kalna, and N. Seoane, "FinFET versus gate-all-around nanowire FET: Performance, scaling, and variability," *IEEE J. Electron Devices Soc.*, vol. 6, pp. 332–340, 2018. doi: 10.1109/JEDS.2018.2804383.

132. C. C. Hobbs *et al.*, "Fermi-level pinning at the polysilicon/metal oxide interface-Part I," *IEEE Trans. Electron Devices*, vol. 51, no. 6, pp. 971–977, 2004. doi: 10.1109/TED.2004.829513.

133. C. C. Hobbs *et al.*, "Fermi-level pinning at the polysilicon/metal-oxide interface - Part II," *IEEE Trans. Electron Devices*, vol. 51, no. 6, pp. 978–984, Jun. 2004. doi: 10.1109/TED.2004.829510.

134. X. Wang, G. Roy, O. Saxod, A. Bajolet, A. Juge, and A. Asenov, "Simulation study of dominant statistical variability sources in 32-nm high- $κ$/ metal gate CMOS," *IEEE Electron Device Lett.*, vol. 33, no. 5, pp. 643–645, 2012. doi: 10.1109/LED.2012.2188268.

135. A. R. Brown, N. M. Idris, J. R. Watling, and A. Asenov, "Impact of metal gate granularity on threshold voltage variability: A full-scale three-dimensional statistical simulation study," *IEEE Electron Device Lett.*, vol. 31, no. 11, pp. 1199–1201, 2010. doi: 10.1109/LED.2010.2069080.

136. H. Dadgour *et al.*, "Statistical modeling of metalgate work-function variability in emerging device technologies and implications for circuit design," *Proc. IEEE/ACM Int. Conf. Comput. Des.*, pp. 270–277, 2008. doi: 10.1109/ICCAD.2008.4681585.

137. H. F. Dadgour, K. Endo, V. K. De, and K. Banerjee, "Grain-orientation induced work function variation in nanoscale metal-gate transistors - Part I: Modeling, analysis, and experimental validation," *IEEE Trans. Electron Devices*, vol. 57, no. 10, pp. 2504–2514, Oct. 2010. doi: 10.1109/TED.2010.2063191.

138. G. S. Lujan, S. Kubicek, S. De Gendt, M. Heyns, W. Magnus, and K. De Meyer, "Mobility degradation in high-k transistors: The role of the charge scattering," in *European Solid-State Device Research Conference*, 2003, pp. 399–402. doi: 10.1109/ESSDERC.2003.1256898.

139. J. F. Yang, Z. L. Xia, G. Du, X. Y. Liu, R. Q. Han, and J. F. Kang, "Coulomb scattering induced mobility degradation in Ultrathin-body SOI MOSFETs with high-k gate stack," in *ICSICT-2006: 2006 8th International Conference on Solid-State and Integrated Circuit Technology, Proceedings*, 2006, pp. 1315–1317. doi: 10.1109/ICSICT.2006.306146.

140. K. Goto *et al.*, "Technology booster using strain-enhancing laminated SiN (SELS) for 65nm node HP MPUs," in *Technical Digest - International Electron Devices Meeting, IEDM*, 2004, pp. 209–212. doi: 10.1109/iedm.2004.1419111.

141. V. Moroz, L. Smith, X.-W. Lin, D. Pramanik, and G. Rollins, "Stress-aware design methodology," in *7th International Symposium on Quality Electronic Design (ISQED'06)*, 2006, pp. 6–812. doi: 10.1109/ISQED.2006.124.

142. T. Das and J. H. Ahn, "Development of electronic devices based on two-dimensional materials," *FlatChem*, vol. 3. Elsevier B.V., pp. 43–63, Jun. 01, 2017. doi: 10.1016/j.flatc.2017.05.001.

143. R. Courtland, "Plotting a Moore's law for flexible electronics [News]," *IEEE Spectr.*, vol. 54, no. 7, pp. 7–8, 2017. doi: 10.1109/MSPEC.2017.7951706.

144. L. Xiang, X. Zeng, F. Xia, W. Jin, Y. Liu, and Y. Hu, "Recent advances in flexible and stretchable sensing systems: From the perspective of system integration," *ACS Nano*, vol. 14, no. 6. American Chemical Society, pp. 6449–6469, Jun. 23, 2020. doi: 10.1021/acsnano.0c01164.

145. T. Sekitani and T. Someya, "Stretchable organic integrated circuits for large-area electronic skin surfaces," *MRS Bull.*, vol. 37, no. 3, pp. 236–245, Mar. 2012. doi: 10.1557/mrs.2012.42.

146. Y. Sun and J. A. Rogers, "Inorganic semiconductors for flexible electronics," *Advanced Materials*, vol. 19, no. 15. pp. 1897–1916, Aug. 03, 2007. doi: 10.1002/adma.200602223.

147. W. Lu *et al.*, "Recent progress of multilayer composite transparent conductive film," *Kexue Tongbao/Chinese Science Bulletin*, vol. 62, no. 5. Chinese Academy of Sciences, pp. 372–384, Feb. 15, 2017. doi: 10.1360/N972016-01133.

148. W. Shi, Y. Guo, and Y. Liu, "When flexible organic field-effect transistors meet biomimetics: A prospective view of the internet of things," *Advanced Materials*, vol. 32, no. 15. Wiley-VCH Verlag, Apr. 01, 2020. doi: 10.1002/adma.201901493.

149. S. Mhedhbi *et al.*, "First power performance demonstration of flexible AlGaN/GaN high electron mobility transistor," *IEEE Electron Device Lett.*, vol. 37, no. 5, pp. 553–555, May 2016. doi: 10.1109/LED.2016.2542921.

150. K. Fukuda, T. Suzuki, D. Kumaki, and S. Tokito, "Reverse DC bias stress shifts in organic thin-film transistors with gate dielectrics using parylene-C," *Phys. status solidi*, vol. 209, no. 10, pp. 2073–2077, Oct. 2012. doi: 10.1002/pssa.201228219.

151. C. D. Sheraw, D. J. Gundlach, and T. N. Jackson, "Spin-on polymer gate dielectric for high performance organic thin film transistors," in *Materials Research Society Symposium - Proceedings*, 2000, vol. 558, pp. 403–408. doi: 10.1557/proc-558-403.

152. S. Yang *et al.*, "Environmentally stable transparent organic/oxide hybrid transistor based on an oxide semiconductor and a polyimide gate insulator," *IEEE Electron Device Lett.*, vol. 31, no. 5, pp. 446–448, May 2010. doi: 10.1109/LED.2010.2043334.

153. R. Wilson *et al.*, "A 460MHz at 397mV, 2.6GHz at 1.3V, 32b VLIW DSP, embedding FMAX tracking," in *2014 IEEE International Solid-State Circuits Conference Digest of Technical Papers (ISSCC)*, 2014, pp. 452–453. doi: 10.1109/ISSCC.2014.6757509.

154. M. L. Alles, "Thin-film SOI emerges," *IEEE Spectr.*, vol. 34, no. 6, pp. 37–45, Jun. 1997. doi: 10.1109/6.591663.

155. S. Cristoloveanu, D. Munteanu, and M. S. T. Liu, "A review of the pseudo-MOS transistor in SOI wafers: Operation, parameter extraction, and applications," *IEEE Trans. Electron Devices*, vol. 47, no. 5, pp. 1018–1027, 2000. doi: 10.1109/16.841236.

156. G. K. Celler and S. Cristoloveanu, "Frontiers of silicon-on-insulator," *J. Appl. Phys.*, vol. 93, no. 9, pp. 4955–4978, May 2003. doi: 10.1063/1.1558223.

157. G. G. Shahidi, "SOI technology for the GHz era," *IBM J. Res. Dev.*, vol. 46, no. 2–3, pp. 121–132, 2002. doi: 10.1147/rd.462.0121.

158. K. Oshima *et al.*, "Advanced SOI MOSFETs with buried alumina and ground plane: Self-heating and short-channel effects," *Solid. State. Electron.*, vol. 48, no. 6, pp. 907–917, Jun. 2004. doi: 10.1016/j.sse.2003.12.026.

159. F. Gamiz *et al.*, "Electron mobility in extremely thin single-gate silicon-on-insulator inversion layers," *J. Appl. Phys.*, vol. 86, pp. 62–69, 1999. doi: 10.1063/1.371684.

160. S. Mhedhbi, "Develpoment de composants flexibes en technologies heterogene (GaN et Graphene) pour des applications hautes frequences," PhD Thesis, Université de Lille, 2017.

161. K. Izumi, M. Doken, and H. Ariyoshi, "C.m.o.s. devices fabricated on buried sio2 layers formed by oxygen implantation into silicon," *Electron. Lett.*, vol. 14, no. 18, pp. 593–594, Aug. 1978. doi: 10.1049/el:19780397.

162. P. L. F. Hemment *et al.*, "Novel dielectric/silicon planar structures formed by ion beam synthesis," *Nucl. Inst. Methods Phys. Res. B*, vol. 21, no. 1–4, pp. 129–133, Jan. 1987. doi: 10.1016/0168-583X(87)90811-1.

163. M. Kimura, K. Egami, M. Kanamori, and T. Hamaguchi, "Epitaxial film transfer technique for producing single crystal Si film on an insulating

substrate," *Appl. Phys. Lett.*, vol. 43, no. 3, pp. 263–265, Aug. 1983. doi: 10.1063/1.94320.

164. W. P. Maszara, "Silicon-on-insulator by wafer bonding: A review," *J. Electrochem. Soc.*, vol. 138, no. 1, pp. 341–347, Jan. 1991. doi: 10.1149/1.2085575.

165. M. Bruel, "Silicon on insulator material technology," *Electron. Letts.*, vol. 31, no. 14, pp. 1201–1202, Jul. 1995. doi: 10.1049/el:19950805.

166. M. Bruel, B. Aspar, and A. J. Auberton-Hervé, "Smart-cut: A new silicon on insulator material technology based on hydrogen implantation and wafer bonding," *Japanese J. Appl. Physics, Part 1 Regul. Pap. Short Notes Rev. Pap.*, vol. 36, no. 3 SUPPL. B, pp. 1636–1641, Mar. 1997. doi: 10.1143/jjap.36.1636.

167. W. Kohn and L. J. Sham, "Self-consistent equations including exchange and correlation effects," *Phys. Rev.*, vol. 140, pp. A1133–A1138, 1965. doi: 10.1103/PhysRev.140.A1133.

168. J. M. Luttinger and W. Kohn, "Motion of electrons and holes in perturbed periodic fields," *Phys. Rev.*, vol. 97, no. 4, pp. 869–883, 1955. doi: 10.1103/PhysRev.97.869.

169. P. Hohenberg and W. Kohn, "Inhomogeneous electron gas," *Phys. Rev. B*, vol. 136, no. 3B, pp. 864–871, 1964. doi: 10.1103/PhysRev.136.B864.

170. J. C. Phillips, "Energy-band interpolation scheme based on a pseudopotential," *Phys. Rev.*, vol. 112, no. 3, pp. 685–695, Nov. 1958. doi: 10.1103/PhysRev.112.685.

171. J. Ihm *et al.*, "Momentum-space formalism for the total energy of solids," *J. Phys. C Solid State Phys.*, vol. 12, no. 21, p. 4409, 1979. doi: 10.1088/0022-3719/12/21/009.

172. M. C. Payne, M. P. Teter, D. C. Allan, T. A. Arias, and J. D. Joannopoulos, "Iterative minimization techniques for ab initio total-energy calculations: Molecular dynamics and conjugate gradients," *Rev. Mod. Phys.*, vol. 64, no. 4, pp. 1045–1097, Oct. 1992. doi: 10.1103/RevModPhys.64.1045.

173. P. J. H. Denteneer and W. van Haeringen, "The pseudopotential-density-functional method in momentum space: details and test cases," *J. Phys. C Solid State Phys.*, vol. 18, no. 21, p. 4127, 1985. doi: 10.1088/0022-3719/18/21/010.

174. P. Blaise *et al.*, "Nanoscale FET: How to make atomistic simulation versatile, predictive, and fast at 5nm node and below," in *2020 International Conference on Simulation of Semiconductor Processes and Devices (SISPAD)*, 2020, pp. 249–252. doi: 10.23919/SISPAD49475.2020.9241651.

Chapter 2

Fabless Intelligent Manufacturing

Arup Ratan Saha
Synopsys India, Hyderabad 500032, Andhra Pradesh, India
saha.arup2020@gmail.com

The modern era marks the beginning of the fourth industrial revolution (Industry 4.0), made possible by the integration of the computer network and the physical world (cyber-physical system), commonly known as the internet of things (IoT) [1–4]. Artificial intelligence (AI), big data, machine learning, and deep learning are the other emerging technologies constituting the majority of technological developments soon [5, 6]. Sensors are the cornerstone of all these technologies allowing the continuous transmission of data from the physical world to the computer network. Institute of Audiovisual and Telecommunications in Europe (IDATE) considers that there are around 15 billion connected objects at the moment against only 4 billion in 2010. With an annual growth of 17.5%, the number of connected objects in the world will exceed 36 billion by 2030. These data show an increased integration of electronics in our daily lives. Indeed, the purpose of this chapter is to lay a strong

Fabless Semiconductor Manufacturing: In the Era of Internet of Things
Edited by Chinmay K. Maiti

ISBN 978-981-4968-27-0 (Hardcover), 978-1-003-31497-4 (eBook)
www.jennystanford.com

foundation for a large and interdisciplinary discussion about the distinguishing characteristics, technologies, and enabling factors of smart semiconductor manufacturing in the research and industrial communities [7].

All the technologies for smartphones, automotive, virtual reality, and AI, which seemed close to science fiction a few years ago, are the result of the incredible technical, economic, and industrial boom that began in the second half of the twentieth century with microelectronics. The growth of this industry has increased in recent years with the arrival of connected objects [8] or even the ever-greater empowerment of cars. In the coming years, the demand in emerging economies should contribute significantly to this industry growth. The purpose of this chapter is to illustrate the magnificence of the fabless semiconductor ecosystem [9]. In a few decades, microelectronics has established itself as an important economic driver by the technological innovations that it brings to many fields: automotive, space, telephony, and information technology. The significant investments made by the industry are a financial important windfall and a source of direct and indirect employment for the regions hosting manufacturing units (see Figure 2.1). The arrival of the first circuits containing more than one billion transistors was shown in the 2000s. For example, A11 processor Bionic, which powers Apple's latest smartphones, has 4.3 billion transistors.

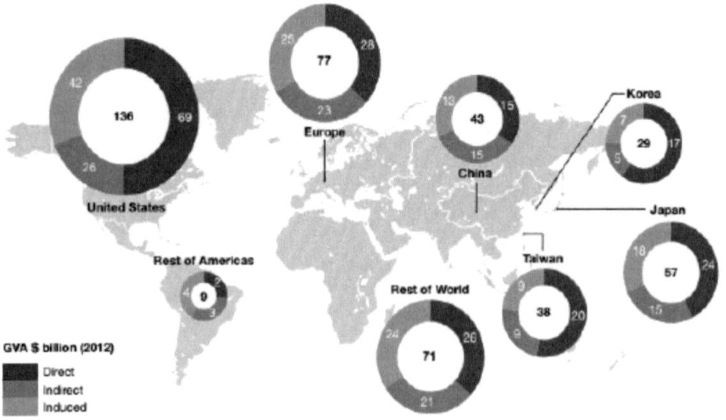

Figure 2.1 Influence of the semiconductor industry on GDP of different regions of the world [10].

The economics of designing a chip and getting it manufactured is becoming more and more complex. It takes teams of hundreds of design engineers to design an integrated circuit (IC), and a complex ecosystem of software, components, and services to make it happen. Getting an IC designed and the "fabs" that physically manufacture the ICs is incredibly expensive. Manufacturing has progressed, becoming more automated, computerized, and complex. However, IC-based electronics need to be cheap and pervasive, although they are not easy to make. For the last 60 years, the cost per transistor has decreased steadily. There are many reasons for this cost reduction, and the fabless semiconductor business model is among the most important of those reasons over the past three decades. Smart manufacturing is a new type of production that combines sensors, computing platforms, communication technology, control, simulation, data-intensive modeling, and predictive engineering with today's and tomorrow's manufacturing assets [11]. In the area of computer processors, AMD and Intel are now launching processors exceeding 10 billion transistors. The manufacture of chips containing a billion transistors has become a technological and economic challenge.

A revolution in manufacturing systems is underway, with significant recent investment going into the creation of smart manufacturing systems that can react in real-time to changes in consumer demands, as well as supply chain and factory conditions. Smart manufacturing is a key component of Industry 4.0, and it relies on the development of a bridge between digital and physical environments via the IoT technologies, as well as improvements to those digital environments via cloud systems, data analytics, and machine learning. Though semiconductor technology computer-aided design has been studied and improved for more than 30 years, it can complement or be used in combination with machine learning algorithms in the solution-finding process. In the IC industry, using machine intelligence to forecast systems and process output is a new approach. In IC technology, where cascaded and repeated process steps are normal, an adaptive weighting neural network model that combines the benefits of statistical machine learning and physical TCAD models may be useful [12].

The increasingly competitive market necessitates superior product performance (e.g., timing and power) as well as novel design approaches. The new proposed approach, which is based on process tuning, is a valuable resource in this regard. It enables even fabless companies to calculate the final product's sensitivity to process parameters and tailor them to the particular VLSI application, with the goal of performance optimization for large-scale productions. While interactions are inevitable in a real-world production environment, TCAD's benefits in clearly separating the impact of changes in each process parameter are obvious. To secure a new growth driver, the manufacturing sector is attempting to increase productivity by combining cutting-edge ICT technologies. Smart manufacturing, also known as Industry 4.0 and a new paradigm, is a series of cutting-edge technologies that enable efficient and precise engineering decision-making in real-time through the integration of various ICT technologies and the convergence of established manufacturing technologies [13].

From a technological and business standpoint, we first trace the history of the semiconductor industry. Since the mid-1980s, the development of the fabless business model has been a key enabler of semiconductor growth. We concentrate on the evolution of the electronics industry because, as much as technology, business models are keeping us excited about new gadgets year after year. We shall look at the industry trends that led to their success (both in terms of technology and business models) and how they are driving the semiconductor industry forward. We also talk about several phases of development in the semiconductor industry and use the commonly used following terms to describe the companies and technologies that define a particular business model.

An IC, also called a chip, is a set of electronic circuits, including transistors and other components, on a silicon substrate. A systems company makes a consumer product from chips that other companies have designed. Semiconductor manufacturing companies, also called integrated device manufacturers (IDMs), like Intel and Samsung, design and manufacture standard ICs that are used by systems companies in their products. They control both the design and manufacture of their chips. Since mid-1980, this trend has changed gradually, and now there are only a few (Intel and Samsung notably) IDMs. All other chip makers outsource the manufacturing of their designs to a foundry.

A business that is a dedicated semiconductor fabrication facility does not design its own ICs. The term "fab" refers to any semiconductor fabrication plant, whether run as part of an IDM (like Intel) or as a foundry (like TSMC). An application-specific IC refers to a chip that is custom-designed for a specific application, rather than for a general-purpose application. This type of company that developed in the 1980s performed the physical design and manufacturing of these application-specific ICs for other semiconductor or systems companies. An IC that integrates all components of a computer or other electronic system into a single chip is known as a system-on-chip (SOC). On a single substrate, it may contain digital, analog, mixed-signal, and sometimes radio-frequency functions. A company that designs its chip but outsources the manufacturing to a third party, either a foundry or an IDM is called the fabless company. This is the prevailing business model today. Electronic design automation (EDA) companies make the software that is used to design all modern semiconductor devices. The three dominant EDA companies today are Synopsys, Cadence Design Systems, and Mentor Graphics.

2.1 Semiconductor Industry

In this section, we discuss the history of the semiconductor industry, including the invention of the basic building block of all modern digital devices and transistors, the invention of the IC, and the businesses that developed around them. The processes used to make semiconductor devices must be precisely regulated down to the atomic level. Characterization is required for understanding and optimizing semiconductor devices, such as local morphological, chemical, and structural characterization to understand the origins of device behavior and to optimize manufacturing processes and device design. Electrical and optical characterization of finished devices is used to check their performance.

The rapid development of new technologies in recent years, high demands in international markets, and high profitability by multinational companies have led to very strong competition at the global level. Several avenues are being explored by companies to reduce production costs without impacting the quality of the product. Among the different tracks, one of the main factors is the controls during the manufacturing of electronic circuits. Indeed, the size of the electronic circuits or chips to be manufactured becomes so small (of the order of a nanometer) that several controls are necessary to ensure that manufacturing processes are performed correctly and that the product meets the customer's specifications. Hence, the challenge for companies to get better control, distribute, and limit this number of non-compulsory checks without increasing the risk (i.e., the potential loss) on production.

Various sampling techniques exist and have been developed by companies and in the literature to find the best compromise between the number of controls and the risk tolerated within the production. Between the static and dynamic techniques, dynamic sampling techniques are considered more robust due to their ability to integrate the dynamics of production and variability. The problem arises concerns about the industrialization of dynamic sampling techniques. They require investment (IT resources, training of operators and engineers, production system, etc.), and the complexities are such that most companies prefer to stay on static techniques while the inability of the latter to anticipate potential drifts for multi-product companies has already been demonstrated [14, 15].

The main activity of a semiconductor company is to manufacture electronic components, interconnect them, and thus obtain chips or circuits that are used in many areas of everyday life. The dimensions of the electronic components manufactured are so small that a special environment called a "clean room" is necessary to avoid any contamination and ensure the proper operation of electronic circuits. In this environment, operators, and engineers are covered from head to toe and the ambient air is renewed every 30 seconds. The manufacturing process of electronic chips or circuits can be summed up in two main stages [16]: front-end and back-end. The front-end brings together the manufacturing steps of the basic elements of the chip while the back-end that of the interconnections of the basic elements and the casing. Figure 2.2 shows the different levels involved in front-end processing.

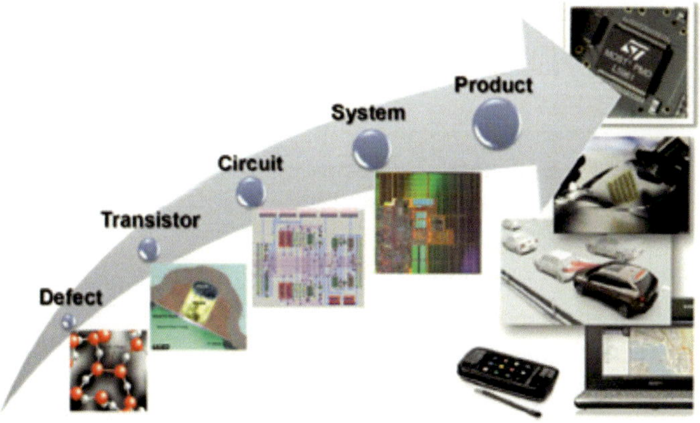

Figure 2.2 Different levels of front-end processing [17].

The main operations of the front-end [16] include oxidation, photoresist deposition, photolithography, development of photoresist and etching, and doping (see Figure 2.3). All these operations carried out during the front-end are repeated several times. Before, during, and after each manufacturing operation, one or more operations control is necessary to check that the process has been carried out and the customer specifications are met.

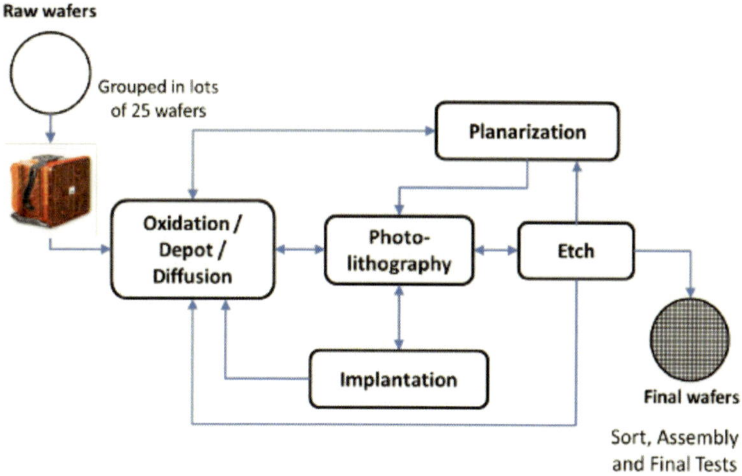

Figure 2.3 The different operations of front-end processing [17].

Several levels of controls exist during the manufacture of electronic components. For each level of control, several types of control exist. Among the main levels of control include:

(a) The parameters related to the environment are checked during production (ambient temperature, contamination, luminosity, liquids, gas, etc.) to guarantee the most optimal possible conditions within production.

(b) Sensors on the control equipment to check the state of the sensors placed on the various production equipment to ensure a correct rise of the data when analyzing the various parameters.

(c) On-line measurements or controls to check throughout the production, the wafers on which the electronic components are made. Several techniques are used (ellipsometry, scatterometry, etc.) giving rise to several types of controls. Most techniques are grouped under advanced process control (APC).

(d) The electrical parameters of the components are analyzed (leakage current, breakdown voltage, threshold voltage of transistors, etc.).

(e) Final or functional tests take place at the end of the front-end. It checks that the circuits produced are functioning correctly before they are packaged (back-end) and sent to the different customers.

(f) Physical characterization of the lifespan of the various components is evaluated under external conditions (temperature, humidity, corrosion, etc.).

The other level of complexity comes from the number of products (more than 200 different products are produced in parallel), technologies (more than 20 technologies), or priorities to be considered (product criticality, customer requirements, delivery times, production costs, the environment, management, etc.) when selecting wafers or batches to control. This constitutes a real challenge for sampling. Indeed, integrating all parameters is quite simply impossible and the major stake, therefore, lies in the ability to use the control capacity as optimally as possible available.

(a) A 100% control is not possible because of the cost that this generates on the final product [18].

(b) A 100% inspection can never guarantee 100% quality in the semiconductor products because of particle size, faults, and reliability of the different control methods. It is, therefore, essential to limit the number of checks during any manufacturing while making sure to do the right controls at the right time.

Many types or methods of sampling exist depending on the objectives sought. The three main techniques are [19]:

(a) **Control of material at risk and management of excursions** [20]: The goal is to select batches to be checked according to a good frequency defined to on the one hand limits the potential loss in the event of a problem, and the other side to be able to detect the various faults as quickly as possible generated during production.
(b) **Integration of new processes and improvement of efficiency** [21]: The goal is to adjust the percentage of batches to be checked to better identify the main detractors for the various technologies and eliminate them as and when. In small factories, we try to adjust the number of lots launched at the start of production to compensate for any losses.
(c) **Statistics and learning** [22]: The aim is to learn about the different types of faults detected as well as their mechanisms.

The objective is twofold viz., limit the number of checks (by sampling) without increasing the material to risk in the event of a problem, and detect the various problems very quickly. There is a necessary compromise to find because, if on the one hand, we focus on the reduction of several checks (which makes it possible to reduce the final cost of the product), we take the risk of not quickly detecting the various faults and therefore of having losses significant in the event of a problem. On the other hand, if we focus only on the rapid detection of faults, there is a risk of increasing the number of checks and rising cost of the end product. Several sampling techniques have been developed in the literature and are classified into three main groups: static, adaptive, and dynamic.

The microelectronics industry has evolved rapidly since the manufacture of the first transistor in 1947 by Bell Telephone Laboratories. Today, this industry is one of the main engines of the world economy. To stay competitive, it must be innovative at all levels. Thus, specialized companies in semiconductor manufacturing are constantly exploring new avenues to reduce manufacturing costs while improving product quality to stay profitable. An IC is mainly composed of chips that contain the elementary electronic components (transistors) made on a wafer circular silicon wafer (see Figure 2.4). The IC production process requires two major steps: (a) the front-end that consists in manufacturing the basic elements of the chip and in testing its functionality, and (b) the back-end that brings together the interconnection stages on the one hand between the chip and the case, and on the other hand the final test of the IC (see Figure 2.5). To reduce costs production, the effort of companies has focused mainly on optimizing the process of front-end, which is the most complex (see Figure 2.3), the longest, and the most expensive step.

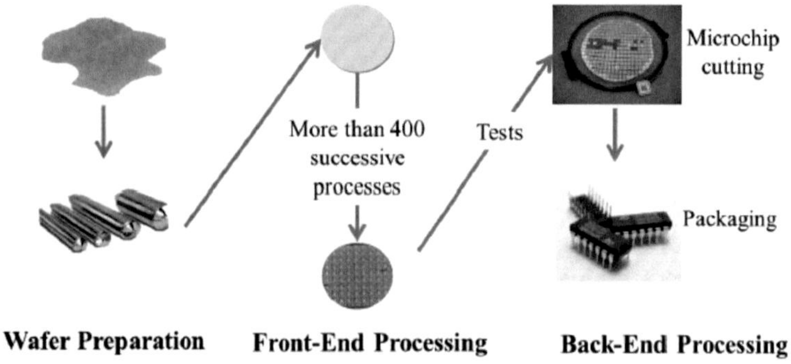

Figure 2.4 Manufacture of ICs [24].

The semiconductor manufacturing process is one of the most complex these days [23]. Silicon is the basic building block in the manufacture of microelectronic components (see Figure 2.4). In particular, it has a conductivity intermediate between those of metals and that of insulators. For chip, manufacturing silicon is crystallized and conditioned to obtain cylindrical bars often 200–300 mm in diameter. These bars are then cut into slices about a millimeter thick and polished. ICs are made on this surface by performing several hundred operating steps. These manufacturing steps are grouped into two main parts: the front-end and the back-end. The main process stages in front-end processing are depicted in Figure 2.3. The procedure is re-entrant rather than linear. A lot of it gets back to the various manufacturing steps, with some process operations being repeated layer after layer on the wafer.

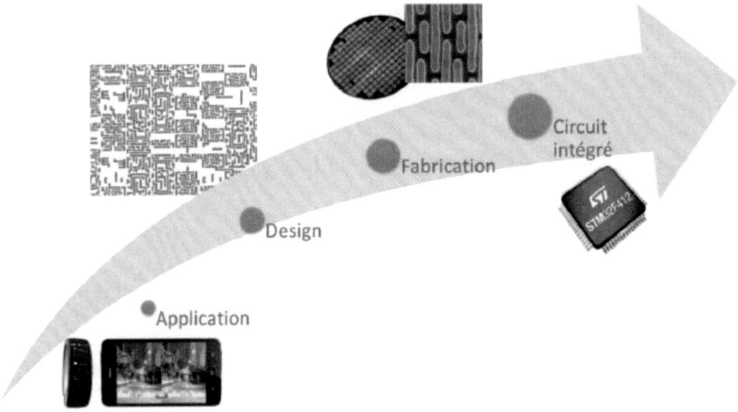

Figure 2.5 The development cycle of an IC [26].

In this step, the wafers are grouped in a container with a maximum capacity of 25 wafers, called a "lot". The evolution of the diameter of the wafers from 200 mm to 300 mm has made it possible to manufacture 50% more chips per wafer. This progress gave birth to a new generation semiconductor manufacturing unit (called a 300 mm fab). The modern unit contains more sophisticated and automated production equipment. The increase in the diameter of the wafer also required the automation of batch handling operations because of its weight. The 200 mm wafer fabs continue to operate at a level low automation and practically manual handling (operators). This explains by the fact that this type of manufacturing unit was initially designed and optimized for mass production. Thus, the evolution in the semiconductor industry from high volume to low volume production and high diversity is a challenge for wafer fabs 200 mm, which lack flexibility [25].

The arrival of the new generation of 300 mm fab wafers has significantly reduced the number of old units to 200 mm. According to market research, 300 mm inserts represented 60% of total sales in 2016. However, recently, due to large investments made by China, the production of 200 mm inserts has been on the rise. Maintaining the competitiveness of 200 mm manufacturing units nevertheless requires modernization of these manufacturing units to face the competition of 300 mm wafer fabs. Thus, several solutions have been adopted by companies to improve their manufacturing processes and increase flexibility such as automation, optimization of planning, and continuous improvement (lean manufacturing). An automated material handling system (AMHS) is considered an efficient solution to gain flexibility and productivity in a manufacturing unit [13].

Front-end processing is the most complex, time-consuming, and costly step in the manufacture of ICs. This process is further divided into two major sub-steps: front-end of line (FEOL) and back-end of line (BEOL). The front-end consists of applying several treatments on a silicon wafer, to manufacture millions of electronic components (FEOL) and then interconnect them (BEOL). The most common elementary operations in the majority of technologies are given below. These steps are repeated several times depending on the technology (complexity) of the final product. Despite the technological evolution of the front-end process, the production environment is still a challenge for several reasons [27].

(a) **The diversity of products**: To make their activities profitable, semiconductor manufacturers are obliged to have a wide range of products to adapt to rapid technological advances in microelectronics.

(b) **The reentrant nature of production flows**: The operating range for manufacturing a product requires returning the wafers several times to the same operation. This leads to a reentrant workflow. Thus, many efforts are oriented toward the optimization of production planning and scheduling.

(c) **A large number of production steps**: The manufacture of a product may require more than 700 operations with very variable processing times (from a few minutes to more than 8 hours). Thus, the cycle time of a product ranges from 4 to 12 weeks.

(d) **The great need for handling**: Transport, storage, and loading/unloading of products during their production are repetitive operations and thus need high attention during handling.

The characteristics presented above form, among other things, strong internal constraints to the front-end manufacturing process. In addition, external factors increase the level of complexity. In fact, on the one hand, great flexibility is required to follow rapid development in the design of new products and adapt to the needs of customers [28]. On the other hand, semiconductor manufacturers are faced with the uncertainty of the market and fluctuating demand [29]. So in this very complex environment and uncertain, the main objective is to optimize the use of production equipment, which is an extremely expensive resource.

During the front-end process, the wafers are grouped in capacity containers maximum of 25 wafers, called "lot". The lots undergo a sequence of several processes (photolithography, chemical mechanical planarization (CMP), and implantation) in the fab wafer, which is broken down into several production workshops.

Figure 2.3 shows schematically the re-entering production flow (the fact of passing several times by the same workshop or the same machine) in a wafer fab. Each production workshop is made up of one or more groups of equipment or bays specialized in well-defined operations. Metrology equipment is also provided in the workshops to carry out several types of controls by measurements. Depending on the diameter of the wafer, there are mainly two types of semiconductor manufacturing units: wafer fabs 200 mm and wafer fabs 300 mm. The transition of the wafer diameter from 200 mm to 300 mm required the automation of handling operations due to the increase in the weight of the lots [30]. However, the 200 mm manufacturing units often remain manual or semi-automatic.

Figure 2.3 also shows that the photolithography step is critical since it is repeated several times. Thus, the photolithography workshop is often the bottleneck of the fab wafer [3]. This is due to three main reasons:

(a) **The reentrance of production flows**: To have the final product (the chip), a wafer can return several times (up to 40 times) to the photolithography workshop. This leads to a high level of work-in-progress (WIP) in the workshop.

(b) **The limited number of production equipment due to high costs**: In particular, high-tech, high-speed equipment, such as steppers, is extremely expensive. Moreover, these systems are often not able to realize all types of treatments. Thus, we are talking about the notion of qualifying a machine for processing a product.

(c) **The use of a reticle as an auxiliary resource** [31]: The reticles are very expensive and often unique. Thus, their management is very complex not only in terms of planning but also in terms of handling. In particular, in the presence of a manual handling system, transport, and storage of a large number of reticles (sometimes more than 5,000 reticles) is a problem that impacts directly the performance of the production process in the photolithography workshop.

Photolithography workshops are often equipped with a production module (called track and stepper), which allows three sub-steps to be linked: depositing the resin, exposure to the light ray using the reticle, and finally the development.

To summarize, the front-end allows the chips to be on the surface of the wafer silicon through a very complex manufacturing process. Then electrical tests (electrical wafer sort [EWS]) are performed to identify faulty chips. Once past the EWS, the pads are sent to the back-end phase.

This step allows the transition from a single chip on the surface of a wafer to a ready-to-use IC use. The stages of the back-end are as follows:

(a) **Cutting**: This involves separating the chips by a high-precision sawing operation.
(b) **Gluing or welding**: The good chips identified by the EWS are welded to a support metallic (grid).
(c) **Wiring**: This step consists of establishing the necessary connections between the chip and the grid for the transmission of electrical signals.
(d) **Encapsulation**: The whole (grid and chips) is encapsulated in an appropriate box to protect it from the external environment.
(e) **Test and marking**: The circuit is tested again, then a marking is made on the case that contains the necessary information (manufacturer's logo and functionality).

2.2 Performance of Manufacturing Process

To ensure the quality of finished products, many types of controls are needed in semiconductor manufacturing. This study focuses on defect inspections, which are used to monitor the process to reduce defects and increase yield. During fabrication, we are interested in controlling and reducing the risk on process tools (i.e., the number of wafers at risk). Inspection operations are conducted on goods to reduce this risk. Since inspection operations have a direct effect on product cycle times, sampling techniques are used to minimize the number of inspected lots while maintaining quality. There are many sampling techniques available, each of which can be categorized based on their ability to deal with factory dynamics. Dynamic sampling techniques, in which lots to inspect are chosen in real-time while taking current production risk into account, have recently been suggested. These methods are much more effective than previous ones, but they are more difficult to plan and execute. A novel approach to selecting the lots to inspect is suggested in this study. Several algorithms have been proposed and validated to efficiently handle defect inspection queues by skipping (i.e., releasing) lots that no longer provide sufficient details. An optimization model for defect inspection capability preparation is also proposed to help strategic and tactical decisions. As the production conditions change, this model calculates the necessary defect inspection ability to ensure the risk limits on process tools are maintained. Without increasing defect inspection capability, industrial results show substantial improvements in terms of risk reduction.

2.3 Cycle Time

This is the total time to make a wafer. It reflects the difference between the date of start of the batch in production in the fab wafer and its release date. Cycle time includes time for operations with and without value addition [32]. Thus, this is a theoretical cycle time that contains the times of processing of production equipment and times of essential inspection operations [33]. Transport, storage, non-quality, and waiting for batches are operations without added values, which considerably degrade the cycle time. The reduction in cycle time allows the launch of new products in the market, to control the level of work in progress as well as to improve the output of the unit manufacturing. To reduce cycle time, the effort is usually directed toward worthless activities added that represent between 50% and 90% of the total cycle time.

2.4 Throughput

This is the number of finished products (wafers) leaving a production workshop or the fab wafer per unit of time. The throughput reflects the level of saturation of the unit of workmanship and the stability of the production process. Indeed, when the launch rate (or start) of the wafers is close to the production rate, the system is stable and generates no lots piling up in front of the machines. However, this hypothesis is far from being realistic because of the variability in the semiconductor manufacturing process, due to breakdowns of machines, mixed products, and processing time. This variability leads necessarily to batch wait times.

2.5 Handling Systems in Wafer Fabs

The many manufacturing steps (more than 600 steps depending on the technology) of a chip are produced on families of specialized machines, which are located in different workshops of the fab wafer. We are talking about a route to follow for each batch according to the product it contains. By definition, the handling system manages the batches between two processing steps on the machines. It allows the physical movement of the batch (transport), its temporary storage (storage), and loading/unloading on wearing equipment. Thus, in general, handling is a process without added value and is only a production service. Given the complexity of the manufacturing process, special attention is paid to the transport and storage of materials in the wafers fabs.

Semiconductor fabrication is one of the most complex manufacturing systems, and multi-cluster tools are a brand new kind of multi-loop coupling structured and automated 300 mm wafer fabrication system. Multi-cluster tools differ from other manufacturing systems in that they have strict time limitations, severe resource conflicts, are expensive, and so on. Solving the multi-cluster tool scheduling problem effectively entails not only the advancement of scheduling theory but also the enhancement of wafer fabrication system production efficiency. Fab costs have risen dramatically in the last decade, and technology offers only marginal benefits, which may not be enough to justify the high financial risk of owning a fab. As fabless technology becomes more common, supply-chain collaboration and innovation are required to lower IC costs and accelerate the semiconductor industry's growth. The fabless/foundry business model is the preferred model for businesses to succeed in the semiconductor industry, with the most flexibility to address dynamic markets, according to an analysis.

For complex technology, capital-intensive, and high value-added features, wafer fabrication is at the heart of the IC product manufacturing process. Because the wafer fabrication system is the most complicated and expensive part of the semiconductor manufacturing process, its scheduling has a huge impact on economic efficiency. In general, every 1% reduction in cycle time will result in tens of millions of dollars in additional annual revenue. The standard of production management, on the other hand, does not correspond to the industrial scale. Low production efficiency, low equipment utilization, and low product yield have often hampered the long-term growth of businesses as scheduling technology has lagged. As a result, advanced scheduling theory is required to guide production.

Combination equipment, also known as a cluster tool, has been extensively used in the 200 mm wafer fab for the past two decades [34, 35]. The cluster tool integrates several processing modules and material handling systems to create a versatile and efficient wafer fabrication environment [36]. The multi-cluster tools, the new integrated, automated, and multi-loop coupling structured wafer fabrication systems, have emerged in recent years. These are typically made up of two or more single cluster tools that are linked together using buffer modules. Several robot transport modules are included in a multi-cluster tool. The multi-cluster tool is fitted with a cassette module to isolate the internal vacuum environment from the outside to meet the standards for cleanliness in wafer fabrication.

In comparison to the single cluster tool, the multi-cluster tool combines the previous loosely coupled discrete wafer processing flow into a direct correlated and tightly coupled discrete processing flow, achieving the goal of increasing the degree of automation and cleanliness of the wafer fabrication system. Furthermore, the multi-cluster tool will integrate the necessary device modules according to the wafer fabrication process's requirements. As a result, the versatility of a multi-cluster tool outweighs the flexibility of a single cluster tool. Currently, the multi-cluster tool is primarily used in the fabrication of 300 mm wafers [29, 37].

2.6 Intelligent Manufacturing

Our contemporary societies are shaped by the massive deployment of information technology. The deployment of the broadband internet, wireless communications networks, and the proliferation of multimedia terminals change significantly the modes of communication, entertainment, consumption, and ways of informing, socializing, or organizing ourselves, up to our relation to knowledge. The generalization of information technologies is intrinsically linked to the constant progress of microelectronics by the nature of the devices on which they are based. Since its invention in the middle of the last century, microelectronics has continued to supply ever more efficient, integrated into ever more efficient microprocessors. Until recently, this progress primarily followed Moore's law. However, Moore's law is now reaching its limits [38]. Reducing the size of the transistors, essential components of microelectronics, has become so important that it now comes up against the physical limits of a matter.

With the announced end of Moore's law, players in the field of microelectronics are looking to new paradigms on which to rely to continue the march of progress. This movement, called "More-than-Moore", was first initiated by the modification of the architectures of the systems with the emergence of multicore processors in the mid-2000s. The parallelization of tasks that this type of architecture allows a performance gain without increasing the frequency, bypassing the self-heating issues that were starting to appear. This reflection on the organization of architectures continues today with the design of systems based on the co-integration and interconnection of specialized chips dedicated to certain functions, especially in the context of 3D integration [39], which is booming today. However, among the "More-than-Moore" strategies currently in development, two of them are particularly disruptive strategies: quantum computing and neuromorphic. The first is to rely on the quantum properties of matter to affect certain types of calculations with an unprecedented performance gain. The second is to build systems that are inspired by the way information is processed in the biological nervous systems and have the particularity of having to be trained in the task to which they are dedicated. These two approaches do not have the same maturity. Quantum computing is still in its infancy, although the first computer-based on this principle was proposed by IBM in 2019.

Neuromorphic systems rely on the execution of algorithms that implement mathematical models of neural networks. Historically, these artificial neural network (ANN) algorithms were executed on standard processors at von Neumann-type architecture. Even today, the best performing ANNs are carried out in this way. However, recent years have seen the emergence of processors whose architectures themselves are modeled on that of the nervous systems-based neural network. For its part, neuromorphism is the basis of many services already available to the general public, such as voice assistants, human language manipulators [40], real-time computer vision, and recognition of content in photographs (OpenCV) [41]. It is by elsewhere at the origin of recent advances in AI have seen an explosion in the number of application outlets. Advanced processors are composed of a multitude of copies of interconnected elementary circuits, which reproduce the behavior of a neuron or a synapse and communicate with each other through the emission of short electrical impulses. This impulse and asynchronous mode of communication drastically reduce the power consumption of these neural processors. Neuromorphism focuses on the study and development of analog pulse neuron circuits with a reduction of the circuit footprint area as a priority. The main motivation for studying and developing neuromorphic systems lies in the energy efficiency of the brain (estimated at 10^{-16} J/operation) compared to that of conventional ICs (10^{-9} J/operation). The architecture of neuromorphic systems differs from von Neumann-type architectures [42] on which the classical calculation systems are based. In particular, the memory units (synapses) are distributed among the computational units (neuron and synapse) within a network, unlike von Neumann architectures where the two types of units are grouped into two separate sets.

The modern era marks the beginning of Industry 4.0, made possible by the integration of the computer network and the physical world. Intelligent manufacturing refers to the use of advanced data analytics in conjunction with physical science to improve device efficiency and decision-making. With the widespread use of sensors and the IoT, there is a growing need for large manufacturing data with high volume, high velocity, and high variety [43]. AI, big data, machine learning, and deep learning are the other emerging technologies constituting the majority of technological developments shortly. Sensors are the cornerstone of all these technologies allowing the continuous transmission of data from the physical world to the computer network. In the following, we provide the reader with some background and a description of the contributions of the state-of-the-art systems that allow one to understand the challenges. After an introduction to the concept of neuromorphism, ANNs are presented through three elements that constitute them: neural models, network topologies, and rules learning. The concept of neuromorphism revolves around the description of the nervous system and biological neurons. The nervous system is a biological system whose role is to regulate the functioning of vital organs and coordinate the physical actions to be performed according to the configuration of the environment as per the needs of the individual. The nervous system is made up of a network of nerve cells, called neurons, and interconnected through terminations, called synapses. The neuron is made up of a body cell, or soma, which contains the nucleus and the connected extensions: dendrites and axons. The axon is subdivided at its end into axonal endings. The neuron is separated from its external environment by its membrane. Axonal endings are connected to dendrites of other neurons through synapses. From a functional point of view, the main role of neurons is to transmit and process information. When a neuron activates, it emits an electrical impulse (the action potential), which propagates along the axon to the axonal endings. The arrival of the action potential at the endings axonal has the effect of releasing neurotransmitters.

Neuromorphic engineering is a concept introduced by professor Carver Mead of Caltech (California, United States) in the late 1980s [38]. Originally, it consists of the use of very-large-scale integration to build information processing systems inspired by the functioning of networks of biological neurons. Nowadays, it designates any analog, digital, or mixed electronic system, or any computer system that implements models of neurons and synapses networked with each other. To develop this concept, Carver Mead relied on the great similarity between characteristic curves of the ionic currents, which circulate through the membrane of neurons as a function of the voltage through it, and similar to the saturation current of the transistors as a function of the gate voltage [39]. In 1958, Franck Rosenblatt introduced the first calculator architecture inspired by networks of biological neurons called "perceptron" [44] based on the model of the formal neuron [45]. The generalization of the perceptron gives rise to the multi-layered perceptron and, more generally, to neuromorphic systems, the perceptron must first be trained to perform a task: internal parameters (synaptic weights and activation thresholds of neurons) must be adjusted during a learning phase. This learning can be supervised or unsupervised, depending on whether the adjustment of the parameters is guided from outside or performed automatically based on network activity. However, the design of the perceptron relies on the work of the psychologist Donald O. Hebb who enunciated in 1949 a general principle on neuronal plasticity, i.e., the way whose synaptic connections change according to the activity of neurons over time [46]. It can be summed up by the fact that neurons that are triggered together bind together [47]. It is the first example of a non-supervised learning rule.

Since the introduction of neuromorphism by Carver Mead, many circuit designs of neurons have been proposed in the literature. Schematically, two different approaches are exploited for their design, one bio-realist and the other simplifying. The bio-realistic approach tends to reproduce the diversity and complexity of the behavior of neurons biological as closely as possible, in particular by modeling the dynamics of ion channels proposed by Hodgkin and Huxley (H-H model) [48]. It leads to circuits of a certain complexity that requires the application of numerous control voltages [49–51] and which suffer from low integration density. It is of great interest for simulation execution in neurobiology.

The simplifying approach seeks to reduce the circuit to a primary function, generally integrate and fire (IF). Circuits designed in this way are not able to reproduce the diversity of behaviors of biological neurons but have a certain simplicity and can be integrated on a larger scale. An example characteristic of this type of circuit is the "Axon-Hillock circuit", originally proposed by Carver Mead, implemented in artificial retinal circuits in particular [52, 53], or neural processors [52], and recently optimized to achieve very low power consumption

[54]. The circuits discussed so far use classic design methods that incorporate standard components (p- and n-MOSFETs and capacitors). These methods exploit the correspondences between the equations, which describe the current-voltage behaviors of components and equations of neural models to be implemented [55–59]. Another approach exploits the specific less conventional physio-electronic mechanisms. It is similar to that which led to the use of resistive memories for the implementation of synapses. The ambition of this strategy is to minimize the size of the neuron circuit. More recently, the electrostatic effect of temporary storage of holes generated by impact ionization in silicon components has been exploited for the manufacture of "neural components", in an NPN structure on optimized SOI [60], in a PDSOI MOSFET [61, 62], and a NIPIN structure [63].

2.7 AI in Smart Manufacturing and Design

Cybernetics boomed as a research field from the 1930s until the 1960s and was driven by an impressive mix of new ideas and results from neuromorphism, control theory, and information theory. Research in neurology had shown that the brain was an electrical network of neurons. This section describes the background and requirements of the next semiconductor design for automation-enabled chips to accelerate system and device performance using neuromorphism technique. With data-driven analytics becoming mainstream, the global demand for dedicated AI and deep learning accelerator chips is rising. These accelerators, designed with densely packed processing elements, are especially vulnerable to manufacturing defects, which are common in the advanced semiconductor process nodes; resulting in significant yield loss. The demand for AI and deep learning is growing at a rapid pace across a wide range of applications, such as self-driving vehicles, image and voice recognition, medical imaging and diagnosis, finance and banking, natural resource exploration, and defense operation. Because of these data-driven analytics and AI growth, demands of AI will also emerge at both the IoT and cloud-computing environments [64–66].

In recent market research, it has been reported that AI-related semiconductors will see a growth of about 18% annually, and by 2025, AI-related semiconductors could account for almost 20% of all semiconductor demand, which would translate into about $67 billion in revenue [67]. Robotics, which is no longer just a look into the future but has been developing for many years, is also happening now rapidly. It will continue to emerge as a part of human life for the unanticipated future. Robots are no longer just a futuristic topic as they were in the early history of AI, they are now a reality and becoming a greater part of everyday life. Advances in robotics are integrally tied to advances in AI.

A conventional definition of intelligence is the ability to develop and utilize existing knowledge from experience and adapt to a new problem-solving environment. In 1956, a watershed conference was hosted by John McCarthy at Dartmouth College (attended by Minsky, Shannon, Herbert Simon, Allen Newell, and others), where people first defined the term AI as we know it today, can be expressed as a superset of human-machine intelligence. Although multiple definitions exist for AI, the most practical one might be like: AI is a program that can sense, act with reasoning, and auto-adapt within a problem environment. AI is a part of computer science that is concerned with designing intelligent computing systems and can exhibit physical characteristics with far better brainpower compared to a human. Moreover, AI provides accurate results with few errors and defects. It can work in complex situations where humans may struggle. Machine learning and deep learnings are a usage-based subset of AI; depending on data handling and training capabilities as shown in Figure 2.6.

The concept of AI has been implemented in the following fields like problem-solving, gaming, natural language understanding, generation and vision system, speech recognition, and advanced robotics. That is why AI produces so much hype in semiconductor manufacturing starting from IoT devices, smart home automation devices, as well as smart auto-driven cars. To scale and go beyond More-than-Moore, smart semiconductor manufacturability is a must. In this context, the reliability study and reverse engineering using TCAD simulation can be driven by AI due to its superiority over human capabilities. Beyond 3–5 nm technology node, yield enhancement and manufacturing with limited DOE time-frame requires AI-driven TCAD engineering [68]. Several applications of AI already exist in the development cycle, e.g.:

- Neural network design for failure analysis on wafer manufacturing,
- Augmented machine learning for reverse engineering, i.e., equipment analysis and failure prediction on the device level, and
- Study of defects in wafer and SEM images using computer vision and deep learning (CNN), etc.

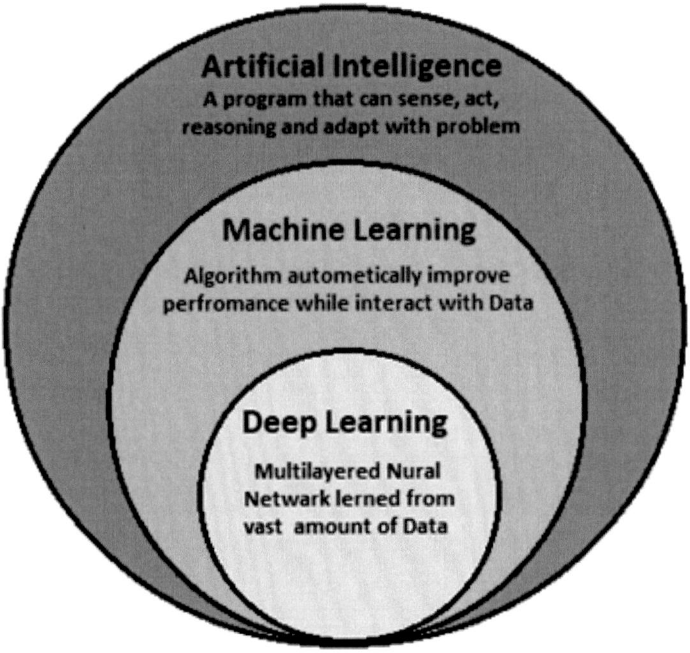

Figure 2.6 The AI paradigm.

It may be noted that there is a tremendous potential of intelligent semiconductor manufacturing for future growth. The smart manufacturing analysis also requires better computing resources to handle such a big amount of data along with futuristic ever-growing data processing capabilities. It is quite apparent that in the future, industry leaders will take help from AI and machine learning-enabled TCAD platforms to build their AI-CHIPs and AI-ICs [69].

2.8 Summary

In this chapter, we have presented the basics of semiconductor manufacturing. First, we described the semiconductor industry, in particular the process of chip manufacturing with an emphasis on the front-end process. This allows identifying the complexity of the production phases. We also stressed the importance of the transport and storage of batches and characterized the different handling systems most commonly used in wafer fabs. We have provided a general description of the old 200 mm wafer fab. The industrial peculiarity affects the current handling system in manufacturing. This is manual transport and storage carried out by operators based on their experiences and very basic rules.

A revolution in manufacturing systems is underway, with significant recent investment going into the creation of smart manufacturing systems that can respond in real-time to changes in consumer demands, as well as supply chain and factory conditions. It is possible that the smart factory (Industry 4.0) can be applied using evolving information technologies, such as IoT, big data, and cloud computing, and AI. Smart manufacturing is a key component of Industry 4.0, and it relies on the development of a bridge between digital and physical environments via the IoT technologies, as well as improvements to those digital environments via cloud systems, data analytics, and machine learning. Smart manufacturing would be the cornerstone of the next industrial revolution if these ideas and innovations were applied.

Neuromorphic computing is an advanced area of study that aims to find new solutions and applications in everyday life. The long evolution of living organisms has given rise to an information processing system based on the networking of neuronal cells through synapses, all of which constitute the nervous system. From the beginnings of computing in the middle of the last century, and thanks to recent discoveries in neurobiology, this method of processing information has guided the manufacture of the first neuro-inspired computers, whose operating principle differs fundamentally that of standard calculators, in particular, by the fact that these calculators must be trained to function. Neuromorphism is one of the new "More-than-Moore" paradigms on which the semiconductor manufacturers now plan to continue for technological progress while Moore's law reaches its limits. Neural processors currently under development are the privileged support due to the energy efficiency that they offer by mimicking the mode of impulse communication of biological nervous systems. They are based on the massive integration of elementary circuits of neurons and interconnected synapses.

References

1. "Industry 4.0: Definition, design principles, challenges, and the future of employment | cleverism," https://www.cleverism.com/industry-4-0/ (accessed Mar. 09, 2021).
2. F. Shrouf, J. Ordieres, and G. Miragliotta, "Smart factories in Industry 4.0: A review of the concept and of energy management approached in production based on the internet of things paradigm," in *IEEE International Conference on Industrial Engineering and Engineering Management*, 2014, vol. 2015-Jan, pp. 697–701. doi: 10.1109/IEEM.2014.7058728.
3. J. Wan *et al.*, "Software-defined industrial internet of things in the context of industry 4.0," *IEEE Sens. J.*, vol. 16, no. 20, pp. 7373–7380, Oct. 2016. doi: 10.1109/JSEN.2016.2565621.
4. M. Wollschlaeger, T. Sauter, and J. Jasperneite, "The future of industrial communication: Automation networks in the era of the internet of things and industry 4.0," *IEEE Ind. Electron. Mag.*, vol. 11, no. 1, pp. 17–27, Mar. 2017. doi: 10.1109/MIE.2017.2649104.
5. D. E. O'Leary, "Artificial intelligence and big data," *IEEE Intell. Syst.*, vol. 28, no. 2, pp. 96–99, Mar. 2013. doi: 10.1109/MIS.2013.39.
6. X. Wu, X. Zhu, G. Q. Wu, and W. Ding, "Data mining with big data," *IEEE Trans. Knowl. Data Eng.*, vol. 26, no. 1, pp. 97–107, Jan. 2014. doi: 10.1109/TKDE.2013.109.
7. S. Mittal, M. A. Khan, D. Romero, and T. Wuest, "Smart manufacturing: Characteristics, technologies and enabling factors," *Proc. Inst. Mech. Eng. Part B J. Eng. Manuf.*, vol. 233, no. 5, pp. 1342–1361, Apr. 2019. doi: 10.1177/0954405417736547.
8. J. Yun, I.-Y. Ahn, S.-C. Choi, and J. Kim, "TTEO (things talk to each other): Programming smart spaces based on IoT systems," *Sensors*, vol. 16, no. 4, 2016. doi: 10.3390/s16040467.
9. S. W. Yoon, B. Petrov, and K. Liu, "Advanced wafer level technology: Enabling innovations in mobile, IoT and wearable electronics," in *2015*

IEEE 17th Electronics Packaging and Technology Conference (EPTC), 2015, pp. 1–5. doi: 10.1109/EPTC.2015.7412320.

10. A. Lakcher, "Nouvelles perspectives de métrologie dimensionnelle par imagerie de microscope électronique pour le contrôle de la variabilité des procédés de fabrication des circuits intégrés," PhD Thesis, Université Grenoble Alpes, 2018.
11. A. Kusiak, "Smart manufacturing," *Int. J. Prod. Res.*, vol. 56, no. 1–2, pp. 508–517, Jan. 2018. doi: 10.1080/00207543.2017.1351644.
12. C. Y. Huang, S. M. Fu, P. Parashar, C. H. Chen, C. Akbar, and A. S. Lin, "Intelligent manufacturing: TCAD-assisted adaptive weighting neural networks," *IEEE Access*, vol. 6, pp. 78402–78413, 2018. doi: 10.1109/ACCESS.2018.2885024.
13. H. S. Kang *et al.*, "Smart manufacturing: Past research, present findings, and future directions," *Int. J. Precis. Eng. Manuf. - Green Technol.*, vol. 3, no. 1, pp. 111–128, Jan. 2016. doi: 10.1007/s40684-016-0015-5.
14. A. Bousetta and A. J. Cross, "Adaptive sampling methodology for in-line defect inspection," in *ASMC (Advanced Semiconductor Manufacturing Conference) Proceedings*, 2005, pp. 25–31. doi: 10.1109/asmc.2005.1438762.
15. M. Purdy, "Dynamic, weight-based sampling algorithm," in *IEEE International Symposium on Semiconductor Manufacturing Conference Proceedings*, 2007, pp. 211–214. doi: 10.1109/ISSM.2007.4446804.
16. M. Pfeffer, H. Richter, R. Altmann, A. Leibold, and A. Bauer, "Enhanced contamination control methods in advanced wafer processing," in *2016 International Symposium on Semiconductor Manufacturing (ISSM)*, 2016, pp. 1–5. doi: 10.1109/ISSM.2016.7934523.
17. A. Benhassain, "Moniteurs de Vieillissement in-situ: Méthodologie d'intégration et application à la gestion dynamique de la fiabilité," PhD Thesis, Universite Grenoble Alpes, 2016.
18. L. Pesotchinsky, "Problems associated with quality control sampling in modern IC manufacturing," *IEEE Trans. Components, Hybrids, Manuf. Technol.*, vol. 10, no. 1, pp. 107–110, 1987. doi: 10.1109/TCHMT.1987.1134700.
19. M. Sahnoun *et al.*, "Optimizing return on inspection trough defectivity smart sampling," *EEE Int. Symp. Semicond. Manuf. ing*, pp. 1–4, 2010.
20. J. N. Munga *et al.*, "Optimized management of excursions in semiconductor manufacturing," *Optim. Manag. excursions Semicond. Manuf.*, pp. 2105–2112, 2011.

21. J. C. Le Denmat *et al.*, "Tracking of design related defects hidden in the random defectivity in a production environment," in *ASMC (Advanced Semiconductor Manufacturing Conference) Proceedings*, 2009, pp. 5–13. doi: 10.1109/ASMC.2009.5155943.
22. A. Ferreira, A. Roussy, and L. Conde, "Virtual metrology models for predicting physical measurement in semiconductor manufacturing," in *ASMC (Advanced Semiconductor Manufacturing Conference) Proceedings*, 2009, pp. 149–154. doi: 10.1109/ASMC.2009.5155973.
23. L. Mönch, J. W. Fowler, S. Dauzère-Pérès, S. J. Mason, and O. Rose, "A survey of problems, solution techniques, and future challenges in scheduling semiconductor manufacturing operations," *J. Sched.*, vol. 14, no. 6, pp. 583–599, Dec. 2011. doi: 10.1007/s10951-010-0222-9.
24. G. L. R. Verjan, "Smart sampling for risk reduction in semiconductor manufacturing," PhD Thesis, Ecole Nationale Supérieure des Mines, 2014.
25. H. Heinrich, G. Schneider, F. Heinlein, S. Keil, A. Deutschländer, and R. Lasch, "Pursuing the increase of factory automation in 200mm frontend manufacturing to manage the changes imposed by the transition from high-volume low-mix to high-mix low-volume production," in *ASMC (Advanced Semiconductor Manufacturing Conference) Proceedings*, 2008, pp. 148–155. doi: 10.1109/ASMC.2008.4529020.
26. J. A. S. Perez, "Risk minimization through metrology in semiconductor manufacturing," PhD Thesis, Université de Lyon, 2017.
27. L. Monch *et al.*, "Production planning and control for semiconductor wafer fabrication facilities: Modeling, analysis, and systems," p. Springer Science & Business Media, 2012.
28. C. A. Mack, "Fifty years of Moore's law," in *IEEE Transactions on Semiconductor Manufacturing*, May 2011, vol. 24, no. 2, pp. 202–207. doi: 10.1109/TSM.2010.2096437.
29. N. Geng and Z. Jiang, "A review on strategic capacity planning for the semiconductor manufacturing industry," *Int. J. Prod. Res.*, vol. 47, no. 13, pp. 3639–3655, 2009. doi: 10.1080/00207540701871051.
30. D. Nazzal and L. F. McGinnis, "Analytical approach to estimating AMHS performance in 300 mm fabs," *Int. J. Prod. Res.*, vol. 45, no. 3, pp. 571–590, 2007. doi: 10.1080/00207540600792333.
31. A. Bitar, S. Dauzère-Pérès, C. Yugma, and R. Roussel, "A memetic algorithm to solve an unrelated parallel machine scheduling problem with auxiliary resources in semiconductor manufacturing," *J. Sched.*, vol. 19, no. 4, pp. 367–376, Aug. 2016. doi: 10.1007/s10951-014-0397-6.

32. E. Akçah, K. Nemoto, and R. Uzsoy, "Cycle-time improvements for photolithography process in semiconductor manufacturing," *IEEE Trans. Semicond. Manuf.*, vol. 14, no. 1, pp. 48–56, Feb. 2001. doi: 10.1109/66.909654.

33. D. P. Martin, "Total operational efficiency (TOE): The determination of two capacity and cycle time components and their relationship to productivity improvements in a semiconductor manufacturing line," in *10th Annual IEEE/SEMI Advanced Semiconductor Manufacturing Conference and Workshop, ASMC 1999*, 1999, pp. 37–41. doi: 10.1109/ASMC.1999.798177.

34. B. H. Zhou, Z. S. Gao, and J. Chen, "Scheduling algorithm of dual-armed cluster tools with residency time and reentrant constraints," *J. Cent. South Univ.*, vol. 21, no. 1, pp. 160–166, Mar. 2014. doi: 10.1007/s11771-014-1927-2.

35. J. Yi, S. Ding, D. Song, and M. T. Zhang, "Scheduling analysis of cluster tools with buffer/process modules," in *Proceedings - IEEE International Conference on Robotics and Automation*, 2007, pp. 985–990. doi: 10.1109/ROBOT.2007.363113.

36. Y. Ma and B. Gong, "Double-layer scheduling strategy of load balancing in scientific workflow," in *Proceedings of the International Conference on Parallel and Distributed Systems - ICPADS*, 2009, pp. 671–678. doi: 10.1109/ICPADS.2009.27.

37. T. Ganesharajah, N. G. Hall, and C. Sriskandarajah, "Design and operational issues in AGV-served manufacturing systems," *Ann. Oper. Res.*, vol. 76, no. 0, pp. 109–154, 1998. doi: 10.1023/a:1018936219150.

38. C. Mead, "Analog VLSI and neural systems," Addison-Wesley, 1989.

39. C. Mead, "Neuromorphic electronic systems," *Proc. IEEE*, vol. 78, no. 10, pp. 1629–1636, 1990. doi: 10.1109/5.58356.

40. "What is natural language processing?" SAS, https://www.sas.com/en_us/insights/analytics/what-is-natural-language-processing-nlp.html (accessed May 14, 2021).

41. "Home - OpenCV," https://opencv.org/ (accessed May 14, 2021).

42. J. Von Neumann and M. D. Godfrey, "First draft of a report on the EDVAC," *IEEE Ann. Hist. Comput.*, vol. 15, no. 4, pp. 27–75, 1993. doi: 10.1109/85.238389.

43. J. Wang, Y. Ma, L. Zhang, R. X. Gao, and D. Wu, "Deep learning for smart manufacturing: Methods and applications," *J. Manuf. Syst.*, vol. 48, pp. 144–156, Jul. 2018. doi: 10.1016/j.jmsy.2018.01.003.

44. F. Rosenblatt, "The perceptron: A probabilistic model for information storage and organization in the brain," *Psychol. Rev.*, vol. 65, no. 6, pp. 386–408, Nov. 1958. doi: 10.1037/h0042519.

45. H. D. Block, "The perceptron: A model for brain functioning. I," *Rev. Mod. Phys.*, vol. 34, no. 1, pp. 123–135, Jan. 1962. doi: 10.1103/RevModPhys.34.123.

46. D. O. Hebb, "The organization of behavior: A neuropsychological theory," Wiley, 1949.

47. S. Löwel and W. Singer, "Selection of intrinsic horizontal connections in the visual cortex by correlated neuronal activity," *Science (80-.).*, vol. 255, no. 5041, pp. 209–212, Jan. 1992. doi: 10.1126/science.1372754.

48. A. L. Hodgkin and A. F. Huxley, "A quantitative description of membrane current and its application to conduction and excitation in nerve," *J. Physiol.*, vol. 117, no. 4, pp. 500–544, Aug. 1952. doi: 10.1113/jphysiol.1952.sp004764.

49. T. Yu and G. Cauwenberghs, "Analog VLSI biophysical neurons and synapses with programmable membrane channel kinetics," in *IEEE Transactions on Biomedical Circuits and Systems*, Jun. 2010, vol. 4, no. 3, pp. 139–148. doi: 10.1109/TBCAS.2010.2048566.

50. S. Saïghi, J. Tomas, Y. Bornat, and S. Renaud, "A conductance-based silicon neuron with dynamically tunable model parameters," in *2nd International IEEE EMBS Conference on Neural Engineering*, 2005, vol. 2005, pp. 285–288. doi: 10.1109/CNE.2005.1419613.

51. M. F. Simoni, G. S. Cymbalyuk, M. E. Sorensen, R. L. Calabrese, and S. P. DeWeerth, "A multiconductance silicon neuron with biologically matched dynamics," *IEEE Trans. Biomed. Eng.*, vol. 51, no. 2, pp. 342–354, Feb. 2004. doi: 10.1109/TBME.2003.820390.

52. M. Azadmehr, J. P. Abrahamsen, and P. Häfliger, "A foveated AER imager chip," in *Proceedings - IEEE International Symposium on Circuits and Systems*, 2005, pp. 2751–2754. doi: 10.1109/ISCAS.2005.1465196.

53. J. A. M. Olsson and P. Häfliger, "Mismatch reduction with relative reset in integrate-and-fire photo-pixel array," in *2008 IEEE-BIOCAS Biomedical Circuits and Systems Conference, BIOCAS 2008*, 2008, pp. 277–280. doi: 10.1109/BIOCAS.2008.4696928.

54. F. Danneville, C. Loyez, K. Carpentier, I. Sourikopoulos, E. Mercier, and A. Cappy, "A Sub-35 pW Axon-Hillock artificial neuron circuit," *Solid. State. Electron.*, vol. 153, pp. 88–92, 2019. doi: 10.1016/j.sse.2019.01.002.

55. C. Toumazou, J. Georgiou, and E. M. Drakakis, "Current-mode analogue circuit representation of Hodgkin and Huxley neuron equations,"

Electron. Lett., vol. 34, no. 14, p. 1376, Jul. 1998. doi: 10.1049/el:19980989.

56. I. Sourikopoulos *et al.*, "A 4-fJ/Spike artificial neuron in 65 nm CMOS technology," *Front. Neurosci.*, vol. 11, no. MAR, p. 123, Mar. 2017. doi: 10.3389/fnins.2017.00123.
57. J. H. B. Wijekoon and P. Dudek, "Spiking and bursting firing patterns of a compact VLSI cortical neuron circuit," in *IEEE International Conference on Neural Networks - Conference Proceedings*, 2007, pp. 1332–1337. doi: 10.1109/IJCNN.2007.4371151.
58. Y. J. Lee *et al.*, "Low power real time electronic neuron VLSI design using subthreshold technique," in *Proceedings - IEEE International Symposium on Circuits and Systems*, 2004, vol. 4. doi: 10.1109/iscas.2004.1329111.
59. E. Chicca, F. Stefanini, C. Bartolozzi, and G. Indiveri, "Neuromorphic electronic circuits for building autonomous cognitive systems," *Proc. IEEE*, vol. 102, no. 9, pp. 1367–1388, 2014. doi: 10.1109/JPROC.2014.2313954.
60. V. Ostwal, R. Meshram, B. Rajendran, and U. Ganguly, "An ultra-compact and low power neuron based on SOI platform," in *International Symposium on VLSI Technology, Systems, and Applications, Proceedings*, Jun. 2015, vol. 2015-June. doi: 10.1109/VLSI-TSA.2015.7117569.
61. S. Dutta, V. Kumar, A. Shukla, N. R. Mohapatra, and U. Ganguly, "Leaky integrate and fire neuron by charge-discharge dynamics in floating-body MOSFET," *Sci. Rep.*, vol. 7, no. 1, pp. 1–7, Dec. 2017. doi: 10.1038/s41598-017-07418-y.
62. M. W. Kwon, H. Kim, J. Park, and B. G. Park, "Integrate-and-fire neuron circuit and synaptic device using floating body MOSFET with spike timing-dependent plasticity," *J. Semicond. Technol. Sci.*, vol. 15, no. 6, pp. 658–663, Dec. 2015. doi: 10.5573/JSTS.2015.15.6.658.
63. B. Das, J. Schulze, and U. Ganguly, "Ultra-low energy LIF neuron using Si NIPIN diode for spiking neural networks," *IEEE Electron Device Lett.*, vol. 39, no. 12, pp. 1832–1835, Dec. 2018. doi: 10.1109/LED.2018.2876684.
64. V. Sze, Y. H. Chen, T. J. Yang, and J. S. Emer, "Efficient processing of deep neural networks: A tutorial and survey," *Proceedings of the IEEE*, vol. 105, no. 12. Institute of Electrical and Electronics Engineers Inc., pp. 2295–2329, Dec. 01, 2017. doi: 10.1109/JPROC.2017.2761740.
65. N. P. Jouppi, C. Young, N. Patil, and D. Patterson, "A domain-specific architecture for deep neural networks," *Commun. ACM*, vol. 61, no. 9, pp. 50–59, Sep. 2018. doi: 10.1145/3154484.

66. J. Lee, J. Lee, D. Han, J. Lee, G. Park, and H. J. Yoo, "7.7 LNPU: A 25.3TFLOPS/W sparse deep-neural-network learning processor with fine-grained mixed precision of FP8-FP16," in *Digest of Technical Papers - IEEE International Solid-State Circuits Conference*, Mar. 2019, vol. 2019-February, pp. 142–144. doi: 10.1109/ISSCC.2019.8662302.
67. G. Batra, Z. Jacobson, S. Madhav, A. Queirolo, and N. Santhanam, "DECEMBER 2018 • HIGH TECH artificial-intelligence hardware: New opportunities for semiconductor companies artificial intelligence is opening the best opportunities for semiconductor companies in decades. How can they capture this value?"
68. C. Teo, K. L. Low, V. Narang, and A. V Thean, "TCAD-enabled machine learning defect prediction to accelerate advanced semiconductor device failure analysis," in *2019 International Conference on Simulation of Semiconductor Processes and Devices (SISPAD)*, 2019, pp. 1–4. doi: 10.1109/SISPAD.2019.8870440.
69. S. Myung *et al.*, "Real-time TCAD: A new paradigm for TCAD in the artificial intelligence era," in *2020 International Conference on Simulation of Semiconductor Processes and Devices (SISPAD)*, 2020, pp. 347–350. doi: 10.23919/SISPAD49475.2020.9241622.

Chapter 3

Simulation Environment

Chinmay K. Maiti
SOA University, Bhubaneswar 751030, Odisha, India
ckmaiti@outlook.com

In the era of the internet of things, the road toward enhancing the performance of integrated circuits comprises the fundamental improvement of devices via new structures, scaling the size, and transistor features, which increase their speed and reduce the power consumption. The combination of new device architectures and the use of stress/strain tuning and high mobility materials have led toward nanoelectronic devices and pushed the limits of CMOS. Computer-aided design (CAD) and simulation of nanoscale transistors for upcoming technology nodes (7 nm and below) are the main focus of this book. Technology computer-aided design (TCAD) tools are widely used in industrial environments for process modeling and predictive device simulation of the electrical characteristics of microelectronics components. This is mainly because TCAD device design and process technology can reduce cost and development

Fabless Semiconductor Manufacturing: In the Era of Internet of Things
Edited by Chinmay K. Maiti

ISBN 978-981-4968-27-0 (Hardcover), 978-1-003-31497-4 (eBook)
www.jennystanford.com

time, and also secure technology choice [1]. A reliable TCAD tool is also required for supporting the physics coherence and performance analysis of highly scaled devices, for detecting operating limits, and also for investigating new device concepts. Nevertheless, the monograph is associated with the recommendations of the IRDS roadmap [2] concerning the efforts made by industry and academic partners in the modeling of future generations of transistors. To create a foundation for the simulation environment and to ensure reproducibility, the used material models and tools are introduced in the beginning.

In this chapter, currently available advanced semiconductor devices/process simulation tools are introduced. The evolution of semiconductor simulation tools goes back to the late 1960s. Starting in the late 1960s, the process and device simulation were predominantly in 1D and 2D. Device simulation is dominantly 2D, due to the nature of devices, became the workhorse in the design and scaling of devices. The transition from n-MOSFET to CMOS technology resulted in the necessity of tightly coupled and full 2D simulators for process and device simulations. Due to the ultra-small size of the state-of-the-art devices, the 3D effects became dominant. To achieve a better understanding of simulated and fabricated device characteristics, the 3D process/device simulation has now been introduced.

Currently, there are several commercial TCAD software platforms for process/device simulation around the world. Most of them were first developed by universities worldwide and later integrated into a suite by TCAD vendors. Since the early 1990s, commercial vendors such as SILVACO and SYNOPSYS have built their graphical user interfaces (GUI) around existing frameworks, making it easier for non-specialists to use process and device simulation tools within a larger engineering environment. The concept of general-purpose process and device simulators became a reality, allowing for versatile simulation of various structures in various technologies. Several TCAD tools will be used extensively in this monograph for the semiconductor device design and characterization including analysis. The design and simulation flow of various tools is briefly described below.

3.1 MINIMOS-NT

MINIMOS-NT is the most versatile discrete device simulator that provides users with the ability to simulate a broad range of fully-featured 2D and 3D devices. It is a general-purpose semiconductor device simulator providing steady-state, transient, and small-signal analysis. In this tool, device simulation is based on solving carrier transport equations in arbitrary semiconductor structures. The device under investigation is partitioned into segments, and for each segment, a material is assigned. The equations that have to be solved are assembled for each segment and the boundaries in between them, depending on the material, active transport equations, and activated physical models [3–5]. A partial differential equation system describes the device behavior using physics-based transport equations and the corresponding model parameters. For the solution of this system, a discretization in space is required, which is accomplished by defining a geometry mesh. The equations are assembled for each grid point. Such an equation system cannot be solved directly. Therefore, iterative solution methods, such as the Newton iteration system, have to be applied. In the mixed-mode device and circuit simulation, numerically simulated devices can be placed in circuits built from compact device models and passive elements. These simulators incorporate advanced physical models and robust numeric methods for the simulation of the most common types of semiconductor devices.

The use of complex physical models, which often add highly non-linear relations between the variables, also increases the complexity of the equation system leading to an increased number of iteration steps and, therefore, longer convergence times. Some setups can also lead to failed conversions. Also, the boundary conditions, e.g., the given contact potentials, influence the convergence. The iterative solution process needs an initial value (called initial guess) to solve the given problem. If no initial value is available, a first initial guess is calculated using thermal equilibrium in the semiconductor. Alternatively, a previous simulation result can be used as an initial guess.

3.2 VSP

VSP is a general-purpose device simulator for arbitrary nanostructures, operating on the Schrödinger-Poisson equation system [6]. VSP uses a finite volume discretization scheme, thus avoiding the weak formulation fundamental to finite elements and relying instead on a formulation based on the conservation of fluxes in each of the finite volumes. As a quantum-electronic simulator, VSP includes quantum mechanical solvers for closed-boundary as well as open-boundary problems. The closed boundary model setup allows one to calculate a self-consistent carrier concentration in FinFET or nanowire cross-sections including quantization effects due to electrical and geometrical confinement. The equation systems are solved on tetrahedral and triangular meshes for arbitrary 2D and 3D geometries. The Schrödinger solver supports effective mass or k.p Hamiltonians including arbitrary strain distribution conditions tensor. The model accounts for arbitrary substrate types and channel orientation. Hetero-structured semiconductor devices can be treated within the closed boundary model for quick estimation of resonant energy levels. The open boundary model allows the evaluation of current-voltage characteristics. Using the experimental 2D non-equilibrium Green's functions solver, nano-MOSFETs in the ballistic operation regime can be investigated. The key features include:

- Arbitrary 1D/2D/3D geometries by using unstructured meshes
- Substrate and channel orientation
- Multi-valley single-band effective mass Hamiltonian
- KdotP Hamiltonian including strain effect
- Subband ladder (Eigen energies) and wave functions (including overlaps)
- Physical modeling for conductivity and mobility
- Based on the Boltzmann equation (Kubo-Greenwood formula)
- Includes crystal orientation and strain effects
- Anisotropic band-structure
- Surface roughness scattering
- Non-polar phonon scattering (ADP, ODP, inter-valley)
- Polar-optical phonon scattering (POP)

- Ionized impurity (Coulomb) scattering (IIS)
- Alloy disorder scattering

The effects of quantum confinement are simulated using the self-consistent coupled Schrodinger Poisson model; however, this model cannot solve transport problems by itself. It is, therefore, used in combination with either the drift-diffusion (DD) mode-space approach or the mode space non-equilibrium Green's function approach [7]. The self-consistent coupled Schrodinger Poisson model is used to solve Schrodinger's equation to give a quantized description of the density of states in the presence of quantum mechanical confinement. Decomposing the simulation domain along the channel direction into 2D slices allows for electronic structure coupling. In every iteration step of MINIMOS-NT, an instance of VSP is invoked on each slice to calculate the confined carrier densities and mobilities. MINIMOS-NT passes the potential and quasi-Fermi energies for electrons and holes. VSP is used to calculate the subband structure and occupancy. Before moving on to the next iteration step, the data on each slice is extruded and interpolated back onto the 3D simulation domain. The self-consistent quantum correction technique ensures a stable and fast convergence. The elastic model is used to quantify mechanical stress in the simulation chain. In VSP, the piezoresistive model is used to calculate the stress in the device. The coupled/uncoupled mode space approach is used to solve the 3D Schrödinger equation, which provides both computational efficiency and high accuracy when compared to direct real-space calculations [8, 9].

3.3 Silvaco TCAD

Silvaco is the world's leading software provider for TCAD, which is a branch of electronic CAD aiming to model the manufacturing of semiconductor devices and their functions. Silvaco offers the virtual wafer fab (VWF), which represents several simulation tools and interactive tools allowing the design and analysis of most semiconductor devices. VWF modules are mainly interactive tools and simulation tools. The interactive tools are:

- DECKBUILD: Interface with Athena and Atlas simulators

- TONYPLOT: 1D and 2D graphical analysis tool for simulation results
- DEVEDIT: The tool allows one to create a component structure from the data provided

The simulation tools are:

- ATHENA: Technology process simulator that simulates different steps carried out during semiconductor manufacturing in clean rooms and generates the structure of components (constituent layers, dimensions, and geometry) with doping profile.
- ATLAS: Semiconductor device simulator that allows one to obtain the device characteristics such as current-voltage and other electrical parameters (static or dynamic). The input and output arguments of ATLAS are shown in Figure 3.1.

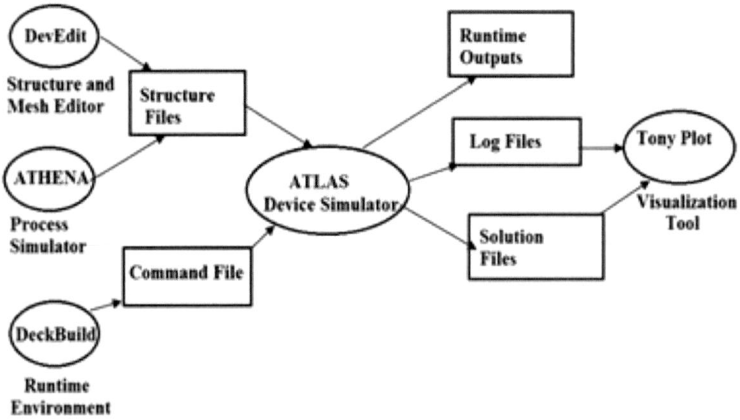

Figure 3.1 Simulation environment (input/output flow) in Silvaco ATLAS [10].

3.4 VictoryProcess

VictoryProcess is a full 3D process simulator that allows users to create a variety of device structures in any shape. The majority of the supported operations correspond to actual processes like etching or deposition, CMP, epitaxy, and others, allowing a process engineer to create a direct link between the manufacturing processes and a VictoryProcess input statement. VictoryProcess is capable of simulating complex physics-based etching and deposition processes. Advanced process simulation can be done quickly and precisely using either "geometrical" mode or very simple physical models. VictoryProcess can simulate complex full physics-based etching and deposition processes that include reactor characteristics (particle flux), shading effects, and/or secondary effects such as etched material redeposition. The etching and deposition process characteristics include:

- Rounded corners
- Tapered sidewall
- Non-conformal epitaxial growth
- Selective geometrical etch
- Selective CMP

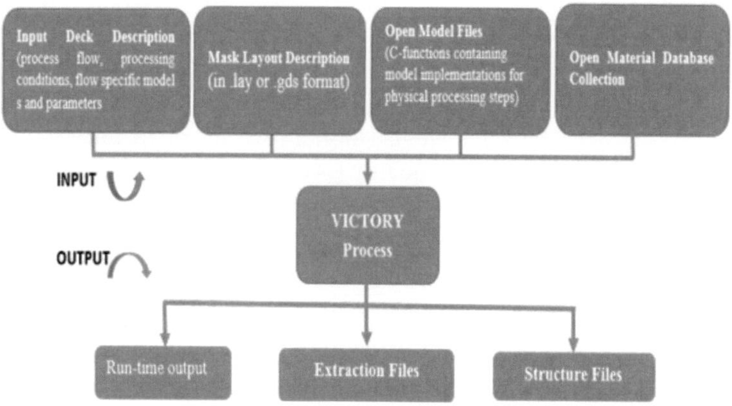

Figure 3.2 The input and output arguments of VictoryProcess [11].

VictoryProcess is capable of simulating complex processes within a feature scale for 3D structures (see Figure 3.2). The technological parameters from the reactor can be included using transport characteristics. Different silicon crystal planes oxidize at different rates; for example, the silicon plane with Miller indices <111> oxidizes 1.7 times faster than the plane with Miller indices <100>. The geometry of a 3D structure may change considerably during an oxidation process, and the silicon/oxide interface may pass through various crystal planes with different oxidation rates. As a result, the geometry of an oxidized structure on this wafer is influenced by the orientation and type of silicon wafer. For the simulation/emulation of an oxidation process, VictoryProcess provides several modes, including

- the analytical oxidation mode,
- the empirical oxidation mode, and
- the full physical oxidation mode.

Since the finite element method provides an approximate solution to the stress analysis problem, the result will have some dependence on the mesh used. First of all, the solution to the problem is calculated in the nodes defined by the mesh. Fewer nodes mean having fewer elements where the solution is calculated. Therefore, it is necessary to include as many elements as possible to improve the accuracy. It is also necessary to have the elements as regularly as possible to enhance accuracy. If the element is too skinny, we will have regions with many closed solutions and regions with solutions spread apart. This makes the resulting system of equations badly conditioned, and in turn, the solution becomes less accurate especially when we use an iterative solver. VictoryProcess provides statistics on mesh quality (average and maximum element aspect ratio) for each material region and gives a warning message when the average element aspect ratio exceeds 1:100 in any region. It is recommended that mesh space in one direction is not too small compared to other directions. VictoryProcess supports multiple meshing algorithms for 3D stress simulation.

3.5 Process Simulation

Process simulation is essential to develop and optimize the technology, and it is especially helpful in the initial phase of technology development. In 1977, the first version of the 1D process simulator SUPREM developed at Stanford was released. Process simulation has continued to improve dramatically since the release of SUPREM. Since then, the process models in SUPREM have developed substantially and versions II and III of SUPREM were released in 1978 and 1983, respectively [12]. The primary objective of process simulation is to accurately predict the physical/structural layers and geometry of devices at the end of a process run, as well as the active dopant/stress distributions. The input to process simulation is a process flow guided by process assumptions and layout/layer masks. The initial wafer/substrate is subject to a variety of process conditions, each of which may involve steps like oxidation, diffusion, implantation, deposition, and etching. Lithography simulation is also performed to accurately capture feature geometries. Process simulation generally uses a finite-element or finite-volume mesh to compute and store the device's dopant and stress profiles. Every geometric change in the simulation domain requires a new mesh that fits the new device boundaries to model the next series of process steps. As device lots become more and more expensive, process modeling is increasingly important. Process simulation and modeling are increasingly sophisticated, but accuracy remains a problem. Hundreds of processing steps, months of processing time, and unexpected relationships between tool performance and yield are common in semiconductor manufacturing. Modeling of front-end processes has been facing challenges caused by a continuous reduction of the implant energies and the thermal budgets of the annealing schemes from soak anneal to the spike anneals, which are now used for production. The complexity of physical models is a major factor that impacts process simulation.

3.6 VictoryDevice

VictoryDevice is a general-purpose 3D device simulator. A tetrahedral meshing engine is used for fast and accurate simulation of complex 3D geometries. VictoryDevice performs DC, AC, and transient analysis for silicon, binary, ternary, and quaternary material-based devices (see Figure 3.3).

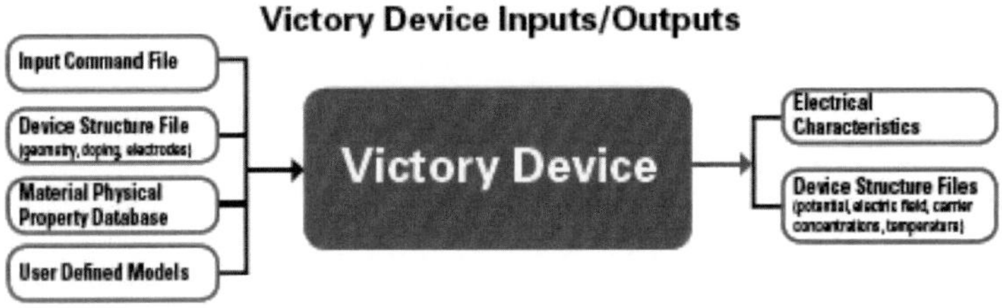

Figure 3.3 Input and output arguments of VictoryDevice [13].

VictoryDevice features include:

- Tetrahedral mesh for accurate 3D geometry representation
- Voronoi discretization for conformal Delaunay meshes
- Advanced physical models with user-customizable material database
- Stress-dependent mobility and bandgap models
- Physical models using the C-interpreter or dynamically linked libraries
- DC, AC, and transient analysis
- DD and energy balance transport equations
- Self-consistent simulation of self-heating effects including heat generation, heat flow, lattice heating, heat sinks, and temperature-dependent material parameters
- Advanced multi-threaded numerical solver library
- Atlas-compatible

3.7 Device Simulation

Transistors have been scaled in tandem with an increase in the number of transistors per chip with each new technology node, on top of device and design developments. The electrical response of the semiconductor device can be clearly understood through different device simulations. DC, AC, transient, and mixed-mode simulations are performed through the device simulator. I_D–V_G and I_D–V_D plots are obtained through DC simulation. C-V characteristics and frequency response are found out by AC analysis. The resulting characteristics are plotted for analysis by using the visualization tool. For the holistic device exploration, either vertical or lateral, various design and technology aspects should be considered. The design and technology should be co-optimized to find the next node device, which would be possible to manufacture, provide sufficient drive to meet application performance requirements, not consume too much energy per switch, and be cost-efficient. The structure of the device determines the method for the simulation of transport properties. Density-Gradient (DG) [14] and Drift-Diffusion (DD) [15] models arise to be the fast and powerful tools to support the predictive technology development for the microelectronics industry, and they require proper calibration. For efficient TCAD purposes, a preliminary identification of the mechanisms responsible for transport degradation, as well as a calibration of empirical TCAD transport models, is mandatory. The model selection hierarchy in device simulation is shown in Figure 3.4.

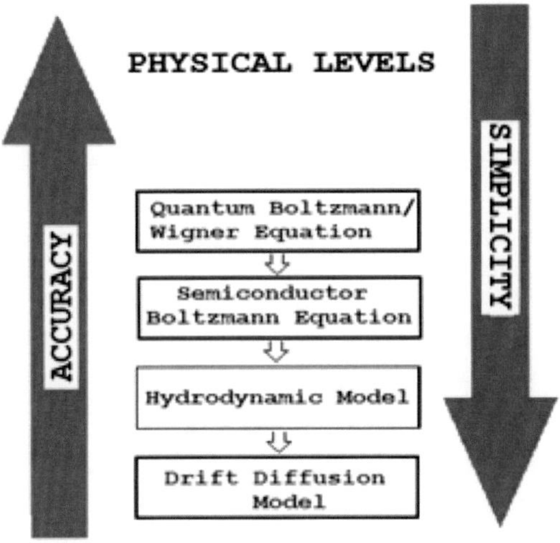

Figure 3.4 Model selection hierarchy in device simulation.

Figure 3.5 shows a list of popular transport models used in TCAD-based simulations of semiconductor devices, starting with the least accurate but the fastest. The size of the devices influences the model that is chosen. In general, the more advanced the transport model, the smaller the system should be. Figure 3.5 shows rough guidelines for when each model should be used. For a long time, the DD model has performed well when tuned to hardware or Monte Carlo simulations. However, models that are more complex than the DD model are needed to accurately predict the behavior of advanced devices, and their complexity grows exponentially with scaling. In general, the device simulation should qualitatively capture the correct device behavior at all the possible device terminal voltage combinations and provide a good convergence for exploratory research.

	Model	Improvements
Semi-classical approaches	Drift-diffusion equations	Good for devices down to 0.5 μm
	Hydrodynamic equations	Velocity overshoot is captured properly
	Boltzmann transport equation	Accurate up to the classical limits
Quantum approaches	Quantum hydrodynamics	All hydrodynamic features + quantum corrections
	Quantum Monte-Carlo methods	All classical features + quantum corrections
	Quantum-Kinetic equation	Accurate up to single particle description
	Green's function methods	Includes correlations in both space and time domain
	Schrödinger equation	Can be solved only for a small number of particles

Figure 3.5 Transport models sorted by complexity, the most computationally intensive and accurate are at the bottom [16].

3.8 Model Selection

To assess the electrical performance, we must first choose the appropriate model for carrier transport. The technique used to simulate the device's transport properties is determined by the device's structure. The device under investigation is divided into segments, with a material assigned to each segment. Each segment, as well as the boundaries between them, has its own set of solved transport equations. The active transport equations and activated physical models have been introduced depending on the material [17]. To solve these types of problems, a discretization in space is required, which is achieved by defining a geometry mesh. Each grid point has its own set of equations. A system of equations like this cannot be solved directly. As a result, the Newton iteration method has been applied. DG and DD models emerge as a quick and powerful tool for predictive technology growth in the microelectronics industry. In this simulation, the DG model was used, along with quantum corrections. The steps in the simulated technique are as follows:

- Computation of band structure of nanosheet using tight-binding model [18]

- Calculation of the charge density using a semiclassical model
- Solution of a 2D Poisson equation in the cross-section of the nanosheet
- Solution to linearized Boltzmann transport (LBT) in nanosheet channel [19]

3.9 Drift-Diffusion Transport Model

DD approach is essentially a low-order approximation to the Boltzmann transport equation. The electron and hole current densities are given by Eqs. (3.1) and (3.2).

$$J_n = q.\mu_n \cdot n\left(\text{grad}\left(\frac{\varepsilon_C}{q} - \psi\right) + \frac{k_B \cdot T_L}{q} \cdot \frac{N_{C,0}}{n} \cdot \text{grad}\left(\frac{n}{N_{C,0}}\right)\right) \tag{3.1}$$

$$J_p = q \cdot \mu_p \cdot p \cdot \left(\text{grad}\left(\frac{\varepsilon_V}{q} - \psi\right) - \frac{k_B \cdot T_L}{q} \cdot \frac{N_{V,0}}{p} \cdot \text{grad}\left(\frac{p}{N_{V,0}}\right)\right), \tag{3.2}$$

where $N_{C,0}$ and $N_{V,0}$ are the effective density of states for electron and hole, respectively, at reference temperature T_0; ε_C and ε_V are the position-dependent band edge energies; and μ_n and μ_p are the mobility of electrons and holes, respectively.

The current density for any carrier type vanishes when the quasi-Fermi potential for that carrier type is taken as constant. The local concentration of a particular carrier type is calculated by quasi-Fermi potential. By considering this, simulation time has been reduced, and the need for solving continuity equations is not required. The quasi-Fermi potential is determined by the contact for which a list of segments is specified. The equation for hole and electron concentration is given by:

$$p = N_V \cdot \exp\left(\frac{q \cdot (\varphi_p - \psi) + \varepsilon_V}{k_B \cdot T_L}\right) \tag{3.3}$$

$$n = N_C \cdot \exp\left(\frac{q \cdot (\varphi_n - \psi) + \varepsilon_C}{k_B \cdot T_L}\right). \tag{3.4}$$

The DG and DD models emerge as a rapid and powerful tool to support the microelectronics industry's predictive technology growth, but they require proper calibration. Preliminary identification of the mechanisms responsible for transport degradation, as well as calibration of empirical TCAD transport models, is required for effective TCAD. Poisson equation and continuity equation are solved to get the following equations for electrons and holes.

$$\text{div}\,(\varepsilon \cdot \text{grad}\,\psi) = q(n - p - C) \tag{3.5}$$

$$\text{div}\,J_n = q \cdot \left(R + \frac{\partial n}{\partial t} \right) \tag{3.6}$$

$$\text{div}\,J_p = q \cdot \left(R + \frac{\partial p}{\partial t} \right), \tag{3.7}$$

where ψ is the electrostatic potential, the electron and hole concentrations are given by n and p, respectively, C denotes the net concentration of the ionized dopants, ε is the dielectric permittivity of the semiconductor, R is the net recombination rate, and J_n is the electron current density.

DD current equations are derived from Boltzmann equations in which basic principles of thermodynamics and the method of the moment are used. Quantum confinement effects are considered in TCAD simulations using the quantum correction included the DD model [20–22], and Fermi-Dirac distribution, Lombardi model, and SRH recombination were used. For all Si and SiGe regions, a bandgap narrowing model was used [23]. To extract various electrical parameters, MINIMOS-NT [4] was used in the simulation of the proposed device.

3.10 Shockley-Read-Hall Recombination

The carriers that are generated in space charge regions and recombination (i.e., in high injection region) have been modeled using Shockley-Read-Hall (SRH) [24] equation:

$$R^{\text{SRH}} = \frac{n \cdot p - n_i^2}{\tau_p \cdot (n + n_1) + \tau_n \cdot (p + p_1)}, \tag{3.8}$$

where n_1 and p_1 are given by:

$$n_1 = N_C(T_L) \cdot \exp\left(\frac{-\varepsilon_C + \varepsilon_T}{k_B \cdot T_L} \right) \tag{3.9}$$

$$p_1 = N_V(T_L)\cdot \exp\left(\frac{-\varepsilon_T + \varepsilon_V}{k_B \cdot T_L}\right), \tag{3.10}$$

where N_C and N_V are the effective density of states of electron and hole, respectively; T_L is the lattice temperature; ε_T is the trap energy level; and τ_n and τ_p are the recombination lifetimes of electron and hole, respectively.

The rate of recombination is found to be maximum when the trap energy level is located in the mid of conduction and valence band (i.e., the value of intrinsic semiconductor). The thermal velocities of the carriers are obtained by the following equations:

$$v_{n,p} = \sqrt{\frac{3\cdot k_B \cdot 300K}{m_{n,p}}} \tag{3.11}$$

$$\tau_{n,300} = \frac{1}{\sigma_{T,n}\cdot N_T \cdot v_{n,300} + S_n/y} \tag{3.12}$$

$$\tau_{p,300} = \frac{1}{\sigma_{T,p}\cdot N_T \cdot v_{p,300} + S_p/y}, \tag{3.13}$$

where N_T is the trap density; $\sigma_{T,n}$ and $\sigma_{T,p}$ are the trap capture cross-section for the electron and hole, respectively; and $\tau_{n,300}$ and $\tau_{p,300}$ are the recombination times of electron and hole, respectively, at 300 K.

The lattice temperature dependence of recombination time for the carriers is given by the following equations:

$$\tau_n(T_L) = \left(\frac{300K}{T_L}\right)^{3/2} \cdot \frac{\tau_{n,300}}{1+\tau_n} \tag{3.14}$$

$$\tau_p(T_L) = \left(\frac{300K}{T_L}\right)^{3/2} \cdot \frac{\tau_{p,300}}{1+\tau_p}. \tag{3.15}$$

3.11 Low-Field Mobility Model

The lattice mobility in Si, which is dependent on temperature, is modeled by a power law. This relation is given by Eq. (3.16) [25, 26]:

$$\mu_v^L = \mu_{v,300}^L \cdot \left(\frac{T_L}{300K}\right)^{\gamma_{0,v}}, \quad v = n, p. \tag{3.16}$$

The reduction in mobility of the carrier due to ionized impurity scattering has been taken by taking the formula of Cauchy and Thomas [25] and the temperature-dependent coefficient. It has been presented below.

$$\mu_v^{LI} = \mu_v^{\min} + \frac{\mu_v^L - \mu_v^{\min}}{1 + \left(\frac{C_I}{C_v^{\text{ref}}} \right)^{\alpha_v}} \tag{3.17}$$

$$\mu_v^{\min} = \begin{cases} \mu_{v,300}^{\min} \cdot \left(\frac{T_L}{300K} \right)^{\gamma_{1,v}} & T_L \geq T_{\text{switch}} \\ \mu_{v,300}^{\min} \cdot \left(\frac{2}{3} \right)^{\gamma_{1,v}} \cdot \left(\frac{T_L}{200K} \right)^{\gamma_{2,v}} & T_L < T_{\text{switch}} \end{cases} \tag{3.18}$$

$$C_v^{ref} = C_{v,300}^{\text{ref}} \cdot \left(\frac{T_L}{300K} \right)^{\gamma_{3,v}} \tag{3.19}$$

$$\alpha_v = \alpha_{v,300} \cdot \left(\frac{T_L}{300K} \right)^{\gamma_{4,v}}, \tag{3.20}$$

where C_1 is denoted by the concentration of ionized impurities.

The surface scattering is modeled by [27]:

$$\mu_v^{\text{LIS}} = \frac{\mu_v^{\text{ref}} + \left(\mu_v^{LI} - \mu_v^{\text{ref}} \right) \cdot (1 - F(y))}{1 + F(y) \cdot \left(\frac{S_v}{S_v^{\text{ref}}} \right)^{\gamma_{6,v}}} \tag{3.21}$$

$$\mu_v^{\text{ref}} = \mu_{v,300}^{\text{ref}} \cdot \left(\frac{T_L}{300K} \right)^{-\gamma_{5,v}}. \tag{3.22}$$

The Lombardi model has been used in this work that accounts for low field and transverse field effects on mobility. The reduction in the mobility of the carriers in the interface was modeled by Lombardi *et al.* [21]. Surface ac-phonon scattering (ac) and surface roughness scattering (sr) were combined with the bulk mobility (b) by Mathiessen's rule. This is expressed as:

$$\frac{1}{\mu_v^{\text{LIS}}} = \frac{1}{\mu_v^b} + \frac{1}{\mu_v^{\text{ac}}} + \frac{1}{\mu_v^{\text{sr}}} \qquad v = n, p. \tag{3.23}$$

The contribution of the two surfaces is given by the following relations:

$$\mu_\nu^{\mathrm{ac}} = \frac{B300}{F_\perp} + \frac{C300 \cdot N_i^{L300}}{F_\perp^{1/3} \cdot T_L} \qquad \nu = n,p \tag{3.24}$$

$$\mu_\nu^{\mathrm{sr}} = \left(\frac{F_\perp^{\alpha}}{D300} + \frac{F_\perp^{3}}{E300} \right)^{-1} \qquad \nu = n,p. \tag{3.25}$$

The dependency of the carrier of the exponent is given by:

$$\alpha = A300 + A_\nu \cdot \frac{\nu}{N_i^{\beta}} \qquad \nu = n,p, \tag{3.26}$$

where N_i is the total doping concentration. The component of the electric field, which is normal to silicon/oxide interface, is denoted by $F_\perp$. The Masetti [28] bulk mobility model is used for the heavy doping range (i.e., from 10^{13} to 4×10^{21}), which is given as follows:

$$\mu_\nu^{\mathrm{LI}} = \mu_{\nu,300}^{\min 1} \cdot \exp\left(\frac{-P_{c,300}}{N_i} \right) + \frac{\mu_\nu^{L} - \mu_{\nu,300}^{\min 2}}{1 + \left(\frac{N_i}{C_{r,300}} \right)^{\alpha_{300}}} - \frac{\mu_{\nu,300}^{\mathrm{LI1}}}{1 + \left(\frac{C_{s,300}}{N_i} \right)^{\beta_{300}}}. \tag{3.27}$$

3.12 Density-Gradient Model

In its most basic form, the DG model [14] can be seen as a direct improvement of the DD model. The standard description can be extended to explain much of the quantum-mechanical behavior exhibited by strong inversion layers by generalizing the electron gas equation of state to include DG dependences. Quantum effects are included, which primarily affect the spatial distribution of the inversion charge in the channel as well as the threshold voltage. The DG extension equations are activated by the DG scheme. The phenomenology technique is used in the MINIMOS-NT simulation tool's DG model, which is the most common approach in quantum mechanical analysis.

$$J_n = q.\mu_n \cdot n \left(\mathrm{grad}\left(\frac{\varepsilon_C}{q} - \psi - \gamma_n \right) + \frac{k_B \cdot T}{q} \cdot \frac{N_{C,0}}{n} \cdot \mathrm{grad}\left(\frac{n}{N_{C,0}} \right) \right) \tag{3.28}$$

$$J_p = q \cdot \mu_p \cdot p \cdot \left(\text{grad}\left(\frac{\varepsilon_V}{q} - \psi - \gamma_p \right) - \frac{k_B \cdot T}{q} \cdot \frac{N_{V,0}}{p} \cdot \text{grad}\left(\frac{p}{N_{V,0}} \right) \right) \tag{3.29}$$

The quantities γ_n and γ_p are defined as the quantum corrected potentials, and these quantities are the summary of a more complex quantum analysis, which, according to a microscopic theory, involves the solving of the Wigner's distribution equation by the method of moments [14]. γ_n and γ_p are given in Eqs. (3.30) and (3.31):

$$\gamma_n = \frac{\hbar}{12 \cdot \lambda_n \cdot m_0} \cdot \text{div grad} \frac{\psi + \gamma_n - \varepsilon_C / q}{k_B \cdot T} \tag{3.30}$$

$$\gamma_p = \frac{\hbar}{12 \cdot \lambda_p \cdot m_0} \cdot \text{div grad} \frac{\psi + \gamma_p - \varepsilon_V / q}{k_B \cdot T}. \tag{3.31}$$

When quantum confinement is taken into consideration, a distance-dependent reduction of the density of state at the Si/SiO_2 interface is modified by the correction factor. The correction factor is given by:

$$h_{\text{corr}} = 1 - \exp \frac{-(z + z_0)^2}{\varsigma^2 \cdot \lambda_{\text{TH}}^2}, \tag{3.32}$$

where z is the distance to the Si/SiO_2 interface, the parameter z_0 moves the whole function relative to the interface, and ς is a newly introduced parameter that enables the variation of λ_{TH} for fitting purposes. The symbol λ_{TH} denotes the thermal wavelength that is given by:

$$\lambda_{\text{TH}} = \frac{\hbar}{\sqrt{2m \cdot k_B \cdot T_L}}. \tag{3.33}$$

The effective density of states with the inclusion of quantum confinement is given with correction factor h_{corr} as follows:

$$N_C = N_{C,0} \cdot h_{\text{corr}}. \tag{3.34}$$

With the inclusion of quantum confinement, band edge energy correction has been introduced by van Dort *et al.* [29], which is given by:

$$\Delta\varepsilon_g = \varepsilon_0 - \varepsilon_c(0) = \frac{13}{9}\cdot\beta\cdot\left(\frac{\varepsilon_{Si}}{4\cdot q\cdot k_B\cdot T_L}\right)^{1/3}\left|E_n\right|^{2/3}, \tag{3.35}$$

where β is 4.1 × 10^{-8} eVcm that is an empirically determined constant, ε_{Si} is the permittivity of silicon, and E_n is the electric field at the Si/SiO_2 interface perpendicular to the interface [27]. When the concentration of impurity is higher, the band formed by the impurity states leads to a reduction in the bandgap of the semiconductor.

3.13 Bandgap Narrowing Model

It has been seen experimentally that shrinkage of the bandgap occurs [30, 31] when the impurity concentration is particularly high. This effect is called the bandgap narrowing effect, which is ascribed to the emerging of the impurity band formed by the overlap impurity states. The change in bandgap is given by:

$$\Delta\varepsilon_g = \varepsilon_{\text{ref}}\cdot\left(\ln\left(\frac{C_I}{N_{\text{ref}}}\right) + \sqrt{\ln^2\left(\frac{C_I}{N_{\text{ref}}}\right) + \frac{1}{2}}\right). \tag{3.36}$$

The shift in conduction and valence band energy is given by:

$$\Delta\varepsilon_C = \left(\frac{\Delta\varepsilon_c}{\Delta\varepsilon_g}\right)\cdot\Delta\varepsilon_g \tag{3.37}$$

$$\Delta\varepsilon_V = \nabla\varepsilon_g - \Delta\varepsilon_C. \tag{3.38}$$

Generally, the semiconductor and metal boundaries are connected by ohmic contact. Dirichlet boundary conditions are applied at the ohmic contacts. The contact potential at the semiconductor boundary is given by:

$$\varphi_s = \varphi_m + \psi_{\text{bi}}, \tag{3.39}$$

where φ_m is the metal quasi-Fermi level or contact voltage. The build-in potential ψ_{bi} is given by [32]:

$$\begin{aligned}\psi_{\text{bi}} &= \frac{k_B\cdot T_L}{q}\cdot\ln\left(\frac{1}{2\cdot C_1}\cdot\left(C + \sqrt{C^2 + 4\cdot C_1\cdot C_2}\right)\right)\\ &= -\frac{k_B\cdot T_L}{q}\cdot\ln\left(\frac{1}{2\cdot C_2}\cdot\left(-C + \sqrt{C^2 + 4\cdot C_1\cdot C_2}\right)\right).\end{aligned} \tag{3.40}$$

The variables C_1 and C_2 are given by:

$$C_1 = N_C \cdot \exp\left(\frac{-\varepsilon_C}{k_B \cdot T_L}\right) \tag{3.41}$$

$$C_2 = N_V \cdot \exp\left(\frac{\varepsilon_V}{k_B \cdot T_L}\right). \tag{3.42}$$

The concentration of the carrier in the semiconductor is given as follows:

$$n_s = N_C \cdot \exp\left(\frac{-\varepsilon_C + q \cdot \psi_{bi}}{k_B \cdot T_L}\right) \tag{3.43}$$

$$p_s = N_V \cdot \exp\left(\frac{\varepsilon_V - q \cdot \psi_{\mathrm{bi}}}{k_B \cdot T_L}\right). \tag{3.44}$$

The carrier temperature is set equal to the lattice temperature:

$$T_n = T_L, T_p = T_L. \tag{3.45}$$

The semiconductor contact potential is defined as the difference of the metal quasi-Fermi level and the metalwork function difference potential.

$$\varphi_s = \varphi_m - \phi_{ms} \quad \text{where} \quad \phi_{ms} = -\frac{\varepsilon_W}{q}, \tag{3.46}$$

where e_w is the work function difference energy.

3.14 Stress/Strain Modeling

Based on the linear elasticity theory, the following procedures are performed to obtain the stress/strain profiles in 3D structures:

(a) The simulation domain and initial material structure are specified. Material properties such as Young modulus and Poisson's ratio, or elasticity constants $C11$, $C12$, $C44$, etc. are specified.

(b) The system of the following equilibrium equation (Eq. (3.47)) is solved in terms of displacement functions (Eq. (3.48)), with the constitutive relations (Eq. (3.49) or Eq. (3.50)) between stress and strain are taken into consideration.

$$\frac{\partial \sigma_{xx}}{\partial x}+\frac{\partial \sigma_{xy}}{\partial y}+\frac{\partial \sigma_{xz}}{\partial z}=0$$

$$\frac{\partial \sigma_{xy}}{\partial x}+\frac{\partial \sigma_{yy}}{\partial y}+\frac{\partial \sigma_{yz}}{\partial z}=0$$

$$\frac{\partial \sigma_{xz}}{\partial x}+\frac{\partial \sigma_{yz}}{\partial y}+\frac{\partial \sigma_{zz}}{\partial z}=0 \qquad (3.47)$$

The system of the following equilibrium equation (Eq. (3.47)) is solved in terms of displacement functions given as (Eq. (3.48)):

$$\varepsilon_{xx}=\frac{\partial u_x}{\partial x}, \varepsilon_{yy}=\frac{\partial u_y}{\partial y}, \varepsilon_{zz}=\frac{\partial u_z}{\partial z},$$

$$\varepsilon_{yz}=\frac{\partial u_z}{\partial y}+\frac{\partial u_y}{\partial z}, \varepsilon_{xz}=\frac{\partial u_z}{\partial x}+\frac{\partial u_x}{\partial z}, \varepsilon_{xy}=\frac{\partial u_y}{\partial x}+\frac{\partial u_x}{\partial y}, \qquad (3.48)$$

with the constitutive relations (Eq. (3.49) or Eq. (3.50)) between stress and strain are taken into consideration as:

$$\sigma = C(\varepsilon - \varepsilon_0) + \sigma_0, \qquad (3.49)$$

where ε_0 and σ_0 are the initial strain and stress, written in vector form for arbitrary matrix C. In the generic case, Eq. (3.49) could be presented as the following system of equations:

$$\begin{pmatrix} \sigma_{xx} \\ \sigma_{yy} \\ \sigma_{zz} \\ \sigma_{yz} \\ \sigma_{xz} \\ \sigma_{xy} \end{pmatrix} = C \times \begin{pmatrix} \varepsilon_{xx}-\varepsilon_{xx}^0 \\ \varepsilon_{yy}-\varepsilon_{yy}^0 \\ \varepsilon_{zz}-\varepsilon_{zz}^0 \\ \varepsilon_{yz}-\varepsilon_{yz}^0 \\ \varepsilon_{xz}-\varepsilon_{xy}^0 \\ \varepsilon_{xy}-\varepsilon_{xy}^0 \end{pmatrix} + \begin{pmatrix} \sigma_{xx}^0 \\ \sigma_{yy}^0 \\ \sigma_{zz}^0 \\ \sigma_{yz}^0 \\ \sigma_{xz}^0 \\ \sigma_{xy}^0 \end{pmatrix}. \qquad (3.50)$$

For isotropic materials, elasticity properties (Eq. (3.51)) are used:

$$\sigma_{xx}=\frac{E(1-\nu)}{(1+\nu)(1-2\nu)}\left(\varepsilon_{xx}-\varepsilon_{xx}^0\right)+\frac{E\nu}{(1+\nu)(1-2\nu)}\left(\varepsilon_{yy}-\varepsilon_{yy}^0\right)$$
$$+\frac{E\nu}{(1+\nu)(1-2\nu)}\left(\varepsilon_{zz}-\varepsilon_{zz}^0\right)+\sigma_{xx}^0$$

$$\sigma_{yy} = \frac{Ev}{(1+v)(1-2v)}\left(\varepsilon_{xx} - \varepsilon_{xx}^{0}\right) + \frac{E(1-v)}{(1+v)(1-2v)}\left(\varepsilon_{yy} - \varepsilon_{yy}^{0}\right)$$
$$+ \frac{Ev}{(1+v)(1-2v)}\left(\varepsilon_{zz} - \varepsilon_{zz}^{0}\right) + \sigma_{yy}^{0}$$

$$\sigma_{zz} = \frac{Ev}{(1+v)(1-2v)}\left(\varepsilon_{xx} - \varepsilon_{xx}^{0}\right) + \frac{Ev}{(1+v)(1-2v)}\left(\varepsilon_{yy} - \varepsilon_{yy}^{0}\right)$$
$$+ \frac{E(1-v)}{(1+v)(1-2v)}\left(\varepsilon_{zz} - \varepsilon_{zz}^{0}\right) + \sigma_{zz}^{0}$$

$$\sigma_{yz} = \frac{Ev}{2(1+v)}\left(\varepsilon_{yz} - \varepsilon_{yz}^{0}\right) + \sigma_{yz}^{0}$$

$$\sigma_{xz} = \frac{Ev}{2(1+v)}\left(\varepsilon_{xz} - \varepsilon_{xz}^{0}\right) + \sigma_{xz}^{0}$$

$$\sigma_{xy} = \frac{Ev}{2(1+v)}\left(\varepsilon_{xy} - \varepsilon_{xy}^{0}\right) + \sigma_{xy}^{0}. \qquad (3.51)$$

3.15 VictoryStress

Accurate simulation of mechanical stress/strain distributions generated during device fabrication is an important part of technology and device design. In many cases, stress effects should be taken into account to predict better manufacturability and to increase the reliability of semiconductor devices. In recent years, stress simulation has become a critical issue in TCAD, due to advances in stress engineering. Stress engineering could be described as a collection of device optimization methods based on the deliberate introduction of stresses into the device cell structure. These methods include dual stress liners (DSL), strain SiGe channels, SiGe (SiC) S/D pockets, the hybrid orientation of n-MOSFET and p-MOSFET devices, and an etch stop layer. All these methods are used to improve device performance by altering carrier mobility in specific areas of the device structure. VictoryStress is a generic 3D stress simulator (see Figure 3.6) that allows accurate prediction of stresses generated during semiconductor fabrication as well as assists users in all aspects of stress engineering. The three main modules in VictoryStress are input file generation, 3D VictoryStress simulation, and output generation. Program flow in VictoryStress is shown in Figure 3.7.

Figure 3.6 VictoryStress information flow [33].

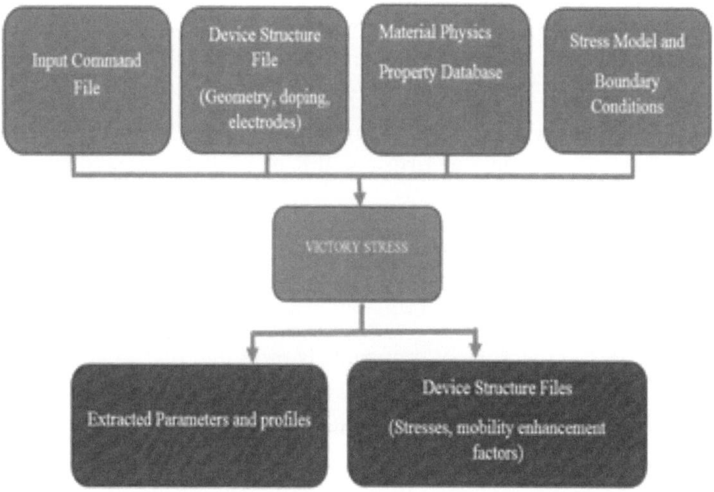

Figure 3.7 Program Flow in VictoryStress [33].

3.15.1 VictoryStress Features and Capabilities

VictoryStress provides a comprehensive set of models and capabilities covering various aspects of stress simulation and stress engineering:

- Stress analysis of arbitrary 3D device structure
- Import of device structures from VictoryProcess, VictoryCell, and ATLAS3D

- Export of 3D stress distributions to VictoryDevice and ATLAS3D
- Models for various sources of strain and stress
- Thermal mismatch between material layers
- Local lattice mismatch due to doping
- Initial intrinsic stress in specified regions
- Hydrostatic stress from capping layers
- Stress/strain generated in the previous processing step (e.g., oxidation)
- Stress simulation for various crystalline (e.g., Si, SiGe, GaAs) and isotropic (e.g., silicon nitride and oxide) materials
- Generic 3D anisotropic stress simulation accounts for wafer orientation and arbitrary wafer flat rotation
- Estimation of mobility enhancement factors (p- and n-type) by the use of piezoresistivity model devices
- Used with VWF for designing of experiments to analyze stress dependence on the process and geometrical parameters of semiconductor

3.16 Piezoresistivity

The electronic structure of materials is modified by the application of stress. Thus, the resistivity of the material depends on its state of deformation commonly known as piezoresistivity. In the framework of Si and Ge, this phenomenon was demonstrated and formalized by Smith in 1954 [34]. Piezoresistivity is an anisotropic property: the increase or decrease of resistivity depends on the orientation of the stress [35, 36]. Although there is a consensus on the value of the piezoresistance coefficients in solid materials, their values require calibration. Indeed, the values piezoresistance coefficients will depend on the nature of the transport in the architecture of the device. The calibration of these coefficients is carried out on experimental measurements, for well-determined stress values, or even on the results of advanced simulation transportation. The piezoresistance model is commonly used in DD simulators for its ease of use and efficiency.

3.17 Modeling Approaches

The basic concept of modeling is the process of solving physical problems through adequate simplification of reality. The definition of modeling varies depending on the application. Modeling typically entails creating a mathematical model with appropriate assumptions for corresponding physical problems, which can take the form of differential or algebraic equations. These mathematical models cannot be solved analytically in many engineering cases, necessitating a numerical solution. Modeling usually entails the creation of an adequate physical model, as well as careful calibration and validation against previously collected data and/or analytical results. For the microelectronics field, the term "device modeling" refers to a set of physical models and methodologies that describe carrier transport and other physical effects in the devices [1].

The Poisson equation, which is largely based on Gauss's law [37], is used to calculate the system's electrostatic solution. For a given material, the Poisson equation defines a relationship between the electrostatic potential and the electric charge density. The continuity equation, which describes charge conservation, can be used to write carrier transport models. The expressions used to calculate the carrier's current density vary between carrier transport models. DD is one of the simplest transport models, and it has been widely used in the industry primarily due to its ease of use, and it will presumably continue to be used in the future, assuming that the simulation parameters are correctly calibrated [38]. DD, which considers carrier transport as diffusion and drift processes, is one of the transport models available in traditional TCAD tools. The carrier movement in the drift component is caused by an applied electric field, while the carrier displacement in the diffusion component is caused by spatial charge variation. The electron gas is believed to be in thermal equilibrium with the lattice temperature in the DD technique. Nonetheless, nanoscale devices are frequently exposed to a strong electric field, which causes carriers to gain energy from the field, raising their temperature and making the device non-uniform [39]. In comparison to the DD model, the hydrodynamic (HD) model has additional driving forces that account for energy distribution on carrier transport. In hetero-structure devices, the current density has contributions from the carrier temperature gradient and the spatial variation of the effective mass, in addition to contributions from the spatial variation of electrostatic potential and the gradient of carrier concentration. Non-uniform temperatures and changes in effective masses along the device are no problem for the HD model.

Furthermore, as device sizes shrink into the nanometer range, carrier transport becomes quasi-ballistic, resulting in non-local effects like velocity overshoot [40]. More advanced transport models, such as HD, have been proposed in an attempt to capture these phenomena, and can also be found in TCAD solvers. The DD and HD models are both approximations of the Boltzmann transport equation, which is a rigorous way of modeling carrier transport in semiconductors. Different models, such as spherical harmonics expansion (SHE), that account for greater moments of the BTE can be found in the literature [41–43]. Furthermore, more complex models must be used to account for additional characteristics in thin-layer structures, such as stress, non-parabolic bands, and geometric quantization [44], and all of these models require proper calibrations.

To calculate the electron concentration in the presence of the confinement effect, one must solve the Poisson and Schrodinger equations consistently [45]. When DG is required, however, the self-consistent Poisson-Schrödinger solution can be used to properly calibrate the model [46]. However, due to velocity overshoot phenomena, the driving electric field in a practical short-channel device is no longer constant along the current path, and the velocity saturation concept becomes somewhat irrelevant [47]. For standard TCAD simulations, the saturation regime of MOSFET devices is a major concern.

3.18 Summary

In this chapter, the requirements of TCAD tools are discussed. Different types of TCAD tools are described briefly. Along with it, the models used to explain the device behavior are given. TCAD has been indicated by the IRDS, as one of the enabling methodologies that can support the advancement of technology progress at the remarkable pace of Moore's law by reducing development cycle times and costs involved in the semiconductor industry. TCAD approaches yielding accurate physical insight and useful predictive results for real-world semiconductor applications are described using multidimensional (2D and 3D) simulations. In this chapter, a brief background on past and present simulation development activities has been given. An overview of currently available commercial TCAD tools, viz., GTS Framework, Silvaco, and Synopsys, has been presented. Nevertheless, the reader should be aware that several other comprehensive simulation programs are in existence today. Developer groups in academia and industry have developed their own set of simulation programs. Indeed, several university groups have developed programs and released them for public use. Each of these programs has certain advantages over the other regarding specific physical models, precise mathematical equations solved, numerical algorithms used, and pre- and post-processing handling of data.

References

1. G. A. Armstrong and C. K. Maiti, *TCAD for Si, SiGe and GaAs Integrated Circuits*, The Institution of Engineering and Technology (IET), UK, 2008.
2. IEEE, International Roadmap for Devices and Systems, 2018.
3. S. Selberherr, *Analysis and Simulation of Semiconductor Devices*, Springer-Verlag, Vienna, 1984.
4. GTS Framework, MINIMOS-NT User Manual, 2020.
5. MINIMOS-NT User's Guide, 1998.
6. Z. Stanojevic *et al.*, "Nano device simulator: A practical subband-BTE solver for path-finding and DTCO," *IEEE Trans. Electron Devices*, pp. 1–7, 2021. doi: 10.1109/TED.2021.3079884.
7. O. Baumgartner, Z. Stanojevic, K. Schnass, M. Karner, and H. Kosina, "VSP - A quantum-electronic simulation framework," *J. Comput. Electron.*, vol. 12, no. 4, pp. 701–721, Dec. 2013. doi: 10.1007/s10825-013-0535-y.
8. D. A. Lemus, J. Charles, and T. Kubis, "Mode-space-compatible inelastic scattering in atomistic nonequilibrium Green's function implementations," *J. Comput. Electron.*, 2020. doi: 10.1007/s10825-020-01549-8.
9. A. Abramo, "A general purpose 2D Schrodinger solver with open/closed boundary conditions for quantum device analysis," in *SISPAD '97. 1997 International Conference on Simulation of Semiconductor*

Processes and Devices. Technical Digest, 1997, pp. 105–108. doi: 10.1109/SISPAD.1997.621347.

10. Silvaco International, ATLAS Device Simulator, 2018.
11. S. International, VictoryProcess User Manual, 2018.
12. R.W. Dutton and Z. Wu, "Technology-oriented CAD," *Technology CAD — Computer Simulation of IC Processes and Devices*, Springer US, pp. 1–36, 1993. doi: 10.1007/978-1-4615-3208-8_1.
13. Silvaco International, VictoryDevice User Manual, 2018.
14. A. Wettstein *et al.*, "Quantum device-simulation with the density-gradient model on unstructured grids," *IEEE Trans. Electron Devices*, vol. 48, no. 2, pp. 279–284, Feb. 2001. doi: 10.1109/16.902727.
15. N. A. Zakhleniuk, "Nonequilibrium drift-diffusion transport in semiconductors in presence of strong inhomogeneous electric fields," *Appl. Phys. Lett.*, vol. 89, no. 25, p. 252112, Dec. 2006. doi: 10.1063/1.2420787.
16. D. Yakimets, "Vertical Transistors: A slippery path towards the ultimate CMOS scaling," PhD Thesis, KU Leuven, 2017.
17. H. F. Dadgour, K. Endo, V. K. De, and K. Banerjee, "Grain-orientation induced work function variation in nanoscale metal-gate transistors part II: Implications for process, device, and circuit design," *IEEE Trans. Electron Devices*, vol. 57, no. 10, pp. 2515–2525, 2010. doi: 10.1109/TED.2010.2063270.
18. S. Datta *et al.*, "High mobility Si/SiGe strained channel MOS transistors with HfO_2/TiN gate stack," in *IEEE International Electron Devices Meeting 2003*, 2003, pp. 28.1.1–28.1.4. doi: 10.1109/IEDM.2003.1269365.
19. E. P. Gusev, D. A. Buchanan and E. Cartier, "Ultrathin high-K gate stacks for advanced CMOS devices," in *IEEE IEDM Tech. Dig.*, 2001, pp. 451–454. doi: 10.1109/IEDM.2001.979537.
20. L. Shen, S. Di, L. Yin, X. Liu, and G. Du, "Calibration of drift-diffusion model in quasi-ballistic transport region for FinFETs," *Sci. China Inf. Sci.*, vol. 61, no. 6, Jun. 2018. doi: 10.1007/s11432-017-9315-4.
21. C. Lombardi *et al.*, "A physically based mobility model for numerical simulation of nonplanar devices," *IEEE Trans. Comp. Aided Des.*, vol. 7, no. 11, pp. 1164–1171, 1988. doi: 10.1109/43.9186.
22. D. Vasileska, S. Goodnick, and G. Klimeck, "Computational electronics: Semiclassical and quantum device modeling and simulation," *Comput. Electron. Semiclassical Quantum Device Model. Simul.*, 2017. doi: 10.1201/b13776.

23. C. K. Maiti and G. A. Armstrong, *Applications of Silicon-Germanium Heterostructure Devices*, Inst. of Physics Publishing, Bristol, 2001.
24. A. Schenk, "A model for the field and temperature dependence of shockley-read-hall lifetimes in silicon," *Solid-State Electron.*, vol. 35, no. 11, pp. 1585–1596, Nov. 1992. doi: 10.1016/0038-1101(92)90184-E.
25. D. M. Caughey and R. E. Thomas, "Carrier mobilities in silicon empirically related to doping and field," in *Proc. IEEE*, 1967, vol. 55, pp. 2192–2193. doi: 10.1109/PROC.1967.6123.
26. S. Selberherr *et al.*, "The evolution of the MINIMOS mobility model," *Solid-State Electron.*, vol. 33, no. 11, pp. 1425–1436, Nov. 1990. doi: 10.1016/0038-1101(90)90117-W.
27. S. Selberherr, "MOS device modeling at 77 K," *IEEE Trans. Electron Dev.*, vol. 36, pp. 1464–1474, 1989. doi: 10.1109/16.30960.
28. G. Masetti *et al.*, "Modeling of carrier mobility against carrier concentration in arsenic-, phosphorous-, and boron-doped silicon," *IEEE Trans. Electron Dev.*, vol. 30, no. 7, pp. 764–769, 1983. doi: 10.1109/T-ED.1983.21207.
29. M. J. van Dort *et al.*, "Lifetime calculations of MOSFETs using depth-dependent non-local impact ionization," *Microelectron. J.*, vol. 26, no. 2–3, pp. 301–305, 1995. doi: 10.1016/0026-2692(95)98932-H.
30. J. W. Slotboom and H. C. de Graaff, "Measurements of bandgap narrowing in Si bipolar transistors," *Solid State Electron.*, vol. 19, no. 10, pp. 857–862, Oct. 1976. doi: 10.1016/0038-1101(76)90043-5.
31. J. Eberhardt and E. Kasper, "Bandgap narrowing in strained SiGe on the basis of electrical measurements on Si/SiGe/Si hetero bipolar transistors," *Mater. Sci. Eng.*, vol. B89, no. 1–3, pp. 93–96, Feb. 2002. doi: 10.1016/S0921-5107(01)00808-X.
32. J. Slotboom *et al.*, "Measurements of band-gap narrowing in Si bipolar transistor," *Solid State Electron.*, vol. 19, no. 10, pp. 857–862, Oct. 1962. doi: 10.1016/0038-1101(76)90043-5.
33. Silvaco International, VictoryStress User Manual, 2018.
34. C. S. Smith, "Piezoresistance effect in germanium and silicon," *Phys. Rev.*, vol. 94, no. 1, pp. 42–49, 1954. doi: 10.1103/PhysRev.94.42.
35. F. Payet, "Modélisation et intégration de transistors a canal de silicium contraint pour les noeuds technologiques CMOS 45nm et en deçà," 2005.
36. T. Guillaume, "Influence des contraintes mécaniques non-intentionnelles sur les performances des transistors MOS à canal ultra-court," 2005.

37. J. C. Maxwell, *A treatise on electricity and magnetism*, Dover Publ., 1873.

38. M. Lundstrom, "Drift-diffusion and computational electronics-still going strong after 40 years!," *Simul. Semicond. Process. Devices Conf.*, 2015. doi: 10.1109/SISPAD.2015.7292243.

39. T. Grasser *et al.*, "A review of hydrodynamic and energy-transport models for semiconductor device simulation," *Proc. IEEE*, vol. 91, pp. 251–274, 2003. doi: 10.1109/JPROC.2002.808150.

40. G. Sasso, N. Rinaldi, G. Matz, and C. Jungemann, "Analytical models of effective DOS, saturation velocity and high-field mobility for SiGe HBTs numerical simulation," in *2010 International Conference on Simulation of Semiconductor Processes and Devices*, 2010, pp. 279–282. doi: 10.1109/SISPAD.2010.5604505.

41. B. H. Hong *et al.*, "Possibility of transport through a single acceptor in a gate-all-around silicon nanowire PMOSFET," *IEEE Trans. Nanotechnol.*, vol. 8, no. 6, pp. 713–717, 2009. doi: 10.1109/TNANO.2009.2021844.

42. C. Jungemann *et al.*, "Stable discretization of the Boltzmann equation based on spherical harmonics, box integration, and a maximum entropy dissipation principle," *J. Appl. Phys.*, vol. 100.2, p. 024502, 2006. doi: 10.1063/1.2212207.

43. T. Grasser, R. Kosik, C. Jungemann, B. Meinerzhagen, H. Kosina, and S. Selberherr, "A non-parabolic six moments model for the simulation of sub-100 nm semiconductor devices," *J. Comput. Electron.*, vol. 3, no. 3–4, pp. 183–187, 2004. doi: 10.1007/s10825-004-7041-1.

44. O. Penzin, L. Smith, A. Erlebach, and K. H. Lee, "Layer thickness and stress-dependent correction for ingaas low-field mobility in TCAD applications," *IEEE Trans. Electron Devices*, vol. 62, no. 2, pp. 493–500, Feb. 2015. doi: 10.1109/TED.2014.2383377.

45. J. Colinge *et al.*, "Quantum`mechanical effects in trigate SOI MOSFETs," *IEEE Trans. Electron Devices*, vol. 53, no. 5, pp. 1131–1136, 2006. doi: 10.1109/TED.2006.871872.

46. N. Pons *et al.*, "Density gradient calibration for 2D quantum confinement: Tri-gate SOI transistor application," in *2013 International Conference on Simulation of Semiconductor Processes and Devices (SISPAD)*, 2013, pp. 184–187. doi: 10.1109/SISPAD.2013.6650605.

47. G. Hiblot *et al.*, "Impact of short-channel effects on velocity overshoot in MOSFET," in *Conference Proceedings - 13th IEEE International NEW Circuits and Systems Conference, NEWCAS 2015*, Aug. 2015, pp. 1–4. doi: 10.1109/NEWCAS.2015.7182061.

Chapter 4

Nanowire Transistors

Suprava Dey
Council of Scientific and Industrial Research—Institute of Minerals and Materials Technology, Bhubaneswar 751013, Odisha, India
supravadey24@gmail.com

Owing to the demands generated by the internet of things (IoT) applications on having low power and low leakage devices, nanowire devices are becoming very important. The field-effect transistors (FETs) are today the cornerstone of microelectronics and the component becomes ubiquitous in everyday life. Efforts are being made today on reducing the power consumed as well as on the development of smart devices. The microelectronics industry is facing difficult challenges in decreasing the physical dimensions, due to the precision of the manufacturing processes, materials, and physical phenomena [1, 2] at the nanoscale (Figure 4.1). Intel marketed transistors at the 10 nm technology node in 2017, and Global Foundries is actively working on making transistors at the 7 nm node. This extreme-scale involves going beyond many barrier techniques, to make operational chips composed of billions of transistors.

Fabless Semiconductor Manufacturing: In the Era of Internet of Things
Edited by Chinmay K. Maiti

ISBN 978-981-4968-27-0 (Hardcover), 978-1-003-31497-4 (eBook)
www.jennystanford.com

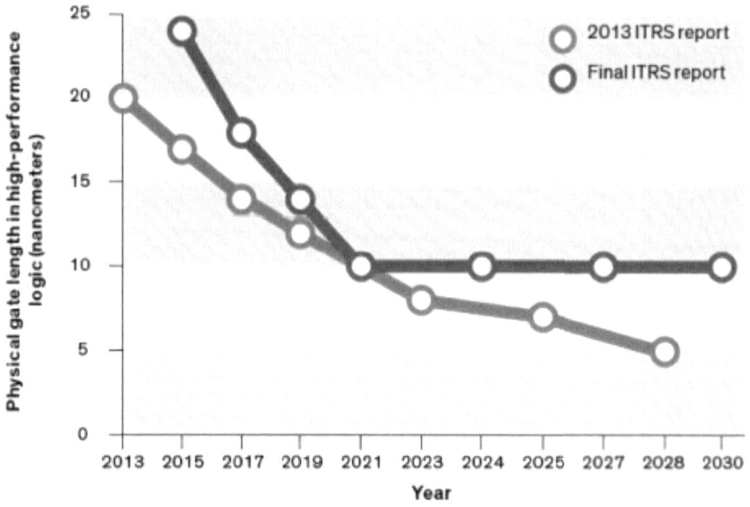

Figure 4.1 Evolution of transistor gate length for high-performance logic planned for future technology nodes by ITRS 2013 [3].

Gate-all-around (GAA) transistors are based on nanowire and nanosheets. Nanowires and nanosheets have the advantage of having a smaller channel and a high surface-to-volume ratio as transistor sizes shrink. Furthermore, these transistors offer several advantages over the FinFET technology. The GAA architecture resembles the FinFET architecture in some ways. The nanowire channel in a GAAFET is surrounded by gates in all directions. As a result, electrostatic gate control is excellent and is an excellent candidate for pushing the scalability down to the sub-7 nm node regime, ushering in a new era of ultra-scaled transistors. Reducing the dimension of the transistors also means reducing the thickness of the gate oxide to help maintain optimum electrostatic control of the conduction in the channel. At the nanoscale, the doped zones, which define the source and the drain of the transistor, have non-negligible so-called short channel effects (SCE). As a result, the potential of the channel does not reach its maximum value in the neutral zones of the channel with a lowering of the threshold voltage. This lowering of the threshold voltage has a negative impact on the performance of the transistor. The SCE and the drain-induced barrier lowering (DIBL) shift the current characteristics toward low voltages while increasing their slope below the threshold. The off current, I_{OFF} increases, and therefore the passive consumption of the component rises. These effects can be associated with electrostatic integrity (EI), which allows quantifying their impacts on the electrical behavior of the transistor. The FDSOI architecture making it possible to remove this technological barrier of miniaturization has been adopted in the industrial environment targeting low consumption applications.

The thickness of gate oxide (SiO_2) needed to control correctly the short channel is today about 1 nm. This thickness is so small that it causes a tunnel leakage phenomenon through the gate oxide [4]. Leakage current increases exponentially with the decrease in oxide thickness. Silicon oxide is not a satisfactory gate dielectric material to fulfill the role of gate insulator because the thickness required to maintain good electrostatic control becomes too small [5]. This SCE phenomenon reduces the potential barrier between the source and the drain, which virtually amounts to modifying channel doping or threshold voltage. DIBL reduces further the threshold voltage. The DIBL is thus defined by the variation of threshold voltage (V_T) when the drain voltage (V_D) is changed (when normalized) may be given as:

$$\mathrm{SCE} = \frac{\varepsilon_{\mathrm{Si}}}{\varepsilon_{\mathrm{ox}}} \frac{t_{\mathrm{ox}}}{L_{\mathrm{EL}}} T_{\mathrm{Dep}} L_{\mathrm{EL}} \times \left(1 + \frac{X_j^2}{L_{\mathrm{EL}}^2}\right) \times \frac{k_B T}{q} \ln\left(\frac{N_{\mathrm{ch}} N_{\mathrm{sd}}}{\mathrm{ni}^2}\right)$$

$$\mathrm{DIBL} = \frac{\varepsilon_{\mathrm{Si}}}{\varepsilon_{\mathrm{ox}}} \frac{t_{\mathrm{ox}}}{L_{\mathrm{EL}}} \frac{T_{\mathrm{Dep}}}{L_{\mathrm{EL}}} \times \left(1 + \frac{X_j^2}{L_{\mathrm{EL}}^2}\right) \times V_D$$

$$V_{\mathrm{Tshort}} = V_{\mathrm{Tlong}} - \mathrm{SCE} - \mathrm{DIBL},$$

where $\varepsilon_{\mathrm{Si}}$ is the relative permittivities of the materials, t_{ox} is the gate oxide thicknesses, T_{Dep} is the depletion, L_{EL} is the electrical length, X_j is the depth of the junction, N_{ch} is the doping of the channel, and N_{sd} is the source-drain doping. The equations show that to combat the effects of short channels, several solutions are possible. First of all by reducing the depth of the junctions X_j either by playing on the implantation depth of dopants, or by adopting the FDSOI structure that naturally limits the latter. Second, by decreasing the depth of desertion T_{Dep}, a criterion also fulfilled by the fine silicon channel architecture and finally by reducing the oxide thickness.

To compensate for the effects of DIBL and tunnel leakage in the barrier, the oxide has been changed to oxide with a higher dielectric constant. Indeed, a dielectric with a larger constant will give the same electrical capacity as the SiO_2 but with a greater physical thickness. This thickness can be large enough to reduce leakage by tunnel effect through the gate dielectric. An equivalent oxide thickness (EOT) is defined by calculating the ratio of dielectric coefficients between SiO_2 and high-k oxide. Selecting high-k oxide is complex because it has to meet several physical criteria, such as it must have a large bandgap and it must be stable enough to be used during the process for manufacturing transistors. It must also have a good interface with the silicon channel, otherwise, carrier mobility in the channel and therefore the current in the ON-state would be greatly reduced by the gate/channel oxide interface defects. In practice, it is hafnium oxide (HfSiO (N) or HfO (N)) that is in use to replace silicon oxide [6].

4.1 Silicon on Insulator (SOI) Transistors

Today transistors are fabricated SOI substrates. These substrates are made using SmartCut technology, which involves a thin layer of high-quality monocrystalline silicon on a substrate of lower quality thus reducing the production costs, and on the other hand, results in a new architecture with an insulator that separates the bulk substrate and the transistor channel [7]. The isolation of the channel and the substrate offers many advantages in terms of electrostatic control of the transistor. This removes particular the effects of leakage by the silicon substrate (punch-through) that appear for transistors with a low gate length. On the other hand, it allows having additional electrostatic control of the channel. Indeed, the SOI structure is equivalent to an additional gate located on the backside of the channel. This additional gate can also be used to control the threshold voltage of the transistor dynamically. SOI technology also allows the introduction of strong deformations in the active layer of silicon by transferring stresses from an initial relaxed SiGe substrate [8]. The study of strain engineering on the properties of the SOI transistors will be discussed later.

4.2 Multi-Gate Architecture

Beyond the so-called planar technology, for which the substrate and the channel are stacked, the insulator then the gate, it is possible to manufacture transistors with a surrounded gate and channel partially or totally (see Figure 4.2). To do this, we anneal the silicon film to obtain a narrower silicon wire, which we will then oxidize on certain surfaces according to the number of gates required [9, 10]. We can thus differentiate several architectures (see Figure 4.3). These particular architectures make it possible to improve greatly the electrostatic control of the limiting channel as well as the effects of short channels, which have been described previously. This improvement of electrostatic control is proportional to the gate area that covers the channel. GAA architecture, therefore, represents the ultimate architecture in terms of transport control. In addition, these architectures also make possible high-density integration

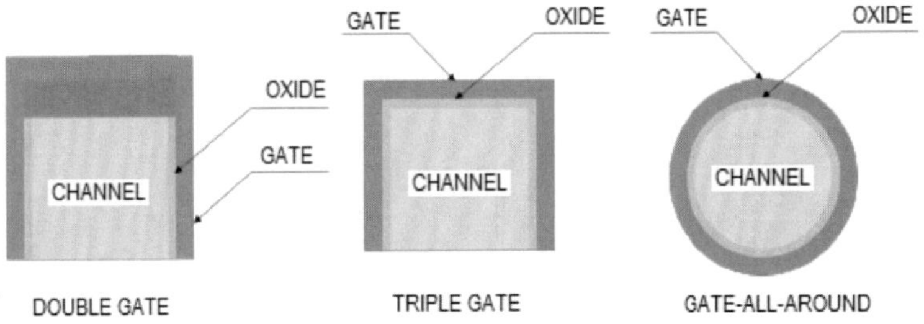

Figure 4.2 Diagram illustrating different multi-gate stacks. From left to right: double gate, triple gate, and GAA [14].

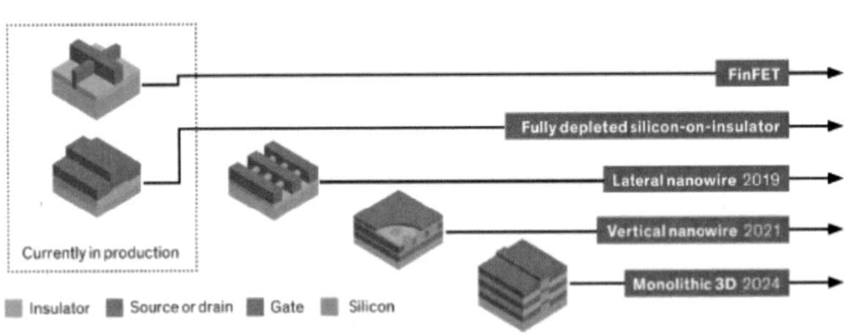

Figure 4.3 Evolution of the architectures of transistors envisaged for the technological nodes to come [14].

strategies for transistors since they make it possible to build stacks of transistors and thus to take advantage of the vertical dimension of a chip contrary to the standard design, which occupies only the horizontal plane [11, 12]. It is thus possible to stack transistors completely or partially encapsulating gates, which allow multiplying the density of transistors (see Figure 4.4) [13]. TriGate nanowire architecture will be described in more detail later.

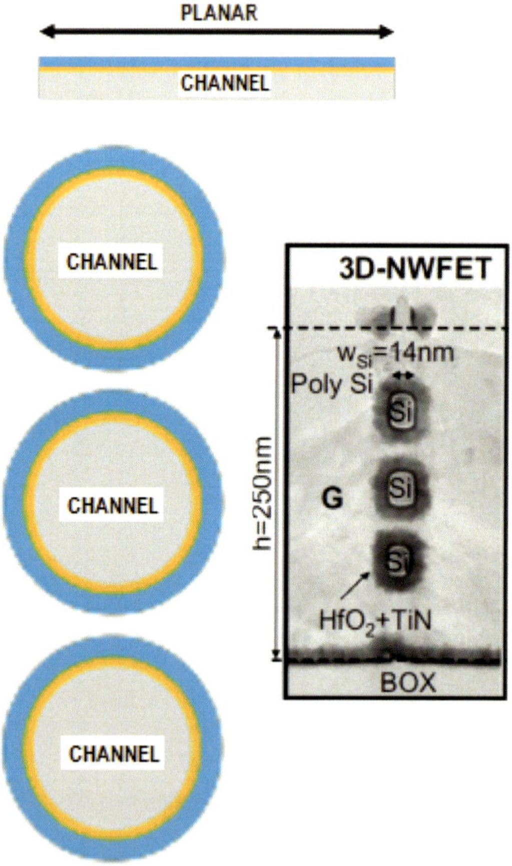

Figure 4.4 (Left) Comparative diagram between a planar design and a stacked GAA design illustrating the density gain of transistors for stacked nanowires. (Right) Example of a structure with stacked nanowires realized on SOI substrate [14].

4.3 Ultimate Device: GAA Nanowire Transistor

The performance of integrated circuits has improved significantly through scaling during the last six decades. The downscaling of MOS transistors has been limited by the SCE. To overcome the SCE, the strain was introduced. Below the 90 nm technology node, strain engineering appears to be one of the reliable technologies for the improvement of overall carrier mobility in planar transistors. Similarly, high-k/metal gate stacks are in use since 45 nm technology node to overcome gate leakage and static power dissipation [15]. It was not sufficient for further scaling of the devices. In this regard, the introduction of triple gate structures (mainly in FinFETs) has emerged as a defending contender to overcome the SCE along with further miniaturization. The channel electrostatics was improved by taking triple gate structures with scaling of gate length down to 15 nm at 7 nm technology node [16]. The performance of triple gate structures is degraded below 7 nm technology node [17]. The scaling of gate length can be continued in sub-10 nm node with improved electrostatics using GAA nanowire transistors [18]. The best electrostatic control of the gate over the channel can be achieved only when it can be controlled from all sides of the channel.

The focus of this chapter is to understand the effect brought by extreme reduction of the channel and gate lengths in multigate transistors. The ultimate miniaturization of the silicon channel leads to the production of nanowire devices. At the nanoscale, the devices guarantee to better control the conduction channel when the gate surrounds the channel. Different methods exist to manufacture such devices, and small dimensions can be defined by lithography on Si or SiGe [19, 20]. The release of the nanowire structure is generally permitted by etching the BOX under a structure manufactured on SOI [21]. This method allows one to define nanowire diameters up to 3 nm in manufacturing [22]. Several studies have shown the interest to realize stacked nanowire transistors [23–25]. Using a multilayer stacking with minimal bulk, excellent currents have been reported [26]. With the increase in the number of channels, the definition of the gate becomes critical since access to materials is limited during etching processes that define surface structures only. It is thus necessary to define the gates in a self-aligned way without etching process. The use of the damascene process (deposit of cavity filling gate) was proposed on nanowire devices [27, 28]. The integration is complex and requires in particular two stages of etching to define the gate. The method makes it possible to define a nanowire transistor with a small diameter and 30 nm gate length. To circumvent the difficulty of self-aligned surrounding gate fabrication, an 'ideal' process has been proposed by Monfray *et al.* [29]. This innovative process proposes the use of a single lithography step to define the gate under the channel at the same time as the upper gate (Figure 4.5). This process thus allows us to get closer to standard process sequences.

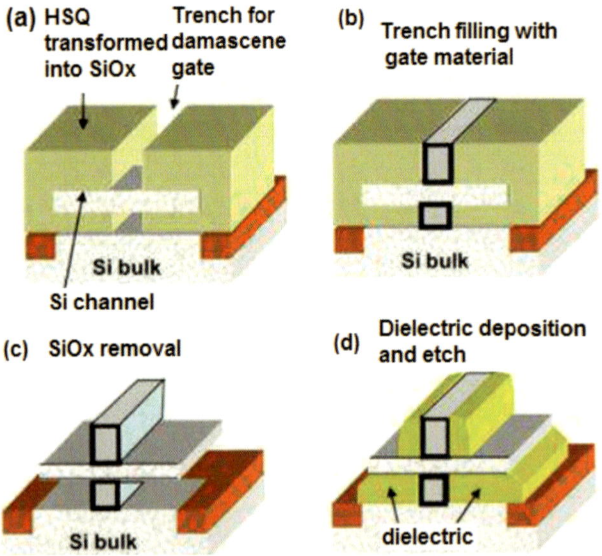

Figure 4.5 Process for manufacturing nanowire transistor. The sequence of key process steps of a GAA self-aligned gate isolated by a buried dielectric [14].

4.4 Design of Nanowire Transistors

The nanowire transistor is one of the most promising candidates because GAA nanowires have better gate control than FinFETs. The nanowire's cross-section is smaller than that of the FinFET. As a consequence, the quantization model requires the solution of the Schrödinger equation. At the 5 nm technology node, transport is quasi-ballistic and should be physically modeled. To determine the best efficiency, design, and manufacturability trade-offs, a subband Boltzmann transport equation (subband-BTE) simulation [30] of the nanowire transistor is required. The phonon scattering and surface-roughness scattering models are used in the transport simulation. To perform 3D analysis of 5 nm Si nanowires, the following features were included in simulation: electronic band structure, ellipsoidal model calibrated to two-band k.p model for electrons, and six-band k.p model for holes. Scattering mechanisms included in simulations are three g-type and three f-type intervalley inelastic processes and one intravalley acoustic scattering for electrons. Optical phonon scattering and acoustic phonon scattering are for holes. Surface roughness scattering: the exponential model of the power-spectrum density functions for both electrons and holes.

The distance of the subband energy widens as the cross-section of the nanowire shrinks. As a result, fewer subbands would be involved in transport. Since the number of subbands (the number of Boltzmann equations and the inter-subband scattering matrix) determines the simulation time, it is important to choose the right number of subbands to complete the simulation in a reasonable amount of time.

GAA nanowire transistors are now considered to be the most suitable and feasible alternative structures for ongoing rapid scaling even below 7 nm technology nodes [31]. Today's emphasis on low-power and low-voltage operation in industries and academies requires an understanding of the nanodevices at ultimate scaling [32]. Silicon nanowire (SiNW) FET has received serious attention since long as biological [33] and chemical sensors [34] and as thermoelectric [35] photovoltaic [36], and nanoelectronic devices [37, 38]. For all of these applications, performance improvements are necessary for nanowire devices with scaling. According to the report from the International Roadmap for Devices and Systems [31], the technology generation of logic devices in the industry would reach 5 nm and 3 nm in the years 2021 and 2024, respectively.

IRDS considers nanowire as the ideal transistor [31]. The design of the semiconductor device and optimization of its key parameters are dependent on accurate device simulation. The physical behavior of the device and its electrical characteristics can be easily understood from the simulation. The behavior of the device can be modeled by using the technology computer-aided design (TCAD) tools [39, 40]. TCAD can be "predictive" when it has a very high level of accuracy. But, proper calibration is required for predictive TCAD. This approach is very expensive and time-consuming. Also, it is very difficult to calibrate a set of TCAD simulators. Suppose, if the threshold voltage does not match the experimental value, one needs to check the assumptions in both the process and the device simulators, as well as the electrical measurements and the test structure layout. Thus, such calibration is performed by incorporating the exact device geometry and materials along with other relevant device parameters, as reported in the experimental work. Continual calibration of model parameters is also required. TCAD calibration has been performed for an experimental 22 nm Si GAANW [21].

4.5 Process Simulation of Nanowire Transistors

The more effective suppression of the short-channel effect will be needed as CMOS technology progresses toward the 5 nm technology node [41, 42]. The nanowire transistor is one of the most promising candidates because GAA nanowires have better gate control than FinFETs. The nanowire's cross-section is smaller than that of the FinFET. As a consequence, the quantization model requires the solution of the Schrödinger equation. The 5 nm technology node's transport is quasi-ballistic and should be physically modeled. To determine the best efficiency, design, and manufacturability trade-offs, a subband-BTE simulation of the nanowire transistor is required.

The TCAD calibration involves constructing 3D Si nanowire structures, simulating source/drain doping diffusion, and simulating the subband-BTE with the Schrödinger–Poisson solver. The phonon scattering and surface-roughness scattering models are used in the transport simulation. Device engineers working on cutting-edge transistor architecture are the expected customers. The project involves creating 3D Si nanowire structures, simulating source/drain (S/D) doping diffusion, and simulating the subband-BTE with the Schrödinger–Poisson solver. The phonon scattering and surface-roughness scattering models are used in the transport simulation.

For both n-type and p-type applications, the drain current-gate voltage (I_D–V_G) curve of 3D circular and trapezoidal nanowires was simulated. 3D device simulations were performed using the TCAD software tool, Sentaurus Device, with calibrated models for carrier transport and quantum mechanical effects described in [43]. On-state drive current, off-current, subthreshold slope (SS), and threshold voltage are all significant system characteristics. Furthermore, it gives 3D carrier density and electrostatic potential profiles, as well as carrier density, potential, and current for each subband, spectral current, and density-of-states.

The simulated device is similar to Samsung's technology [44] with a 15-nm gate length. This technology uses process steps that are similar to those used in traditional CMOS but are combined to create a two nanowire structure. The simulation detail is outlined and the simulation results are discussed in the following sections. Structure generation, diffusion dynamics, and implantation simulations are all included in this 3D process simulation. Some of the most important parameters of the device are gate length 15 nm, nanowire radius 5 nm, and oxide thickness 3.5 nm. Sentaurus Process [45] and Sentaurus Structure Editor library are used to simulate 3D NWFET devices. The electrical characteristics of the device are simulated in 3D using Sentaurus Device. The process simulation has been performed using TCAD Sentaurus Version A-2007.12.

The first procedure that involves a new mesh in the current process flow is the implantation of a lightly doped drain (LDD). The arsenic doping profile after 3D implantation is shown in Figure 4.6. Doping distribution in NWFET; the trench, gate, and spacer regions are transparent; and a quarter of the device is shown. The nanowire channel shape after emulation of rounding is presented in Figure 4.7, along with the XX-stress distribution profile.

By setting the contacts and building the device simulation mesh suitable for device simulation, the stand-alone Sentaurus Structure Editor is used to generate the structure for device simulation. The simulation mesh is created using the meshing tool, Noffset3D. Noffset3D enables high-quality meshes to be generated in curved interface areas like the nanowire channel. The use of a boundary-conforming mesh in the nanowire channel region is critical for accurate simulation of the device's electrical characteristics. Sentaurus Device was used to perform the device simulations. The set of transport models used for all simulations includes the drift-diffusion (DD) transport model, Philips unified mobility model, high-field mobility degradation, mobility degradation at the silicon-oxide interface, density-gradient (DG) quantum corrections, and Egley stress-induced mobility model. The electron current profile, electric field, hole mobility, and electron mobility are shown in Figures 4.8–4.11.

Figure 4.12 shows the drain current vs. gate voltage characteristics for the device. Two different drain biases are simulated: 0.05 V and 1.0 V. The findings are consistent with those reported in the literature. The simulated SS and maximum transconductance are both in very good agreement with the ones reported in reference [46].

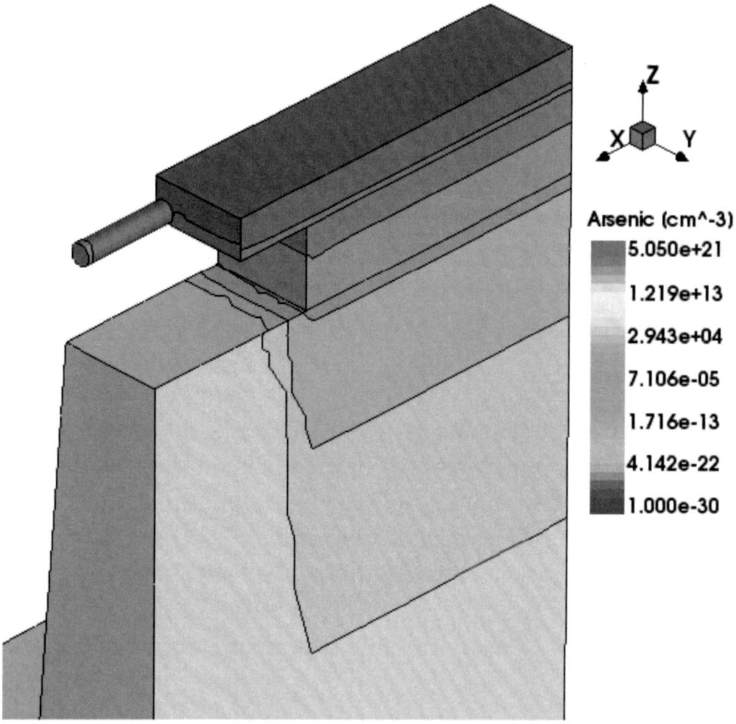

Figure 4.6 Arsenic doing distribution after LDD implantation.

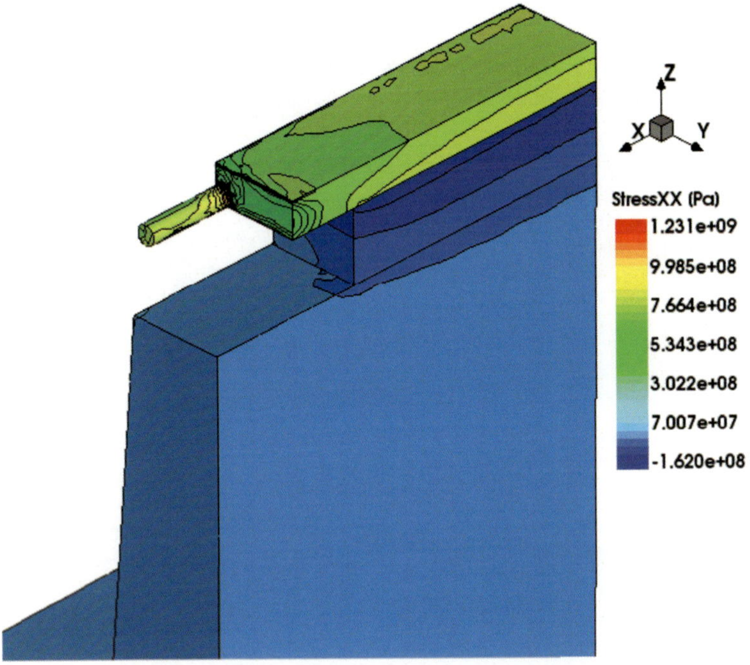

Figure 4.7 Nanowire channel shape after emulation of rounding; XX-stress distribution is shown.

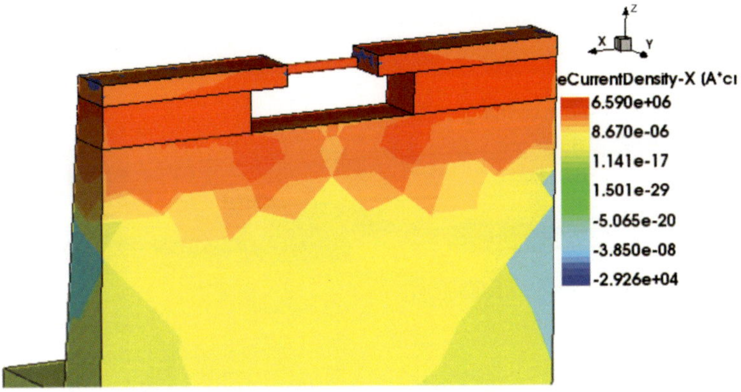

Figure 4.8 Electron current profile.

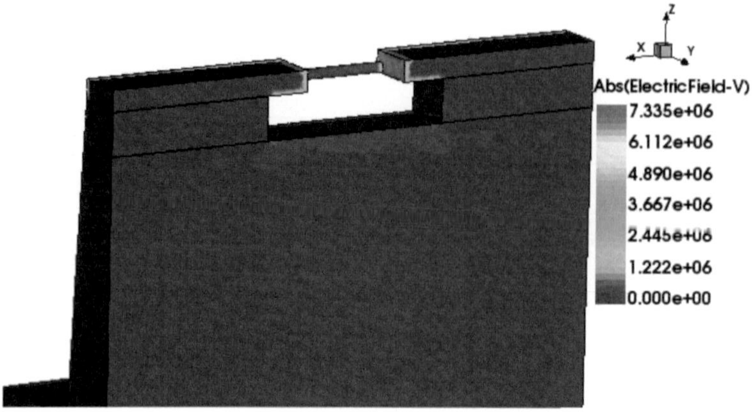

Figure 4.9 Electric field.

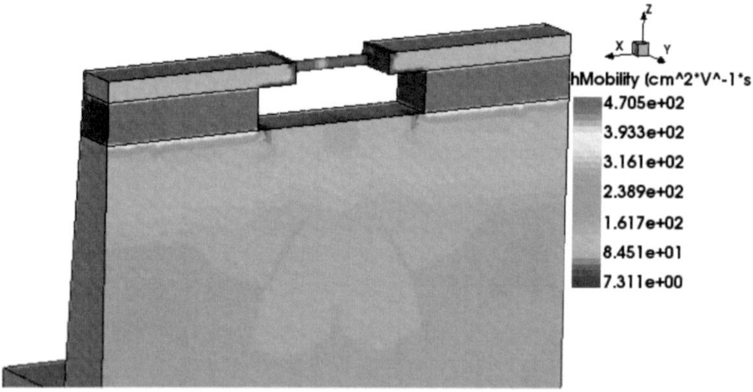

Figure 4.10 Hole mobility.

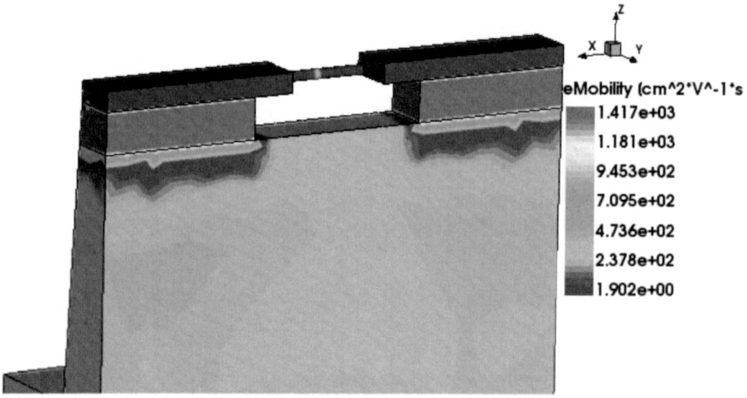

Figure 4.11 Electron mobility.

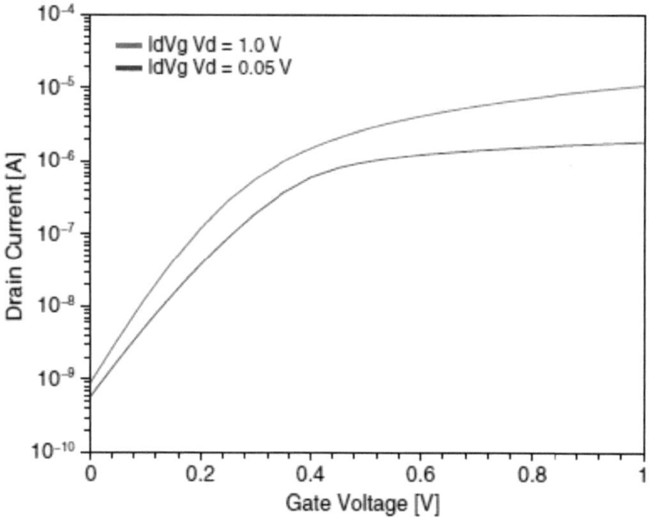

Figure 4.12 Simulation of I_D–V_G characteristics at V_D of 0.05 V and 1.0 V.

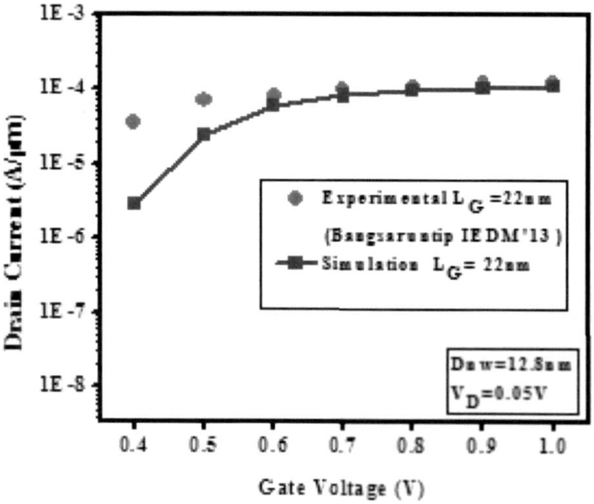

Figure 4.13 TCAD calibration. Comparison of simulation and experimental I_D–V_G characteristics at V_D of 0.05 V for a 22 nm GAANW [21].

In the following, toward device simulation calibration, we have chosen the experimental GAA SiNW reported by Bangsaruntip *et al.* [21]. The authors have demonstrated a manufacturing method capable of producing a GAA SiNW with a pitch of 30 nm and a scaled gate pitch of 60 nm that is compatible with the CMOS scaling requirements of the 10 nm node and beyond. The ensemble of simulated devices consists of n-type and p-type nanowire transistors, both with a circular cross-section, 12.8 nm in diameter, with an EOT of 1.5 nm. The gate length was set to 22 nm. Figure 4.13 shows a comparison of simulation and experimental I_D–V_G characteristics at a V_D of 0.05 V. At the electrical calibration stage, the mobility models implemented in the TCAD device simulation tool have been tuned to experimental data in conjunction with the published mobility models. Note that at a low drain bias of 50 mV and gate biases greater than 0.4 V, the simulated current match is good with the experimental data for 22 nm gate length.

4.6 Design at Nanoscale

Various design and technological factors should be considered for device design. The design and technology should also be co-optimized to predict the next node device that is feasible to build, has enough drive to meet application performance requirements, consumes minimal energy, and is cost-effective [47]. The uniqueness of this design and simulation experiment stems from the fact that we should conduct benchmarking comprehensively, incorporating design methodologies and technical limits that are expected at the 5 nm node. The SCE immunity of nanowire FET can be quantitatively analyzed by using the natural length λ and is used to analyze the immunity toward the SCE of nanowire FET. It has been derived from Poisson's equation as [48]:

$$\lambda = \sqrt{\frac{\varepsilon_{\mathrm{Si}}}{8\varepsilon_{\mathrm{ox}}} t_{\mathrm{Si}}^2 \ln\left(1 + \frac{2t_{\mathrm{ox}}}{t_{\mathrm{Si}}}\right) + \frac{t_{\mathrm{Si}}{}^2}{16}}, \tag{4.1}$$

where the permittivity of silicon is presented by $\varepsilon_{\mathrm{Si}}$, the permittivity of oxide is presented by $\varepsilon_{\mathrm{ox}}$, the thickness of silicon is presented by t_{Si}, and the oxide thickness is presented by t_{ox}.

The value of λ is found to be 4.9, which is less than half of the gate length. Hence, the impact of SCE can be reduced by designing the device. The device (i.e., Figure 4.14) is designed at a 5 nm technology node using the parameters, which are listed in Table 4.1. First, the mesh of the device was created (i.e., shown in Figure 4.14a). The different region was assigned with material (see Figure 4.14b). The net concentration of different regions is shown in Figure 4.14c. The schematic of the device is shown in Figure 4.14d.

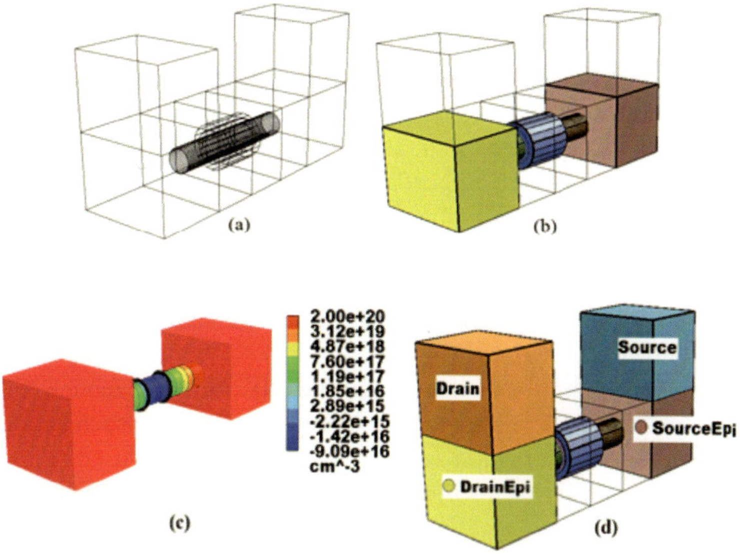

Figure 4.14 Nanowire FET device with a gate length of 10 nm.

Table 4.1 Device geometry and other parameters are used in the simulation

Parameter	Value	Unit
Number of Wire(s)	1	-
Nanowire Diameter	5	nm
Channel Length	10	nm
High-k (HfO_2) Thickness	1.2	nm
Oxide (SiO_2) Thickness	1	nm
Contact Width	20	nm
Contact Length	15	nm
Spacer Length	10	nm
LDD Doping	1e18	nm
Channel Doping	1e17	cm^{-3}
S/D Doping	2e20	cm^{-3}

Nanowire FET design using Si as channel material is considered first. With scaling down the thickness of SiO_2 reduces which leads to an increase in leakage in the gate. The minimization of gate leakage is possible by using high-k along with SiO_2 [49]. So, HFO_2 has been used along with SiO_2. Here, TiN has been used as the metal electrode. The simulation has been carried out by using GTS TCAD tools [50]. Simulation of nanoscaled devices needs different modeling approaches and selection of suitable transportation models, as followed by several research groups. Among them, full quantum models based on the tight-binding Hamiltonian provide a detailed physical description of nanostructures, while semi-classical models based on the solution of the Boltzmann transport equation (e.g., by Monte Carlo techniques) have been recently extended to describe nanowire inversion layers for both the low field and far from equilibrium quasi-ballistic regime where conventional TCAD tools become inaccurate [51].

In the classical model, such as the DD transport model, the quantum effect has not been considered. The DG approach is one of the quantum correction approaches, which is widely used. When the devices are scaled down into the nanometer range, the quantum effect plays an important role in device characteristics. Thus, the quantum effects need to be considered in simulation [52–54]. Generally, in classical theory, the position of the particle can be determined precisely, but in quantum mechanics, the position of the particle is represented in terms of probability. To describe the position of the electron, the probability density function has been used and the behavior of an electron inside the crystal has been represented by the wave function $\psi_w(x, y)$. The wave function has been written in a complex form whose magnitude $R(x, y)$ and phase $S(x, y)$, which is given by:

$$\psi_w(x,y) = R(x,y)\exp\left[\frac{iS(x.y)}{\hbar}\right], \tag{4.2}$$

where $\hbar$ is the reduced Planck's constant.

The simulated structure is shown in Figure 4.14. The simulation has been carried out using carrier recombination mobility models along with tunneling effects for precise device simulation. Typical physics-based models described in Chapter 3, such as concentration and field-dependent model for mobility, Shockley-Read-Hall

(SRH) recombination model, non-local band to band tunneling model, bandgap narrowing model, and Fermi-Dirac statistics, are implemented in simulation. The transverse field-dependent mobility model of Si (Lombardi CVT) along with the SRH model considering both doping and temperature has been applied. The suitability of the DG model, which originally was introduced with the idea of simplifying the derivation of transport equations from the Boltzmann equation, including quantum corrections over the DD model, is shown through simulation. The impact of quantum corrections on the electron concentration in the nanowire channel with different drain biases is shown in Figures 4.15 and 4.16. As discussed before, simulations of the device structure (shown in Figure 4.14) have been performed for device performance comparison using both the DD and DG models. The implemented framework permits decoupling of the drift and diffusion components of the current density allowing a broader understanding and manipulation of the carrier transport mechanisms. At first, the distribution of electron density in the nanowire, which has been cut perpendicularly is shown in Figures 4.15 and 4.16 for two different

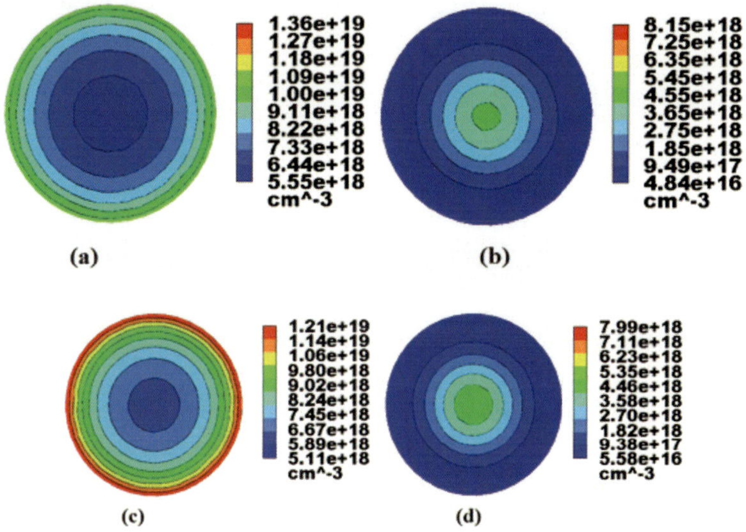

Figure 4.15 Electron concentration along the radial direction in single NW at $V_D = 0.05$ V using (a) DD model, (b) DG model and at $V_D = 0.7$ V in (c) DD model, and (d) DG model at a fixed $V_G = 0.05$ V.

gate voltages 0.05 V and 1 V, where the drain voltage has been kept at 0.05 V and 0.7 V, respectively. The electron density and spatial distribution are calculated via the self-consistent solution of coupled Schrodinger and Poisson equations in the nanowires with a circular cross-section. It is known that the spatial distribution of the carriers depends on the density of surface states and the nanowire width. In Figures 4.15 and 4.16, different configurations can be distinguished, particularly the localization of electrons in the center of the wire, or uniform tubular distribution in the nanowire.

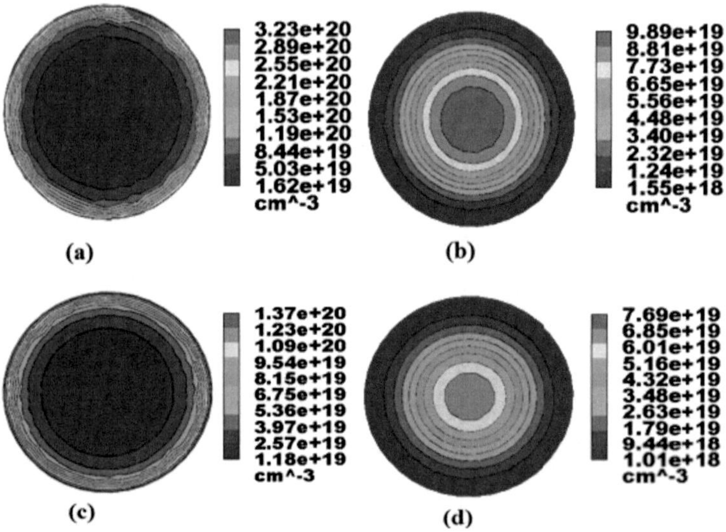

Figure 4.16 Electron concentration along the radial direction in single NW at $V_D = 0.05$ V using (a) DD model, (b) DG model and at $V_D = 0.7$ V in (c) DD model, and (d) DG model at a fixed $V_G = 1$ V.

The transfer characteristics of nanowire transistors using (a) DD model and (b) DG model are presented in Figure 4.17 (for the linear region) and Figure 4.18 (for saturation region), respectively. As the DD approach is essentially a low-order approximation to the Boltzmann transport equation, it gives rise to a higher current compared to the DG model. Also, as the number of wires (channel) increases, the drain current increases. The output characteristics are shown in Figure 4.19. Similar features are observed in all cases and the drain current is more for the DD model, and current increases (x2) as the number of channels increases. In addition to that, the effects of both the models on other devices, parameters have been summarized in Table 4.2. From the table, the value of V_T is 17.54% and 18.03% lower in 1-wire and 2-wire devices, respectively, at lower drain bias (i.e., $V_D = 0.05$ V) using the DD model. The quantum confinement effect leads to an increase in threshold voltage [55]. I_{ON} is found to be 32.29% and 32.07% higher for 1-wire and 2-wire devices, at lower drain bias while using DD model as the carriers are mostly confined in the center at low drain bias while taking quantum confinement effect. The I_{ON} is lower while considering the DG model in both 1-wire and 2-wire transistors.

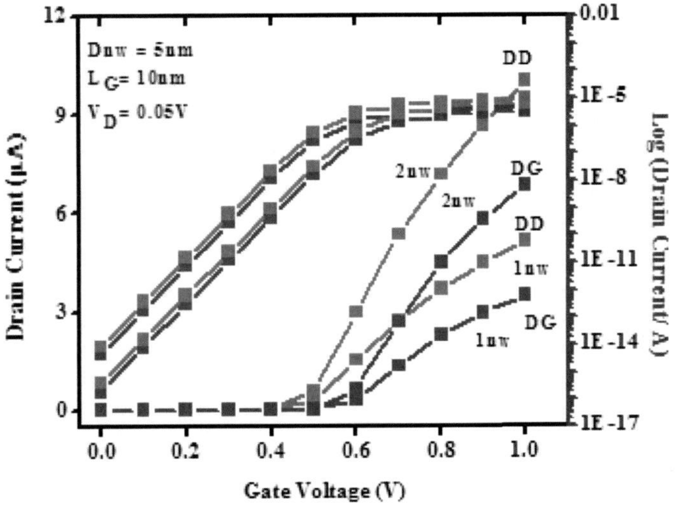

Figure 4.17 I_D–V_G plot of nanowire FET at (a) $V_D = 0.05$ V using DD and DG models.

The I_{OFF} is found to be 94.90% and 94.87% lower in 1-wire and 2-wire devices, respectively, while including the DG model. The inclusion of quantum effects leads to an increase in the bandgap, which ultimately reduces I_{OFF} [34, 55]. The value of SS is found to be 4.55% and 8.81% higher in 1-wire and 2-wire devices, respectively, while including the DG model. The electrons are more concentrated at the center than at the Si/SiO_2 interface in the DG model. So some parasitic capacitance may have been introduced increasing EOT [54]. So, the SS value is found to be higher in the DG model. The value of DIBL is found to be 45.86% and 21.79% higher in 1-wire and

2-wire devices, respectively, with the DG model. To get improved, the geometrical performance variation in nanowire is discussed in the next section.

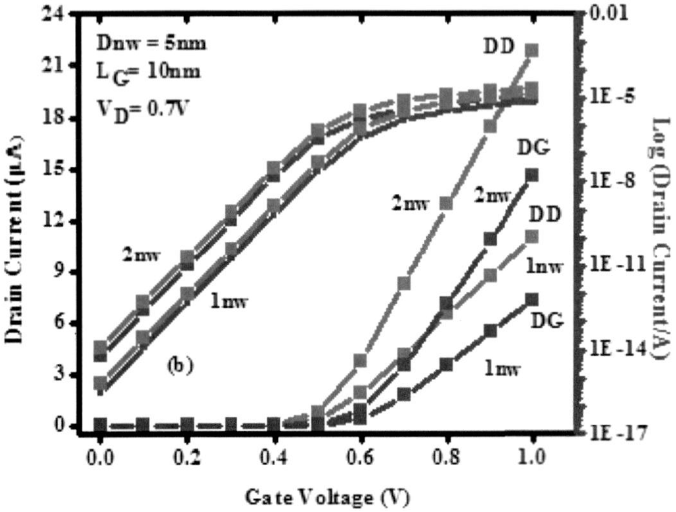

Figure 4.18 I_D–V_G plot of nanowire FET at (a) V_D = 0.7 V using DD and DG models.

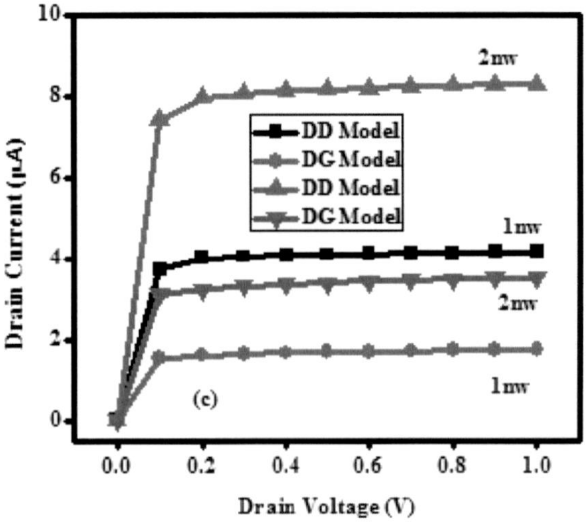

Figure 4.19 I_D–V_G plot of nanowire FET at V_G = 0.7 V using DD and DG models.

Table 4.2 Device parameters extracted using different models

	II. DD Model		III. DG Model	
I. Parameter	**1NW**	**2NW**	**1NW**	**2NW**
V_T_lin (mV)	429.16	407.35	520.50	496.96
V_T_sat (mV)	420.67	398.52	508.12	486.21
I_{ON}_lin (μA)	5.14	10.07	3.48	6.84
I_{ON}_sat (μA)	10.9	21.82	7.27	14.56
I_{OFF}_lin (pA)	0.00429	0.0086	0.00019	0.00039
I_{OFF}_sat (pA)	0.00589	0.0117	0.000305	0.000624
SS_lin(mV/dec)	71.75	68.94	75.02	71.04
SS_sat (mV/dec)	72.22	68.90	75.55	70.61
DIBL (mV/V)	13.06	13.58	19.05	16.54

4.7 Effects of Gate Length Variation

To increase the packing density, the size of the device should be small. For that, the gate length of the device should be reduced. In Figure 4.20, the impact of the reduction in gate length on the I_D–V_G plot of the device has been shown. The figure shows the I_D–V_G plot for different gate lengths in the subthreshold region (i.e., at V_D = 0.05 V). The gate length has been varied from 5 nm to 20 nm with a step of 5 nm. The left axis (i.e., black color) of each figure represents the drain current in the linear scale, and the right axis (i.e., blue color) represents the drain current in the logarithmic scale.

From Figure 4.20, it may be noted that the drain current is increased with the reduction in gate length. As the gate length reduces the impact of source and drain on the electrostatic potential of the channel increases. It leads to an increase in the potential at the central part of the channel. Larger potential in the channel results in a larger electron concentration. The distribution of potential is shown in Figure 4.21 for different gate lengths. Two extreme cases are taken (i.e., (a) 5 nm and (b) 20 nm). From Figure 4.21, it can be observed that the potential is higher at the center for the 5 nm L_G case than the other case (i.e., 20 nm L_G). In the 5 nm L_G case, the potential is found to be 0.0852 V, whereas it is about 0.0542 V in the 20 nm L_G case. The increase in potential leads to an increase in electron concentration, which increases drain current.

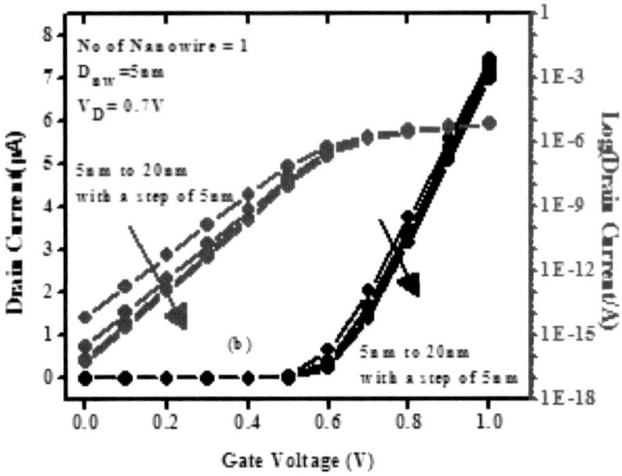

Figure 4.20 I_D–V_G plot of nanowire FET for different gate lengths at a drain bias of 50 mV in linear and logarithmic scale.

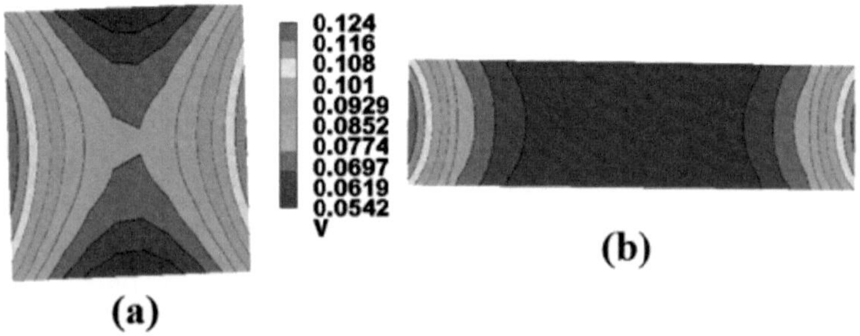

Figure 4.21 Potential distribution in single NW that is cut in XY plane at $V_G = 0.0$ V and $V_D = 0.05$ V (a) for 5 nm gate length and (b) 20 nm gate length.

The potential distribution in NWFET is given by [56]:

$$\psi(r,z) = \psi_0(r) + \psi_1(r,z). \tag{4.3}$$

$\psi_0(r)$ is the gradual channel approximation part can be found out by using:

$$\psi_0(r) = \psi_c - \frac{2 \cdot k_B \cdot T}{q} \ln\left[1 - \frac{r^2}{R^2} e^{-q(\psi_c - \psi_{\text{cm}})/k_B T}\right], \tag{4.4}$$

where ψ_c is the central potential (i.e., the potential at the center of the nanowire), and R is the radius of the nanowire.

The impact of the reduction in gate length on the other performance metrics (i.e., I_{ON}, I_{OFF}, SS, and DIBL) is shown below. Threshold voltage and DIBL are shown in Figure 4.22. The blue color represents the DIBL for different gate lengths, whereas the black color represents the threshold voltage. It can be observed from Figure 4.22 that the threshold voltage in the subthreshold region is 6.5% higher than the ON region at 5 nm gate length. In addition, the threshold voltage in the subthreshold region is 1% higher than the ON region at 20 nm gate length. It can be seen that with the reduction in gate length, the variation in threshold voltage between the subthreshold region and ON region increases, which is represented in the dotted circle. The DIBL at 20 nm gate length is about 88.8% lower than the DIBL at 5 nm gate length. DIBL can be defined as the ratio of absolute change of threshold voltages at linear and saturation drain bias to the change of drain bias. In addition, the variation in DIBL is also increasing with the decrease in gate length. With the reduction in gate length, the carriers of the channel are also controlled by drain voltage as a result of which DIBL is increasing with a reduction in gate length, which can be observed from Figure 4.22.

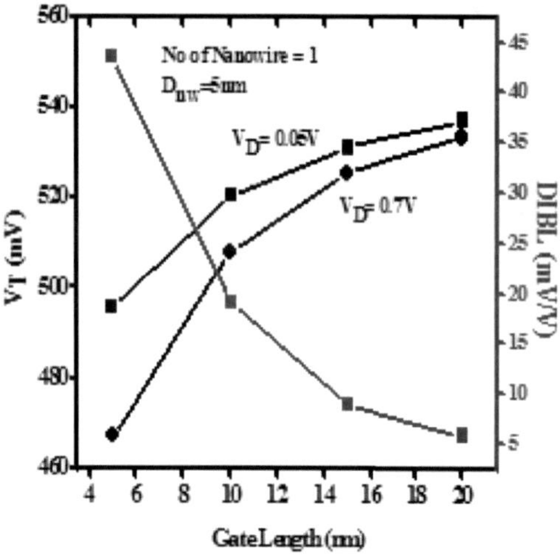

Figure 4.22 Variation in threshold voltage, and DIBL at two different drain biases.

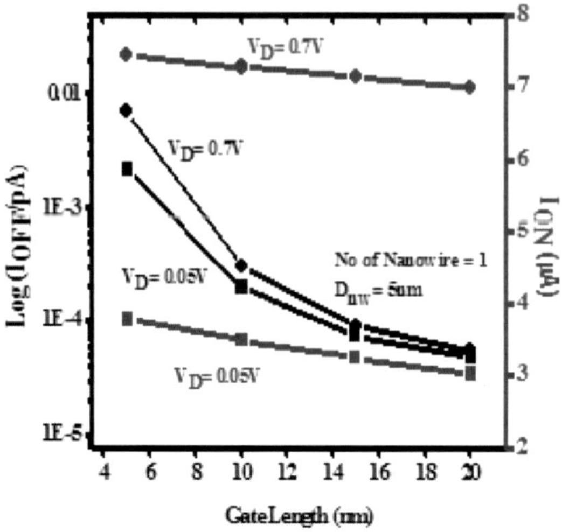

Figure 4.23 Variation in I_{ON}, and I_{OFF} at two different drain biases.

Variation in I_{ON} and I_{OFF} at two different drain biases is shown in Figure 4.23. I_{OFF} is presented in black color, and I_{ON} is presented in blue color. I_{OFF} is presented in the logarithmic scale on the left axis of Figure 4.23, and I_{ON} is given on the right axis. With decreasing in gate length, I_{OFF} increases both in the subthreshold region and ON region. But the value of I_{ON} increases with a reduction in gate length. It may be noted that the ON-current in the subthreshold region is 53.33% lower than the ON region at 5 nm gate length. More numbers of carriers are available for current conduction in the ON region than in the subthreshold region. That is why the ON region current is higher than the subthreshold region current for all the gate lengths (i.e., from 5 nm to 20 nm with a step of 5 nm).

The impact of different gate lengths on the SS is shown in Figure 4.24. The blue color bar represents the SS in the subthreshold region, and the pink color bar represents the SS in the ON region. The SS value has been found using Eq. (4.5).

$$SS = \frac{\partial V_G}{\partial \log(I_D)} \tag{4.5}$$

As SS represents the steepness of the I_D–V_G plot. The higher value of SS represents the reduction in gate control over the channel. This occurs because of the quantum confinement effect. The charge centroid is shifted toward the center. So, the reduction in gate length causes the degradation in the control of gate over channel carriers. This degradation can be improved by reducing the diameter of NWFET [57]. The diameter of the NW has been varied from 3 nm to 7 nm with a step of 1 nm. The impact of different diameters on the drain current is shown in Figures 4.25 and 4.26. These figures present the I_D–V_G plots of NWFET for different diameters of the nanowire in (a) the subthreshold region and (b) the ON region. From Figures 4.25 and 4.26, it can be observed that with a reduction in the diameter of the nanowire, the drain current reduces. It occurs because, with an increase in diameter, the area of current conduction increases. With the reduction in NW diameter, the 2D quantum confinement of current carriers becomes stronger [55] and as a result of which drain current reduces.

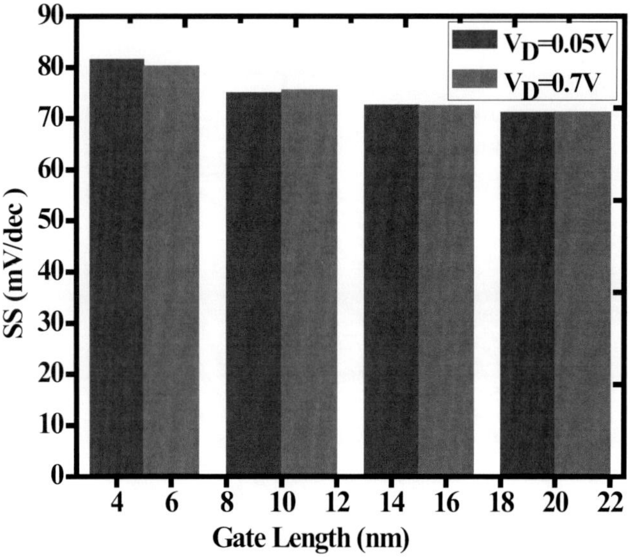

Figure 4.24 Variation in SS at two different drain biases.

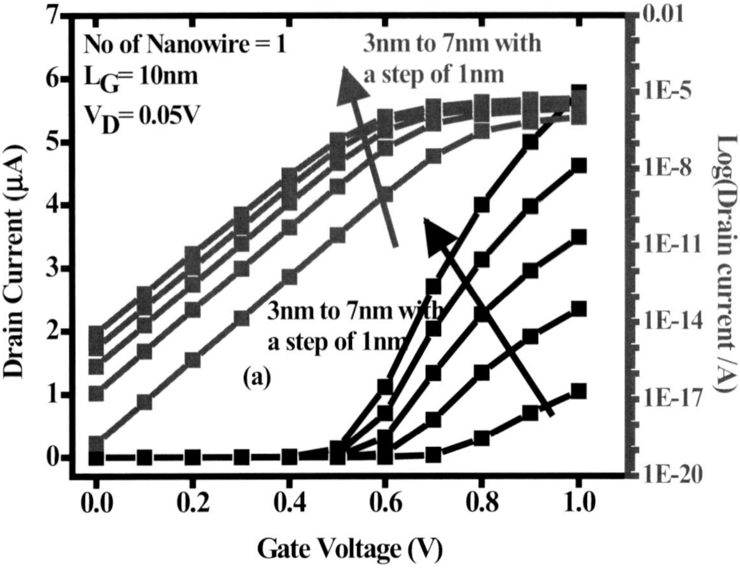

Figure 4.25 I_D–V_G plot of nanowire FET for different diameters of nanowire at a drain bias of 50 mV in linear and logarithmic scale for the subthreshold region.

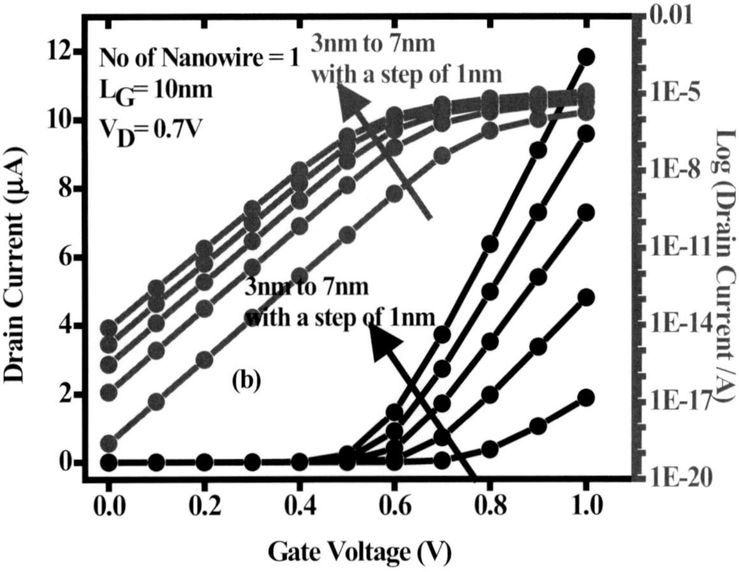

Figure 4.26 I_D–V_G plot of nanowire FET for different diameters of nanowire at a drain bias of 0.7 V in linear and logarithmic scale for the ON region.

4.8 Effects of Nanowire Diameter Variation

The diameter of the NW has been varied from 3 nm to 7 nm with a step of 1 nm. The impact of different diameters on V_T, I_{ON}, I_{OFF}, SS, and DIBL has been extracted. The impact of different diameters on the DIBL is presented on the right axis of Figure 4.27 in blue color. The DIBL at 3 nm diameter is found to be 71.42% lower than the DIBL at 7 nm. Two extreme cases (i.e., 3 nm and 7 nm diameters) have been considered. With a decrease in diameter, DIBL is getting lowered. The control of the gate over the channel carrier becomes stronger with the reduction in the diameter of the NWFET. I_{OFF} and I_{ON} are presented in Figure 4.28. The left axis represents the I_{OFF} on a log scale (i.e., in black color), and the right axis gives the I_{ON} in linear scale (i.e., in blue color) in both subthreshold and ON regions. The ON region is presented by the circle, and the subthreshold region is presented by the rectangle. The impact of different diameters SS is shown in Figure 4.29. With a reduction in diameter, the control of the gate over the channel increases. This leads to an increase in the effective bandgap. Since more carriers are available for current conduction at higher drain bias than the lower drain bias, the penetration of the drain electric field is shielded more by the narrow gate at a higher drain bias than the lower drain bias region. As a result of which better EI is maintained [6]. It leads to the lowering of SS value at higher drain bias.

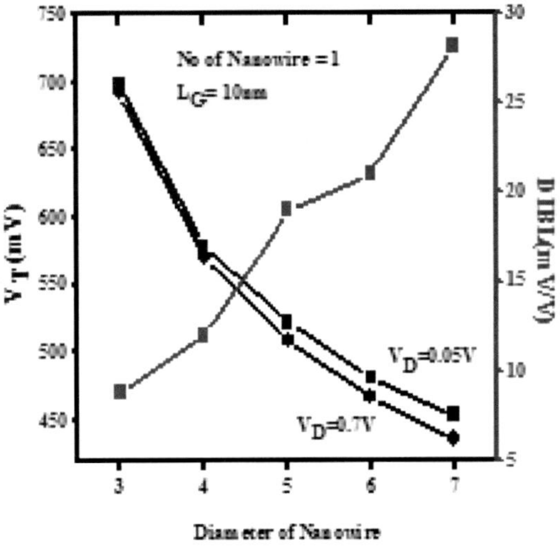

Figure 4.27 Variation in threshold voltage, and DIBL two different drain biases.

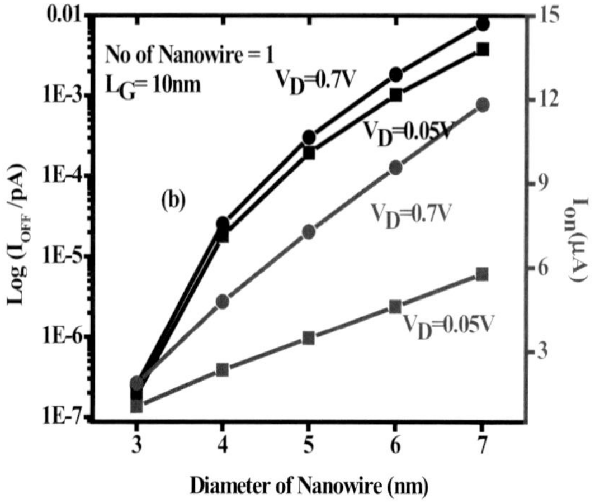

Figure 4.28 Variation in I_{ON}, and I_{OFF} at two different drain biases.

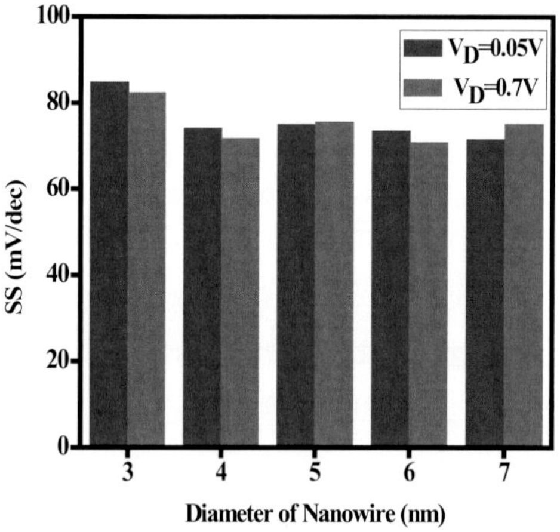

Figure 4.29 Variation in SS at two different drain biases.

The impact of channel orientation on the drain current is shown in Figure 4.30. The I_D–V_G characteristics for nanowire FET with different surface orientations <110> and <100> for NW having a diameter of 4 nm are shown at higher drain bias (0.7 V). The drain current is higher in <100> channel orientation than in <110> as the electron mobility is higher in the (100) plane than in (110). The impact of different orientations on the performance metrics is presented in Table 4.3. In <100> channel orientation the threshold voltage is 0.398% higher than in <110> channel orientation. I_{ON} is 10.784% is higher in <100> orientation than in <110> orientation. SS in <100> is 2.82% lower than <110> orientation. I_{OFF} in <100> orientation is 65.231% lower than <110> orientation.

Table 4.3 Extracted parameters for different orientations

Channel Orientation	Threshold Voltage (mV)	I_{ON} (A)	I_{OFF} (A)	SS slope (mV/dec)
<100>	468.18	1.13E-5	3.15E-15	75.38
<110>	466.32	1.02E-5	9.06E-15	77.57

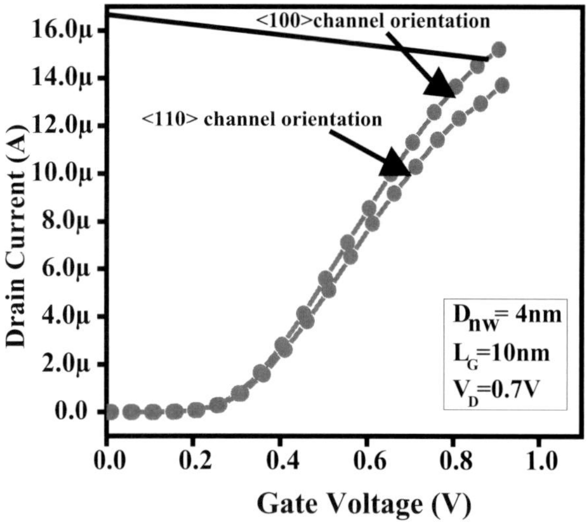

Figure 4.30 I_D–V_G plot of nanowire FET for different substrate orientations.

4.9 Strain-Engineered Nanowire FETs

As there is a lattice mismatch between Si and Ge, the performance of Si NWFET can be improved by using SiGe in the SD extension region. The lattice mismatch introduces stress in the device. In the following, the effects of mechanical stress in Si nanowire transistors with SiGe source/drain extension have been investigated by taking the channel length of 10 nm and 5 nm in diameter, with an EOT of 0.7 nm. The material of the source and drain extension regions is SiGe to introduce compressive stress into the nanowire channel. Linear elasticity model is applied to the segments (only semiconductors) of the nanowire devices, i.e., crystalline materials with elasticity and lattice parameters are used for stress simulation. After the calculation of the stress tensor, we use the output of the VSP tool as the input of the device simulation with MINIMOS-NT. The stress has been modeled by the following approaches:

When mechanical stress is applied to an elastic material, the latter can deform. The deformation is reversible and proportional to the applied stress. The state of deformation of the material can be obtained using the formalism of the theory of linear elasticity, more commonly called Hooke's law. To calculate the band-structure of the strained silicon has been calculated by taking the strain as the starting point in the crystal coordinate system (CCS) that can be expressed as [58]:

$$\varepsilon_{3c} = \begin{pmatrix} \varepsilon_{c,xx} & 2\varepsilon_{c,xy} & 2\varepsilon_{c,zz} \\ 2\varepsilon_{c,xy} & \varepsilon_{c,yy} & 2\varepsilon_{c,yz} \\ 2\varepsilon_{c,zz} & 2\varepsilon_{c,yz} & \varepsilon_{c,zz} \end{pmatrix} \tag{4.6}$$

$$\varepsilon_c = (\varepsilon_{c,xx}, \varepsilon_{c,yy}, \varepsilon_{c,zz}, 2\varepsilon_{c,yz}, 2\varepsilon_{c,xz}, 2\varepsilon_{c,xy}), \tag{4.7}$$

where ε_{3c} is a symmetric three by three matrix while ε_c is the corresponding six-component vector in the reduced (or Voigt) notation. Clearly, ε_{3c} and ε_c consist of the same six scalar quantities $\varepsilon_{c,xx}$, $\varepsilon_{c,yy}$, $\varepsilon_{c,zz}$, $\varepsilon_{c,xy}$, $\varepsilon_{c,yz}$, and $\varepsilon_{c,xz}$.

The stress in the CCS is similarly indicated as T_{3c} while ε_3, ε, T_3, and T are the symbols used for the strain and the stress in the device coordinate system (DCS). In the linear regime, namely for small strain magnitude, ε_c and T_c are linked as:

$$(\varepsilon_c)^T = S(T_c)^T \tag{4.8}$$

where S is the elastic compliance matrix. Cubic crystals have only three independent components S_{11}, S_{12}, and S_{44}. However, one needs a coordinate transformation for tensors of the second rank to calculate the stress (and then the strain) in the CCS [56]. For strain simulation, the evaluation of mobility enhancement is done by taking the piezoresistivity constants for different wafer orientations. The calculation of mobility enhancement is done by taking piezoresistivity factors and the average stress value into account (see Table 4.4) [57].

Table 4.4 Piezoresistance coefficients used in stress simulation for mobility extraction

IV. Si	V. n-type		VI. p-type	
10^{-12} cm^2/dyne	<100>	<110>	<100>	<110>
π_{11}	−102.2	−31.1	6.6	71.8
π_{12}	53.4	−17.5	−1.1	−66.3
π_{13}	53.4	53.4	−1.1	−1.1

The stress/strain profile is calculated under varying SiGe composition and thereby stress conditions, Ge content in the source/drain extension was varied from 10% to 50%. The variation of the stress under tensile stress along the nanowire axis was investigated first, and the results are shown in Figure 4.31a. 1D stress profile across the nanowire channel is found to increase from about 1GPa (at 10% Ge) to about 4GPa for Ge content of 0.5 in the source/drain extension (see Figure 4.31b). The hole mobility in the nanowire is found to be around 100 cm^2/Vs for no-stress conditions, whereas it increases to 250 cm^2/Vs at $V_D = V_G = -0.7$ V for the Ge content of 0.3.

The impact of Ge composition on the I_D–V_G plot is shown in Figure 4.32 (a and b) at two different drain bias regions in both linear and logarithmic scales. The impact of Ge composition on V_T, SS, I_{ON}, and I_{OFF} is shown in Figure 4.32 (c and d). The blue color in Figure 4.32 (a and b) presents the drain current in the linear scale, and the black color presents the drain current in the logarithmic scale. The current/voltage characteristics have been calculated under varying SiGe composition, and thereby stress conditions, the drain and gate voltages are stepped to generate the output characteristics in the subthreshold and the ON region. The variation of the transfer characteristics (I_D–V_G) under tensile stress due to variation in Ge content from 0.1 to 0.5 in the source/drain extension along the nanowire axis is shown in Figure 4.32 for different drain voltage (a) V_D = 0.05 V and (b) V_D = 0.7 V, respectively. With the increase in Ge composition in SiGe, the stress introduced in the material gets enhanced. As a result of which mobility of the carriers is increased, the drain current increases. Most of the simulated data on the output characteristics of NWFETs show excellent gate control.

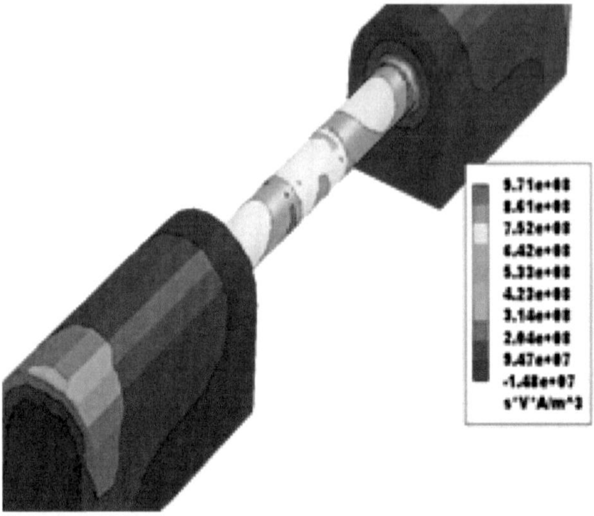

(a)

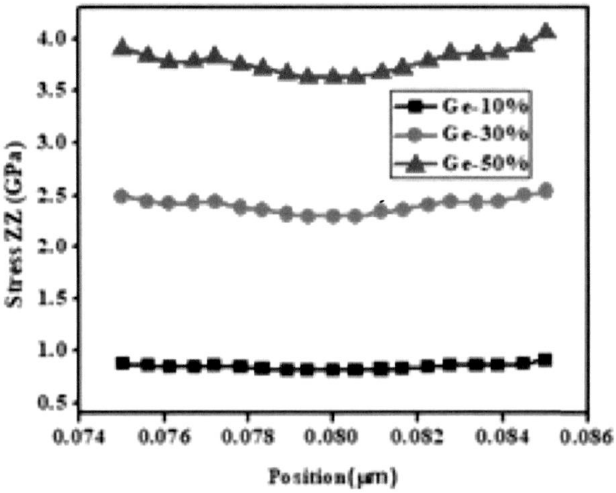

(b)

Figure 4.31 Stress profile: (a) 3D stress mapping for 10% Ge and (b) 1D stress profile across the nanowire channel.

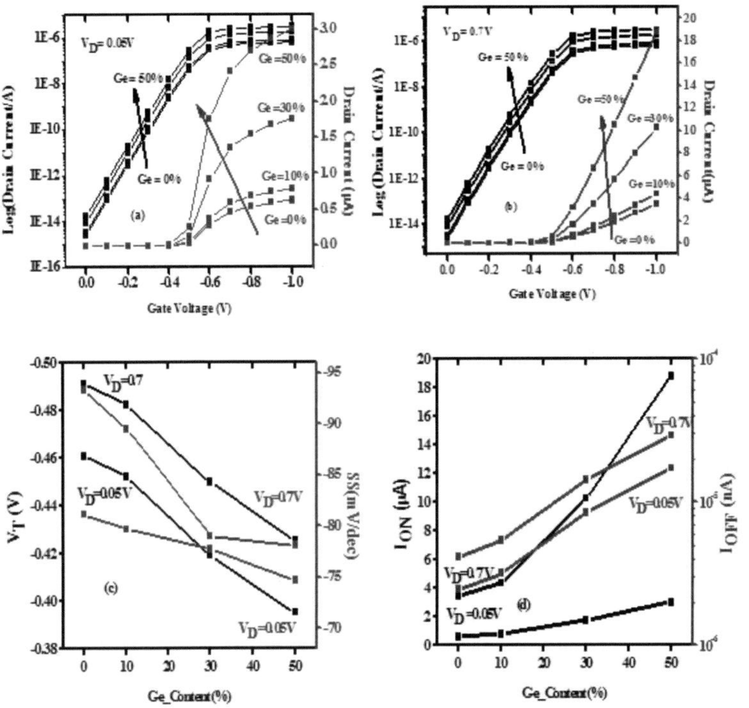

Figure 4.32 I_D–V_G for different drain voltage (a) V_D = 0.05 V and (b) V_D = 0.7 V, variation of V_T, SS, I_{ON}, and I_{OFF} as a function of Ge content in the source/drain extension in (c) and (d), respectively.

In Figure 4.32 (c), the left axis gives the V_T (i.e., black color) and the right axis gives SS (i.e., blue color); and in Figure 4.32 (d), the left axis gives the I_{ON} (i.e., black color), and the right axis gives I_{OFF} (i.e., blue color). For the digital circuits, the ratio between the off- and on-current is of importance. Most of the simulated data on the output characteristics of GAA NWFETs show near-ideal subthreshold behavior, and the subthreshold swing was determined to be 70–95 mV/dec (see Figure 4.32(c)). High I_{ON}/I_{OFF} ratio and high drive current are also observed from most of the simulated data on the output characteristics, for instance, we obtained I_{ON} values of 18 μA/μm for Ge content of 0.5 and lowest I_{OFF} of 20 nA due to improved electrostatic control and electron transport conditions inside the channel. The other way of improving the performance metrics of NWFET is by changing the material of the channel itself.

By using high mobility semiconductors one can increase the drive current of the NWFETs.

4.10 Variability in Nanowire Transistors

The performance of integrated circuits has improved through scaling by five orders of magnitude in the last six decades. One of the main challenges for further scaling is the management of process variation. Advanced CMOS front-end variation sources include random dopant fluctuation (RDF), implants and anneals (pocket implants, rapid-thermal anneal) induced variation, strain-induced variation (wafer-level biaxial, high-stress capping layers), embedded silicon-germanium (SiGe), and gate material granularity induced variation (for poly gates and metal gates) [59]. In recent years, the importance of random variation has increased further to achieve lower minimum operating voltages. It has been predicted that as devices scale below 7 nm technology node, the average number of dopants could be as low as 5. With scaling down of the devices, macroscopically same but microscopically different transistors are obtained. This difference is created by the discrete dopants in the channel, which are considered continuous in long channel devices. In addition, in sub-40-nm flash memory, random discrete dopant effect modulates post-program/erase (P/E) cycling threshold voltage through quick electron de-trapping as well as random telegraph signal [60].

The punch-through effect and the performance of the device have been improved through higher doping density. In scaled NWFET devices, the effective doping density will be influenced by the presence of any undesired impurity dopant in the undoped (or lightly doped) channel. The intrinsic variation, which is caused by random discrete dopants, occurs due to the presence of the undesired dopants and their random position in the channel. The effective density is expressed as the number of impurity atoms divided by the volume. As such, the device characteristics will be influenced by effective doping density. With downscaling, channel volume reduces, and as a result, the presence of a single dopant in the channel will also influence the device characteristics by varying the threshold voltage. The impact of random fluctuation of the number of dopants in the MOSFET channel has been investigated [61]. Random dopant arises due to the diffusion of dopant atoms from the heavily doped source/drain regions into the channel [62]. The impact of random fluctuation in the number of impurity atoms in the depletion region of a FET was reported by Keyes [63]. Takeuchi *et al.* [64] had reported that RDD plays a major role in contributing to threshold voltage variability. Mizuno *et al.* [62] had experimentally demonstrated that the fluctuation in threshold voltage is dependent on channel length, gate oxide thickness, and channel dopants in MOSFET.

The standard derivation in threshold voltage could be minimized by fabricating the fully depleted channel structures on SOI wafers for the reduction of channel dopant concentration. The impact of electron-electron interaction and electron-impurity interaction has to be taken into account while considering MOSFET below 100 nm. In addition, the inelastic interaction between the charges should be taken within an overall self-consistent particle-based device simulation in terms of correction terms. The correction terms were reported by Gross *et al.* [65]. The Columbic interaction was separated into long-range "mesh" interaction and short-range "scattering" interaction. Long-range interaction was included by solving the Poisson equation, and the short-range interaction was considered with a screening model and scattering rate. Asenov *et al.* [66] have reported the intrinsic fluctuation in double-gate MOSFET, which affects the yield and functionality of analog and digital circuits. Alexander *et al.* [67] have reported the impact of transport variation on the current variability and clarified the importance of electrostatic and transport variation. Variability studies using device simulation in FinFETs have been carried out by Dash *et al.* [68, 69]. In the following, we report our results on the study of the impacts of RDD on the performance of GAA NWFET at the 5 nm technology node.

RDF is caused by the random placement of dopant atoms that follow the Poisson distribution in the channel region [70]. The doping of the source/drain region is higher than the channel. The discrete dopants in the source/drain region are modeled by Gaussian distribution in which dopant diffusion occurs by the post-thermal annealing process. The standard deviation σ in the Gaussian doping profile is related to the diffusion length as shown below:

$$\sigma = \sqrt{2 \cdot D \cdot t} \tag{4.9}$$

$$\sigma = \sqrt{2 \cdot D_0 \cdot \exp\left(\frac{E_a}{k_B \cdot T}\right) t}, \tag{4.10}$$

where D is the diffusion coefficient, t is the diffusion time, D_0 is the typical diffusion coefficient of the dopant, E_a is the activation energy of the dopant, k_B is the Boltzmann constant, and T is the temperature. Also, the discrete dopant position influences the device performance, which should be investigated. In this work, we study the effects of discrete random dopants and their positions on the performance of nanowire FETs. The impact of discrete dopants in different regions of nanowires is carried out. The impact of discrete dopants in GAAFET with vertically stacked nanowires as the channel will also be considered.

4.11 Modeling of Random Discrete Dopants

The number and position of dopants may fluctuate randomly from device to device. The discrete dopants are generated randomly in a given segment, and a list of dopants with their distinct positions has also been created in the device. For each grid point, the number of dopants has been determined using a Poisson distribution. The discrete dopant is placed in such a way that it identifies each Si atom position in the 3D NWFET device tetrahedral grid. One random number is rolled to determine the atomic species at each atomic site. The probability of finding the dopant may exceed the random number for that location. If it is the case, one dopant is placed there. After associating the total number of dopant atoms with each node, the dopant concentration is evaluated at the given node enclosed by the control volume.

Several research groups have tackled the issue of modeling nanoscale transistors adopting different approaches and transport models. The simulation has been carried out using 3D quantum transport, which is based on nonequilibrium Green's function (NEGF) formalism [71, 72]. The noncoherent scattering is included in NEGF formalism by introducing self-energy terms [73–75]. By using the self-energy terms, many-body quantum transport can be converted into one-particle Green's function [76–78]. The penetration charge in the oxide has not been considered here. The 3D solution of the Poisson equation has been used in the simulation, where the doping charge is introduced as a discrete randomly placed individual dopant. At low drain voltage, the current is calculated by solving simplified current continuity equations. DG quantum corrections with fine meshing have been employed for the resolution of individual discrete dopants in simulation. This is particularly important when resolving random discrete dopants in the source/drain regions, where artificial charge trapping can introduce an unphysical increase in the access resistance.

The potential distribution is affected by the presence of discrete dopants, which, in turn, depends on the carrier distribution. The carrier crowding around the single dopants has been avoided physically by using the discrete dopant model in combination with the DG model. In this model, the quantum correction potentials γ_n and γ_p for electrons and holes, respectively, are added to the current relation of the DD model. The electron current density in the DG model and the quantum correction potential have been discussed in Chapter 3. In the conventional "atomistic" simulation to introduce the random dopant, the number of dopants in each mesh region (cell) is chosen from a Poisson distribution with a mean equal to the average dopant number in the region by using a continuous doping distribution of the process simulation result for device simulation. Since the dopants are non-uniformly distributed, the distances among the dopants are different. The potential, associated with each dopant, is termed as Coulomb potential, has been divided into long-range and short-range parts. The long-range part of the number density of a single dopant is given by [79]:

$$\rho^{\text{long}}(r) = \frac{k_c^3}{2\pi^2} \cdot \frac{\sin(k_c \cdot r) - (k_c \cdot r)\cos(k_c \cdot r)}{(k_c \cdot r)^3}. \tag{4.11}$$

The electronic potential is given by:

$$\psi^{\text{long}}(r) = -\frac{2}{\pi} \cdot \frac{\sin(k_c \cdot r)}{r}, \tag{4.12}$$

where k_c is the screening length and is given by $k_c = \left(\frac{n_A \cdot e^2}{\varepsilon_{Si} \cdot k_B \cdot T} \right)^{0.5}$ (inverse of screening length).

The long-range and short-range parts of the number density and the electronic potential by a single dopant are given by the following expressions:

$$\rho_{\text{ac}}(r) = -e\sum_{i=1}^{N}\delta(r - r_i) \tag{4.13}$$

$$\rho_{\text{ac}}(r) = -e\sum_{i=1}^{N}\frac{1}{V}\sum_{k<k_c}e^{ik(r-r_i)} - e\sum_{i=1}^{N}\frac{1}{V}\sum_{k>k_c}e^{ik(r-r_i)} \tag{4.14}$$

$$\rho_{\text{ac}}(r) = -e\sum_{i=1}^{N}\rho_{\text{ac}}^{\text{long}}(r - r_i) - e\sum_{i=1}^{N}\rho_{\text{ac}}^{\text{short}}(r - r_i), \tag{4.15}$$

where N represents the number of dopants, and V is the volume. The above relationships are used to compute the potential distribution in the simulation.

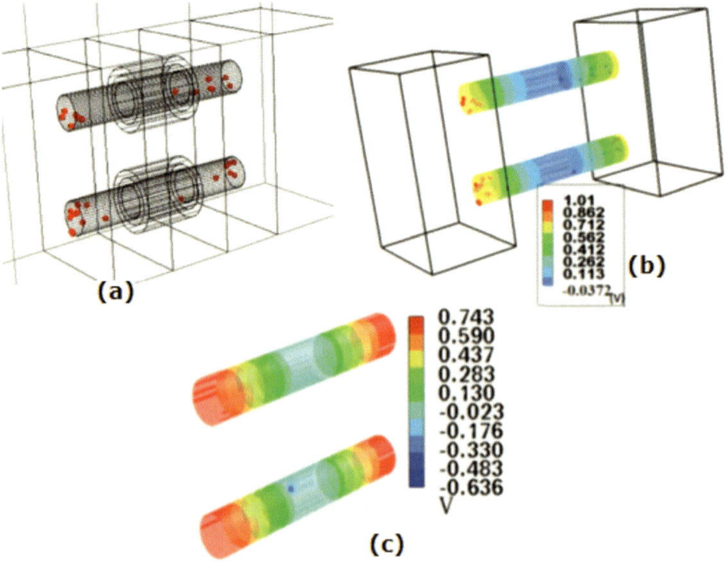

Figure 4.33 (a) Distribution of discrete dopants in the source/drain extension region for a particular configuration. (b) Potential distribution in 2-wire structure at $V_G = 0$ V with discrete dopants in the source/drain extension region. (c) 3D Potential distribution in 2-wire structure at $V_G = 0$ V with discrete dopants in the channel.

Donors are randomly distributed in a 10 nm region from the channel edges toward the source and drain, respectively. Only one discrete dopant is observed in the channel of each nanowire, and donor distribution for one configuration of random dopants. As the discrete dopants trap majority carriers and reduce the mobile electron concentration, a sharp Coulomb potential well is created, resulting in threshold voltage fluctuation. The effects of discrete dopants in the channel are presented in Figure 4.33. The source/drain doping is higher than the channel doping. As the channel is undoped (less doped), the dopants may diffuse from the source/drain region. The effects of one dopant are shown in Figure 4.33(c).

The electrostatic potential and the electron density obtained from a DG solution of the DD equations [80] are used as initial conditions for the Poisson-NEGF solution. After getting a new electron density from the NEGF solver, the quasi-Fermi level is found out using new electron density and the old potential [81]. The electron concentration is updated by the quasi-Fermi level when a non-linear Poisson equation is solved iteratively. When the channel length drops below 20 nm, a high degree of ballisticity is observed in NWFETs [82]. Since the doping of source and drain regions are high, impurity scattering plays a major role. The impact of discrete dopants on the I_D–V_G plot in different regions of NWFET (i.e., channel, Source_ext, and Drain_ext) is shown in Figure 4.34.

From Figure 4.34, it is observed that large variations in the drive current are associated with the position of the discrete dopants due to backscattering, and the drain current can be obtained following references [83–85]. In GAA NWFET, the carrier confinements are considered, and the inversion layer is taken via calibration of quantum correction parameters with a self-consistent Schrodinger-Poisson solver. This analysis is carried out at $V_D = 0.05$ V. $sp^3d^5s^*$ spin-orbit coupled tight binding structure is also taken in the calculation [86]. When the gate bias ($V_G = 0.05$ V) is low, the potential barrier between the source and the channel is high as a result of which the electrons are kept away from the channel by the gate potential. So, the electrons from the source cannot overcome the potential barrier or tunnel through. It leads to lower electron concentration in the channel in comparison to that in the source/drain contacts. The variation in drain current is more with dopants in the S_ext region.

At higher gate voltages ($V_G = 0.7$ V), the potential barrier between the contacts and the channel region decreases, and as a consequence, the electrons can travel above and tunnel through the potential barrier more easily. This results in an increase in the current at higher gate voltages in general. When the discrete dopants are close to the channel, the potential barrier gets lowered, leading to a higher current than the nominal case (i.e., without taking variability sources into account). The degree of backscattering, for a particular RDD configuration, is dependent on the voltage. This scattering is not found at a lower gate voltage. At the lower gate voltage reflection of the carriers is observed by the channel barrier potential.

The standard deviation of threshold voltage distribution due to RDF is modeled as [62]:

$$\sigma_{V_T} = \sqrt[4]{4q^3 \varepsilon_{Si} N_a \varphi_F} \frac{T_{ox}}{\varepsilon_{ox}} \frac{1}{\sqrt{2W * L}} \tag{4.16}$$

From Figure 4.35, it can be observed that the presence of discrete dopants in the S_ext and D_ext leads to more variation in threshold voltage than in the channel. As the channel is lightly doped as compared to the source/drain region, the variation in threshold due to discrete dopants in the channel is less. The reduction in threshold voltage is related to the potential barrier between source and drain. The potential barrier gets lowered by the discrete dopant atom. Secondly, with the increase in gate voltage, the electron concentration around the impurities increases. At the lower gate voltage, the screening of the impurities (which are close to the channel) is poor. The electrons are kept away from the channel by the gate potential, and also affect the carriers in the extension regions. At high gate voltage, the electron concentration in the channel and the screening of the impurities close to the channel increase. In addition, the electrons are injected from the source to drain for the normal functioning of NWFET. With the presence of discrete dopants in the S_ext region, electrons are partially reflected by the impurity potential, decreasing the total transmission. The partial reflection of carriers is called 'coherent scattering'. At lower gate voltage, coherent scattering is less pronounced due to the high potential barrier of the gate. The variation in threshold voltage is less in twin NWFET than in single NWFET.

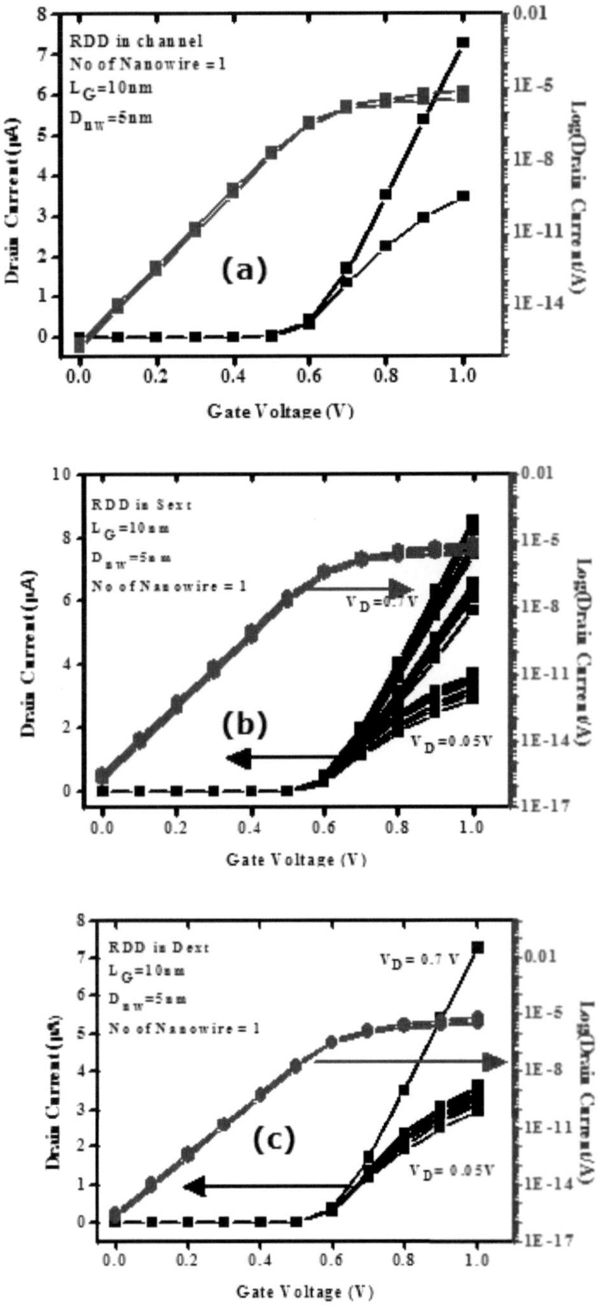

Figure 4.34 (a) I_D–V_G plot of one nanowire FET taking the impact of random discrete dopants in the channel, in S_ext and D_ext, respectively at V_D = 0.05 V and V_D = 0.7 V in both linear and logarithmic scales.

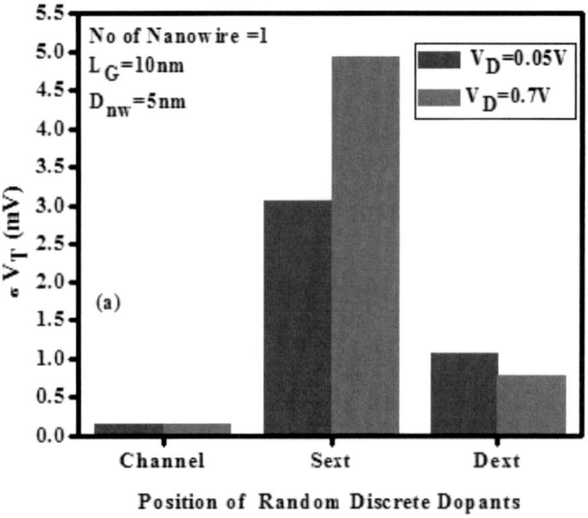

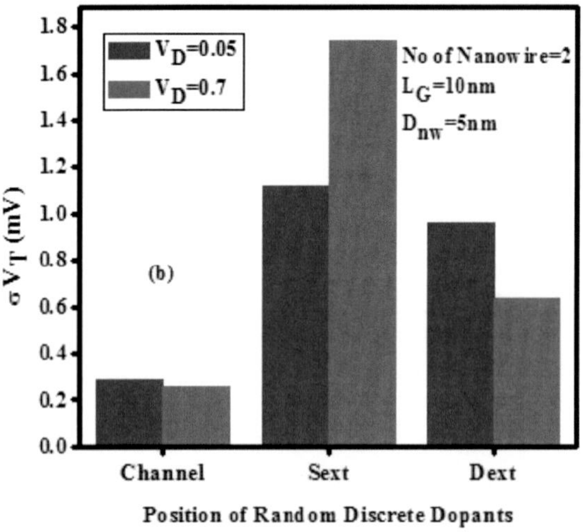

Figure 4.35 Variation of threshold voltage with the position of random discrete dopants in (a) single NWFET, and (b) twin NWFET.

4.12 Summary

In this chapter, the ultimate evolution of the CMOS device, the GAA-type nanowire transistors have been studied. Innovative process integration technique that allows the manufacture of nanowire devices has been discussed. After TCAD calibration with reported experimental devices, NWFETs have been designed with important design parameters at the 5 nm technology node. Then the performance of NWFET has been analyzed by using two models (i.e., DD and DG models). In addition, the effects of stacking multiple nanowires for possible performance improvement of lateral nanowire transistors using predictive simulation have been performed. Improved performance for the 5 nm diameter 10 nm gate length Si-nanowire-based devices has been demonstrated in terms of a higher I_{ON}/I_{OFF} ratio. It has been observed that drain current increases with the reduction in gate length, as expected. For the design of strain-engineered nanowire transistors, stress has been introduced in the channel region by introducing SiGe source-drain extension regions. Toward variability studies, the presence of discrete dopants in different regions of nanowire FETs has been discussed. In addition, the impacts of discrete dopants in different regions (i.e., S_ext, channel, and D_ext) of nanowire FETs on the performance metric of nanowire have been discussed. The stacked twin NWFET structure exhibited lower variation in the performance metrics than in single NWFET. The impact of discrete dopants in the S_ext region is found to be more on the device performance compared to other regions.

References

1. M. Lundstrom, "Device physics at the scaling limit: What matters? [MOSFETs]," in *IEEE International Electron Devices Meeting*, pp. 33.1.1–33.1.4, 2003. doi: 10.1109/IEDM.2003.1269398.
2. T. Skotnicki, J. A. Hutchby, T. J. King, H. S. P. Wong, and F. Boeuf, "The end of CMOS scaling: Toward the introduction of new materials and structural changes to improve MOSFET performance," *IEEE Circuits Devices Mag.*, vol. 21, no. 1, pp. 16–26, Jan. 2005. doi: 10.1109/MCD.2005.1388765.
3. J. Pelloux-Prayer, "Etude expérimentale des effets mécaniques et géométriques sur le transport dans les transistors nanofils à effet de champ," Université Grenoble Alpes, 2017 [Online]. Available: http://www.theses.fr/2017GREAY042/document.
4. J. D. Plummer and P. B. Griffin, "Material and process limits in silicon VLSI technology," *Proc. IEEE*, vol. 89, no. 3, pp. 240–258, 2001. doi: 10.1109/5.915373.

5. M. T. Bohr, R. S. Chau, T. Ghani, and K. Mistry, "The high-k solution," *IEEE Spectr.*, vol. 44, no. 10, pp. 29–35, Oct. 2007. doi: 10.1109/MSPEC.2007.4337663.
6. S. Datta, "Recent advances in high performance CMOS transistors: From planar to non-planar," *Electrochem. Soc. Interface*, vol. 22, no. 1, pp. 41–46, Jan. 2013. doi: 10.1149/2.F04131if.
7. S. Bengtsson, "Wafer bonding and smartcut for formation of silicon-on-insulator materials," *Int. Conf. Solid-State Integr. Circuit Technol. Proc.*, pp. 745–748, 1998. doi: 10.1109/icsict.1998.786103.
8. M. Reiche *et al.*, "Strained silicon-on-insulator: Fabrication and characterization," *ECS Trans.*, vol. 6, no. 4, pp. 339–344, 2019. doi: 10.1149/1.2728880.
9. J. P. Colinge, "FinFETs and other multi-gate transistors," 2007.
10. I. Ferain, C. A. Colinge, and J. P. Colinge, "Multigate transistors as the future of classical metal-oxide-semiconductor field-effect transistors," *Nature*, vol. 479, no. 7373. Nature Publishing Group, pp. 310–316, Nov. 17, 2011. doi: 10.1038/nature10676.
11. K. Ahmed and K. Schuegraf, "Transistors wars," *IEEE Spectr.*, no. November, pp. 50–66, 2011.
12. R. Courtland, "3D transistors for all," *IEEE Spectr.*, pp. 11–12, 2014.
13. M. Y. Wu and M. H. Chiang, "Performance evaluation of stacked gate-all-around MOSFETs at 7 and 10 nm technology nodes," in *Proceedings - International Symposium on Quality Electronic Design, ISQED*, May 2016, vol. 2016-May, pp. 169–172. doi: 10.1109/ISQED.2016.7479195.
14. R. Coquand, "Démonstration de l'intérêt des dispositifs multi-grilles auto-alignées pour les nœuds sub-10nm," Universite Grenoble, Alpes, 2013.
15. ITRS Executive Summary - 2015 Edition., "International technology roadmap for semiconductors 2015," 2015.
16. C. Auth *et al.*, "A 22nm high performance and low-power CMOS technology featuring fully-depleted tri-gate transistors, self-aligned contacts and high density MIM capacitors," in *Digest of Technical Papers - Symposium on VLSI Technology*, 2012, pp. 131–132. doi: 10.1109/VLSIT.2012.6242496.
17. A. Razavieh, P. Zeitzoff, and E. J. Nowak, "Challenges and limitations of CMOS scaling for FinFET and beyond architectures," *IEEE Trans. Nanotechnol.*, vol. 18, pp. 999–1004, 2019. doi: 10.1109/TNANO.2019.2942456.

18. J. Appenzeller, J. Knoch, M. T. Björk, H. Riel, H. Schmid, and W. Riess, "Toward nanowire electronics," *IEEE Trans. Electron Devices*, vol. 55, no. 11, p. 2827, 2008. doi: 10.1109/TED.2008.2008011.
19. P. Hashemi *et al.*, "High performance and reliable strained SiGe PMOS FinFETs enabled by advanced gate stack engineering," in *2017 IEEE International Electron Devices Meeting (IEDM)*, 2017, pp. 37.3.1–37.3.4. doi: 10.1109/IEDM.2017.8268510.
20. P. Hashemi, L. Gomez, and J. L. Hoyt, "Gate-all-around n-MOSFETs with uniaxial tensile strain-induced performance enhancement scalable to sub-10-nm nanowire diameter," *IEEE Electron Device Lett.*, vol. 30, no. 4, pp. 401–403, 2009. doi: 10.1109/LED.2009.2013877.
21. S. Bangsaruntip *et al.*, "Density scaling with gate-all-around silicon nanowire MOSFETs for the 10 nm node and beyond," in *2013 IEEE International Electron Devices Meeting*, 2013, pp. 20.2.1–20.2.4. doi: 10.1109/IEDM.2013.6724667.
22. S. Bangsaruntip *et al.*, "Gate-all-around silicon nanowire 25-stage CMOS ring oscillators with diameter down to 3 nm," in *Digest of Technical Papers - Symposium on VLSI Technology*, 2010, pp. 21–22. doi: 10.1109/VLSIT.2010.5556136.
23. K. Tachi *et al.*, "Experimental study on carrier transport limiting phenomena in 10 nm width nanowire CMOS transistors," in *2010 International Electron Devices Meeting*, 2010, pp. 34.4.1–34.4.4. doi: 10.1109/IEDM.2010.5703476.
24. T. Ernst *et al.*, "3D multichannels and stacked nanowires technologies for new design opportunities in nanoelectronics," *IEEE Int. Conf. ICICDT*, pp. 265–268, 2008. doi: 10.1109/ICICDT.2008.4567292.
25. T. Ernst *et al.*, "Novel 3D integration process for highly scalable nano-beam stacked-channels GAA (NBG) FinFETs with HfO_2/TiN gate stack," in *2006 International Electron Devices Meeting*, 2006, pp. 1–4. doi: 10.1109/IEDM.2006.346955.
26. C. Dupré *et al.*, "15nm-diameter 3D stacked nanowires with independent gates operation: φFET," 2008. doi: 10.1109/IEDM.2008.4796805.
27. S. D. Suk *et al.*, "High performance 5nm radius twin silicon nanowire MOSFET(TSNWFET): Fabrication on bulk Si wafer, characteristics, and reliability," in *Technical Digest - International Electron Devices Meeting, IEDM*, 2005, vol. 2005, pp. 717–720. doi: 10.1109/iedm.2005.1609453.
28. S. D. Suk *et al.*, "Investigation of nanowire size dependency on TSNWFET," in *2007 IEEE International Electron Devices Meeting*, 2007, pp. 891–894. doi: 10.1109/IEDM.2007.4419093.

29. S. Monfray *et al.*, "A solution for an ideal planar multi-gates process for ultimate CMOS?," in *2010 International Electron Devices Meeting*, 2010, pp. 11.2.1–11.2.4. doi: 10.1109/IEDM.2010.5703339.

30. Z. Stanojevic *et al.*, "Phase-space solution of the subband Boltzmann transport equation for nano-scale TCAD," in *2016 International Conference on Simulation of Semiconductor Processes and Devices (SISPAD)*, 2016, pp. 65–67. doi: 10.1109/SISPAD.2016.7605149.

31. IEEE, "International roadmap for devices and systems," 2018.

32. D. Nagy, G. Indalecio, A. J. Garc, A-Loureiro, M. A. Elmessary, K. Kalna, and N. Seoane, "FinFET versus gate-all-around nanowire fet: Performance, scaling, and variability," *IEEE J. Electron Devices Soc.*, vol. 6, pp. 332–340, 2018. doi: 10.1109/JEDS.2018.2804383.

33. Y. Cui, Q. Wei, H. Park, and C. M. Lieber, "Nanowire nanosensors for highly sensitive and selective detection of biological and chemical species," *Science (80-.).*, vol. 293, no. 5533, pp. 1289–1292, Aug. 2001. doi: 10.1126/science.1062711.

34. M. C. Mcalpine, H. Ahmad, D. Wang, and J. R. Heath, "Highly ordered nanowire arrays on plastic substrates for ultrasensitive flexible chemical sensors," *Nat. Mater.*, vol. 6, no. 5, pp. 379–384, 2007. doi: 10.1038/nmat1891.

35. A. I. Boukai, Y. Bunimovich, J. Tahir-Kheli, J. K. Yu, W. A. Goddard, and J. R. Heath, "Silicon nanowires as efficient thermoelectric materials," *Nature*, vol. 451, no. 7175, pp. 168–171, 2008. doi: 10.1038/nature06458.

36. B. Tian *et al.*, "Coaxial silicon nanowires as solar cells and nanoelectronic power sources," *Nature*, vol. 449, no. 7164, pp. 885–889, Oct. 2007. doi: 10.1038/nature06181.

37. N. Singh *et al.*, "Si, SiGe nanowire devices by top-down technology and their applications," *IEEE Trans. Electron Devices*, vol. 55, no. 11, pp. 3107–3118, 2008. doi: 10.1109/TED.2008.2005154.

38. H. H. Hsu, T. W. Liu, L. Chan, C. D. Lin, T. Y. Huang, and H. C. Lin, "Fabrication and characterization of multiple-gated poly-Si nanowire thin-film transistors and impacts of multiple-gate structures on device fluctuations," *IEEE Trans. Electron Devices*, vol. 55, no. 11, pp. 3063–3069, 2008. doi: 10.1109/TED.2008.2005161.

39. C. K. Maiti, *Computer aided design of micro- and nanoelectronic devices*. World Scientific Publishing Co. Pte. Ltd., 2017.

40. G. A. Armstrong and C. K. Maiti, *TCAD for Si, SiGe and GaAs Integrated Circuits*, The Institution of Engineering and Technology (IET), UK, 2008.

41. V. Moroz *et al.*, "Modeling and optimization of group IV and III-V FinFETs and nano-wires," in *2014 IEEE International Electron Devices Meeting*, 2014, pp. 7.4.1–7.4.4. doi: 10.1109/IEDM.2014.7047004.

42. L. Smith *et al.*, "FinFET to nanowire transition at 5nm design rules," in *2015 International Conference on Simulation of Semiconductor Processes and Devices (SISPAD)*, 2015, vol. 2015-Octob, pp. 254–257. doi: 10.1109/SISPAD.2015.7292307.

43. P. Zheng, D. Connelly, F. Ding, and T. K. J. K. Liu, "FinFET evolution toward stacked-nanowire FET for CMOS technology scaling," *IEEE Trans. Electron Devices*, vol. 62, no. 12, pp. 3945–3950, Dec. 2015. doi: 10.1109/TED.2015.2487367.

44. S. D. Suk *et al.*, "High performance 5nm radius twin silicon nanowire MOSFET (TSNWFET) : Fabrication on bulk si wafer, characteristics, and reliability," in *IEEE International Electron Devices Meeting*, 2005, pp. 717–720. doi: 10.1109/IEDM.2005.1609453.

45. Sentaurus Process User Guide, Version A-2007.12., Synopsys Inc., Mountain View, California.

46. K. H. Yeo *et al.*, "Gate-all-around (GAA) twin silicon nanowire MOSFET (TSNWFET) with 15 nm length gate and 4 nm radius nanowires," in *IEEE International Electron Devices Meeting, San Francisco*, 2006, pp. 1–4. doi: 10.1109/IEDM.2006.346838.

47. L. W. Liebmann and R. O. Topaloglu, "Design and technology co-optimization near single-digit nodes," in *IEEE/ACM International Conference on Computer-Aided Design, Digest of Technical Papers, ICCAD*, 2015, vol. 2015-Jan., no. January, pp. 582–585. doi: 10.1109/ICCAD.2014.7001409.

48. M. Luisier and A. Szabo, "Paving the way for ultimate device scaling through nanoelectronic device simulations," in *2013 14th International Conference on Ultimate Integration on Silicon (ULIS)*, 2013, pp. 53–56. doi: 10.1109/ULIS.2013.6523489.

49. H. R. Huff and D. C. Gilmer, *High Dielectric Constant Materials*, vol. 16. Berlin, Heidelberg: Springer Berlin Heidelberg, 2005.

50. Global TCAD Solutions GmbH, "GTS framework," 2018 [Online]. Available: http://www.globaltcad.com/framework.

51. O. Baumgartner, Z. Stanojevic, K. Schnass, M. Karner, and H. Kosina, "VSP - A quantum-electronic simulation framework," *J. Comput. Electron.*, vol. 12, no. 4, pp. 701–721, Dec. 2013. doi: 10.1007/s10825-013-0535-y.

52. G. Baccarani, E. Gnani, A. Gnudi, S. Reggiani, and M. Rudan, "Theoretical foundations of the quantum drift-diffusion and density-gradient

models," *Solid. State. Electron.*, vol. 52, no. 4, pp. 526–532, 2008. doi: https://doi.org/10.1016/j.sse.2007.10.051.

53. M. G. Ancona, "Density-gradient theory: A macroscopic approach to quantum confinement and tunneling in semiconductor devices," *J. Comput. Electron.*, vol. 10, no. 1–2, pp. 65–97, Jun. 2011. doi: 10.1007/s10825-011-0356-9.
54. S. Dey, T. P. Dash, S. Das, E. Mohapatra, J. Jena, and C. K. Maiti, "Gate-all-around Si-nanowire transistors: Simulation at nanoscale," in *2018 IEEE Electron Devices Kolkata Conference (EDKCON)*, 2018, pp. 137–141. doi: 10.1109/EDKCON.2018.8770471.
55. K. Nayak *et al.*, "CMOS logic device and circuit performance of Si gate all around nanowire MOSFET," *IEEE Trans. Electron Devices*, vol. 61, no. 9, pp. 3066–3074, 2014. doi: 10.1109/TED.2014.2335192.
56. D. Esseni, P. Palestri, and L. Selmi, *Nanoscale MOS transistors: Semi-classical transport and applications*, Cambridge University Press, 2011.
57. S. Dey, J. Jena, T. P. Dash, E. Mohapatra, S. Das, and C. K. Maiti, "Performance evaluation of gate-all-around Si nanowire transistors with SiGe strain engineering," in *2019 IEEE Conference on Modeling of Systems Circuits and Devices (MOS-AK India)*, 2019, pp. 29–33. doi: 10.1109/MOS-AK.2019.8902440.
58. C. C. Hobbs *et al.*, "Fermi-level pinning at the polysilicon/metal oxide interface-Part I," *IEEE Trans. Electron Devices*, vol. 51, no. 6, pp. 971–977, 2004. doi: 10.1109/TED.2004.829513.
59. K. J. Kuhn, "Considerations for ultimate CMOS scaling," *IEEE Transactions on Electron Devices*, vol. 59, no. 7. pp. 1813–1828, 2012. doi: 10.1109/TED.2012.2193129.
60. T. Kim, D. He, R. Porter, D. Rivers, J. Kessenich, and A. Goda, "Comparative study of quick electron detrapping and random telegraph signal and their dependences on random discrete dopant in sub-40-nm NAND flash memory," *IEEE Electron Device Lett.*, vol. 31, no. 2, pp. 153–155, Feb. 2010. doi: 10.1109/LED.2009.2036871.
61. H. S. Wong and Y. Taur, "Three-dimensional 'atomistic' simulation of discrete random dopant distribution effects in sub-0.1 μm MOSFETs," in *Technical Digest - International Electron Devices Meeting*, 1993, pp. 705–708. doi: 10.1109/iedm.1993.347215.
62. T. Mizuno, J. I. Okamura, and A. Toriumi, "Experimental study of threshold voltage fluctuation due to statistical variation of channel dopant number in MOSFETs," *IEEE Trans. Electron Devices*, vol. 41, no. 11, pp. 2216–2221, 1994. doi: 10.1109/16.333844.

63. R. W. Keyes, "Special correspondence effect of randomness in the distribution of impurity ions on FET thresholds in integrated electronics," *IEEE J. Solid-State Circuits*, vol. 10, no. 4, pp. 245–247, 1975. doi: 10.1109/JSSC.1975.1050600.

64. K. Takeuchi *et al.*, "Understanding random threshold voltage fluctuation by comparing multiple fabs and technologies," in *Technical Digest - International Electron Devices Meeting, IEDM*, 2007, pp. 467–470. doi: 10.1109/IEDM.2007.4418975.

65. W. J. Gross, D. Vasileska, and D. K. Ferry, "Novel approach for introducing the electron-electron and electron-impurity interactions in particle-based simulations," *IEEE Electron Device Lett.*, vol. 20, no. 9, pp. 463–465, Sep. 1999. doi: 10.1109/55.784453.

66. A. Asenov, A. R. Brown, J. H. Davies, S. Kaya, and G. Slavcheva, "Simulation of intrinsic parameter fluctuations in decananometer and nanometer-scale MOSFETs," *IEEE Trans. Electron Devices*, vol. 50, no. 9, pp. 1837–1852, 2003. doi: 10.1109/TED.2003.815862.

67. C. Alexander, G. Roy, and A. Asenov, "Random-dopant-induced drain current variation in nano-MOSFETs: A three-dimensional self-consistent monte carlo simulation study using -ab initio - ionized impurity scattering," *IEEE Trans. Electron Devices*, vol. 55, no. 11, pp. 3251–3258, 2008. doi: 10.1109/TED.2008.2004647.

68. T. P. Dash, J. Jena, E. Mohapatra, S. Dey, S. Das, and C. K. Maiti, "Stress-induced variability studies in tri-gate FinFETs with source/drain stressor at 7 nm technology nodes," *J. Electron. Mater.*, vol. 48, no. 8, pp. 5348–5362, Aug. 2019. doi: 10.1007/s11664-019-07348-7.

69. J. Jena, T. P. Dash, E. Mohapatra, S. Dey, S. Das, and C. K. Maiti, "Fin shape dependence of electrostatics and variability in FinFETs," *J. Electron. Mater.*, vol. 48, no. 10, pp. 6742–6752, Oct. 2019. doi: 10.1007/s11664-019-07480-4.

70. Y. Ye, F. Liu, M. Chen, S. Nassif, and Y. Cao, "Statistical modeling and simulation of threshold variation under random dopant fluctuations and line-edge roughness," *IEEE Trans. Very Large Scale Integr. Syst.*, vol. 19, no. 6, pp. 987–996, Jun. 2011. doi: 10.1109/TVLSI.2010.2043694.

71. A. Martinez *et al.*, "A self-consistent full 3D real-space NEGF simulator for studying nonperturbative effects in nano-MOSFETs," *IEEE Trans. Electron Devices*, vol. 54, no. 9, pp. 2213–2222, 2007. doi: 10.1109/TED.2007.902867.

72. M. Luisier, A. Schenk, and W. Fichtner, "Three-dimensional full-band simulations of Si nanowire transistors," in *International Electron*

Devices Meeting, 2006 : IEDM '06, 11–13 Dec. 2006, 2006, pp. 563–566. doi: 10.1109/IEDM.2006.346737.

73. R. Lake and S. Datta, "Nonequilibrium Greens-function method applied to double-barrier resonant-tunneling diodes," *Phys. Rev. B*, vol. 45, no. 12, pp. 6670–6685, Mar. 1992. doi: 10.1103/PhysRevB.45.6670.
74. A. Svizhenko and M. P. Anantram, "Role of scattering in nanotransistors," *IEEE Trans. Electron Dev.*, vol. 50, no. 6, pp. 1459–1466, Jun. 2003. doi: 10.1109/TED.2003.813503.
75. S. Jin, Y. J. Park, and H. S. Min, "A three-dimensional simulation of quantum transport in silicon nanowire transistor in the presence of electron-phonon interactions," *J. Appl. Phys.*, vol. 99, no. 12, p. 123719, Jun. 2006. doi: 10.1063/1.2206885.
76. G. Lindblad, "Brownian motion of a quantum harmonic oscillator," *Reports Math. Phys.*, vol. 10, no. 3, pp. 393–406, Dec. 1976. doi: 10.1016/0034-4877(76)90029-X.
77. G. D. Mahan, "Quantum transport equation for electric and magnetic fields," *Physics Reports*, vol. 145, no. 5. pp. 251–318, 1987. doi: 10.1016/0370-1573(87)90004-4.
78. C. Delerue, M. Lannoo, and G. Allan, "Excitonic and quasiparticle gaps in Si nanocrystals," *Phys. Rev. Lett.*, vol. 84, no. 11, pp. 2457–2460, Mar. 2000. doi: 10.1103/PhysRevLett.84.2457.
79. N. Sano *et al.*, N. Sano, K. Matsuzawa, M. Mukai, and N. Nakayama, "Role of long-range and short-range Coulomb potentials in threshold characteristics under discrete dopants in sub-0.1μm Si-MOSFETs," *IEEE Int. Electron devices Meet.*, pp. 275–278, 2000. doi: 10.1109/IEDM.2000.904310.
80. A. Asenov, G. Slavcheva, A. R. Brown, J. H. Davies, and S. Saini, "Increase in the random dopant induced threshold fluctuations and lowering in sub-100 nm MOSFETs due to quantum effects: A 3D density-gradient simulation study," *IEEE Trans. Electron Devices*, vol. 48, no. 4, pp. 722–729, 2001. doi: 10.1109/16.915703.
81. M. P. Anantram and A. Svizhenko, "Multidimensional modeling of nanotransistors," *IEEE Trans. Electron Devices*, vol. 54, no. 9, pp. 2100–2115, 2007. doi: 10.1109/TED.2007.902857.
82. K. H. Cho *et al.*, "Experimental evidence of ballistic transport in cylindrical gate-all-around twin silicon nanowire metal-oxide-semiconductor field-effect transistors," *Appl. Phys. Lett.*, vol. 92, no. 5, p. 052102, Feb. 2008. doi: 10.1063/1.2840187.

83. S. Maheshwaram, S. K. Manhas, G. Kaushal, B. Anand, and N. Singh, "Vertical nanowire CMOS parasitic modeling and its performance analysis," *IEEE Trans. Electron Devices*, vol. 60, no. 9, pp. 2943–2950, 2013. doi: 10.1109/TED.2013.2272651.

84. B. C. Paul, R. Tu, S. Fujita, M. Okajima, T. H. Lee, and Y. Nishi, "An analytical compact circuit model for nanowire FET," *IEEE Trans. Electron Devices*, vol. 54, no. 7, pp. 1637–1644, Jul. 2007. doi: 10.1109/TED.2007.899397.

85. L. Zhang *et al.*, "An analytic model for nanowire MOSFETs with Ge/Si core/shell structure," *IEEE Trans. Electron Devices*, vol. 55, no. 11, pp. 2907–2917, 2008. doi: 10.1109/TED.2008.2007417.

86. K. Nayak *et al.*, "Negative differential conductivity and carrier heating in gate-all-around Si nanowire FETs and its impact on CMOS logic circuits," in *Japanese Journal of Applied Physics*, vol. 53, no. 4 SPEC. ISSUE, 2014. doi: 10.7567/JJAP.53.04EC16.

Chapter 5

Nanosheet Transistors

Eleena Mohapatra
SOA University, Bhubaneswar 751030, Odisha, India
eleenamohapatra@gmail.com

The ever-increasing demand for faster, smaller, and more advanced devices necessitates constant innovation from the micro- and nano-electronics industries. In this chapter, we first present an overview of current and upcoming device structures that enable the continuation of device scaling. A brief overview of the state-of-the-art and next-generation semiconductor logic devices is given first. This paves the way for understanding the challenges involved in designing these devices and how atomic-scale processing techniques can be used to address those challenges.

The introduction of FinFETs has emerged as a defending contender to overcome the short channel effects along with further miniaturization. The channel electrostatics was improved by taking triple gate structures with scaling of gate length down to 7 nm technology node [1]. However, the performance of tri-gate

Fabless Semiconductor Manufacturing: In the Era of Internet of Things
Edited by Chinmay K. Maiti

ISBN 978-981-4968-27-0 (Hardcover), 978-1-003-31497-4 (eBook)
www.jennystanford.com

structures has been found to degrade below 7 nm technology node [2]. As traditional CMOS scaling approaches its physical limits and faces interconnect routing congestion, technologies for reducing cell height by reducing the number of metal tracks are being sought to compensate for the recent trend of more modest pitch scaling and thus help maintain the overall power performance logic roadmap [3–6]. This means that for ultra-scaled cells with fewer metal tracks, fin depopulation from two to one fin per FinFET system will be needed [5, 6], though this will affect device efficiency.

The best electrostatic control of the gate over the channel can be achieved only when it can be controlled from all sides of the channel. The scaling of gate length can be continued in sub 10 nm node with improved electrostatics using gate-all-around (GAA) nanowire transistors [7]. GAA nanowire transistors are now considered to be the most suitable and feasible alternative structures for ongoing rapid scaling even below 7 nm technology nodes [8]. As a result, novel transistor architectures will be needed to maintain scaling gains, with nanowire (NW) [9–15] or nanosheet (NS) [16–21] GAAFETs widely regarded as the most promising FinFET [14, 22–25] replacements for advanced nodes. Furthermore, though lateral NW or NS GAAFET devices can both be considered natural extensions of FinFETs from an integration and layout design standpoint, NSFETs have the advantage of providing more current per layout footprint due to their larger effective width and enhanced electrostatics control [18, 20, 21].

Today's emphasis on low-power and low-voltage operation in industry and academe requires an understanding of nanodevices at ultimate scaling [26]. According to the report from the International Roadmap for Devices and Systems [27], the technology generation of logic devices in the industry would reach 5 nm and 3 nm in 2021 and 2024, respectively. Recently, a GAAFET has been reported by IBM [18]. These devices are intended for being used in the next-generation semiconductor integrated circuits (ICs). Figure 5.1 shows the cross-sectional details of the transistor consisting of nanosheet semiconductor channels [28].

IBM has successfully demonstrated the 5 nm process technology in NS GAAFET, which is a multi-bridge channel field effect transistor (MBCFET). Except for the fact that MBCFET uses nanosheets instead of nanowires as a channel, they are architecturally similar. These devices have many advantages in terms of design and structure. First, the nanosheets can be stacked vertically to increase speed without taking up more space. Second, since they use the same manufacturing process, they are extremely compatible with current FinFET technology. For the same reason, MBCFETs can be used to replace FinFETs in existing designs, making the transition seamless.

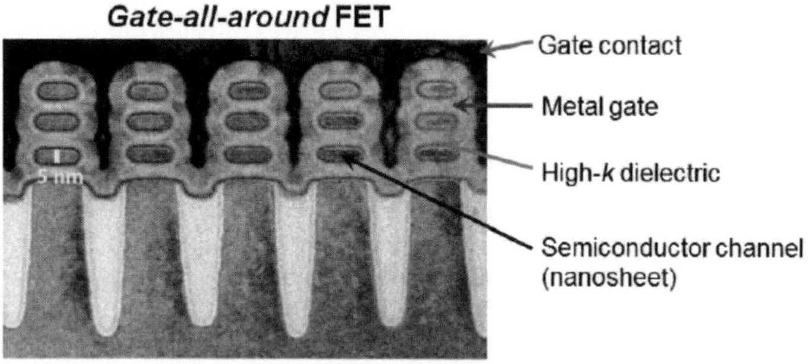

Figure 5.1 A GAAFET cross-sectional details of the transistor consisting of nanosheet semiconductor channels [28].

The design of the semiconductor device and optimization of its key parameters are dependent on accurate device simulation. The physical behavior of the device and its electrical characteristic can be easily known from the simulation. The behavior of the device can be modeled by using the TCAD tools [29]. To be "predictive", the TCAD tools need to have a very high level of accuracy. Also, proper calibration is required for predictive TCAD. However, this approach is very expensive and time-consuming. Also, it is very difficult to calibrate a set of TCAD simulators. Generally, such calibration is performed by incorporating the exact device geometry and materials along with other relevant device parameters, as available from the manufacturer. Continual calibration of model parameters is also required. At first, the TCAD calibration has been performed for an experimental NSFET reported in reference [18]. The ensemble of simulated devices consists of n-type and p-type nanosheet transistors, both with a sheet cross-section (shown in Figure 5.1).

The simulated device NSFET has a 3D structure with a 14 nm gate length, as shown in Figure 5.2. All the critical device geometry parameters are defined in Table 5.1. The width (W_{NS}) and thickness (H_{NS}) of NSFET are 12 nm and 6 nm, respectively. The channel is n-doped (N_D) with a doping concentration of 10^{16} cm$^{-3,}$ and the source/drain (S/D) extension regions are heavily p-doped (N_A) with a doping concentration of 10^{20} cm^{-3}. The gate oxide consists of SiO_2 and HfO_2 of thickness is 0.5 nm and 1.5 nm, respectively, Titanium nitride (TiN) is used as the gate material. Considering the diverse requirements in device simulation, it is natural to ask for the level of confidence at which device simulation tools are useful for device development. To gain confidence in simulations of NSFETs, we first simulate to compare the reported experimental device characteristics of the NSFET. The structure used for 2D numerical simulation is shown in the inset of Figure 5.2, and it is designed to match the actual NSFETs used in this study. Figure 5.2 shows the calibration of the drain current of a three-stack NSFET for V_D = 0.7 V. For calibration, L_g and H_{NS} are selected as 12 nm and 5 nm, respectively. The calibration results obtained from our simulations show an excellent agreement with the reported experimental data. The reproduction of the experimental characteristics of NSFETs by using the physics-based models described below gives the device designers a high confidence level regarding the effectiveness and feasibility of the models used in predictive simulations.

Table 5.1 Device detail of NSFET used in the simulation

Device Parameter	**Value**
Nanosheet Thickness (H_{NS})	4–6 nm
Nanosheet Width (W_{NS})	3–12 nm
Gate Length (L_g)	8–14 nm
High-k (HfO_2) Thickness	1.5 nm
Oxide Thickness (T_{OX})	0.5 nm
Epi-Length	14 nm
Epi-Top-Width	8 nm
Epi-Middle-Width	25 nm
Epi-Bottom-Width	14 nm
Spacer Length (L_{SP})	7 nm
Channel Doping (N_{CH})	10^{16} cm^{-3}
S/D Doping	10^{20} cm^{-3}

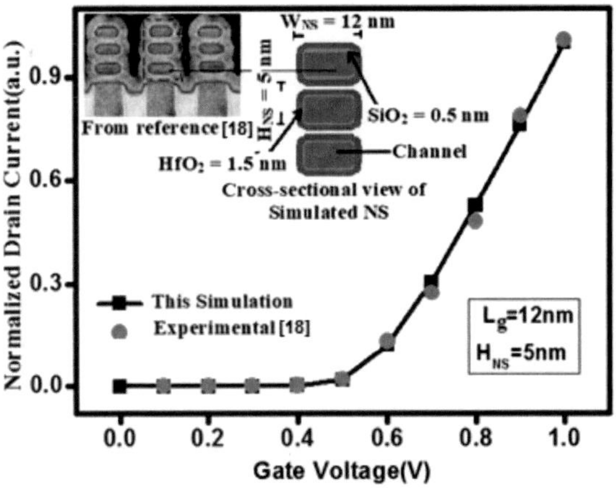

Figure 5.2 Comparison of the simulated transfer characteristics with reported experimental data and the cross-sectional view of three stacked nanosheet semiconductor channels [18].

5.1 GAA Nanowire vs. Nanosheet Transistors

Figure 5.3 shows the cross-sectional view of the nanosheet and nanowire transistors considered for comparison. The ensemble of the simulated device consists of n- and p-type nanowire transistors, both with a circular cross-section, 12.8 nm in diameter, with an EOT of 1.5 nm. The gate length was set to 12 nm. At the calibration stage, the mobility models implemented in the TCAD simulation tools have been tuned to experimental data in conjunction with the published mobility models.

A comparison of electrical performance on-state current (I_{ON}), off-state current (I_{OFF}), SS, and DIBL as a function of channel area for both the shapes is shown in Figure 5.4. In all cases, an increasing trend with the rise in Si-channel area is observed. Even though each structure has the same channel area, the effective channel width is different, i.e., for rectangular shape (NS) by ($2\times H_{NS}+2\times W_{NS}$) and circular by ($\pi\times D_{NW}$). GAA NSFET provides better electrostatic control of the channel, which results in the best subthreshold behavior, followed by the GAA NWFET due to its smallest I_{OFF} and SS. In the case of NS, the I_{ON} exhibits an 18% higher rate than that of the nanowire when the channel area changes from 20 nm^2 to 80 nm^2. Moreover, compared to GAA NWFETs, NSFETs have a larger gate area resulting in higher I_{ON}.

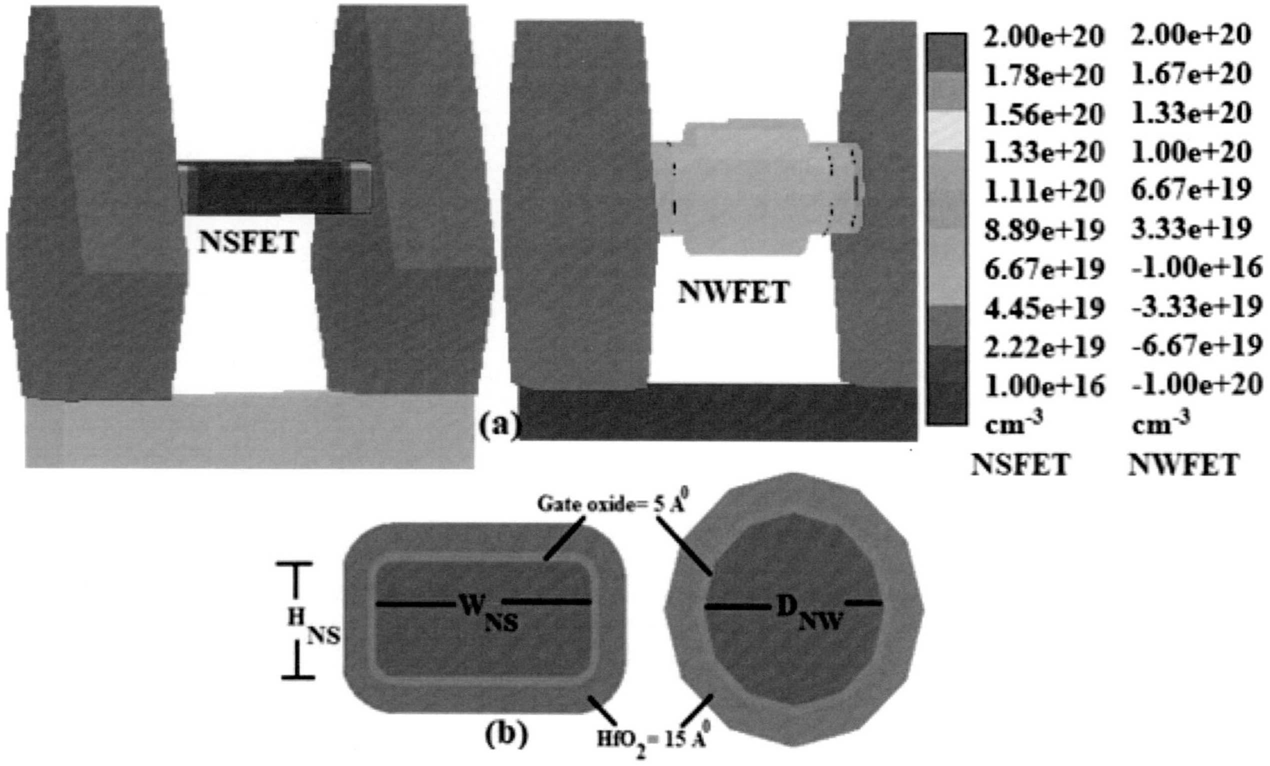

Figure 5.3 (a) Schematic of NSFET and NWFET with net doping concentration (b) cross-sectional view of the nanosheet and nanowire transistors considered for comparison.

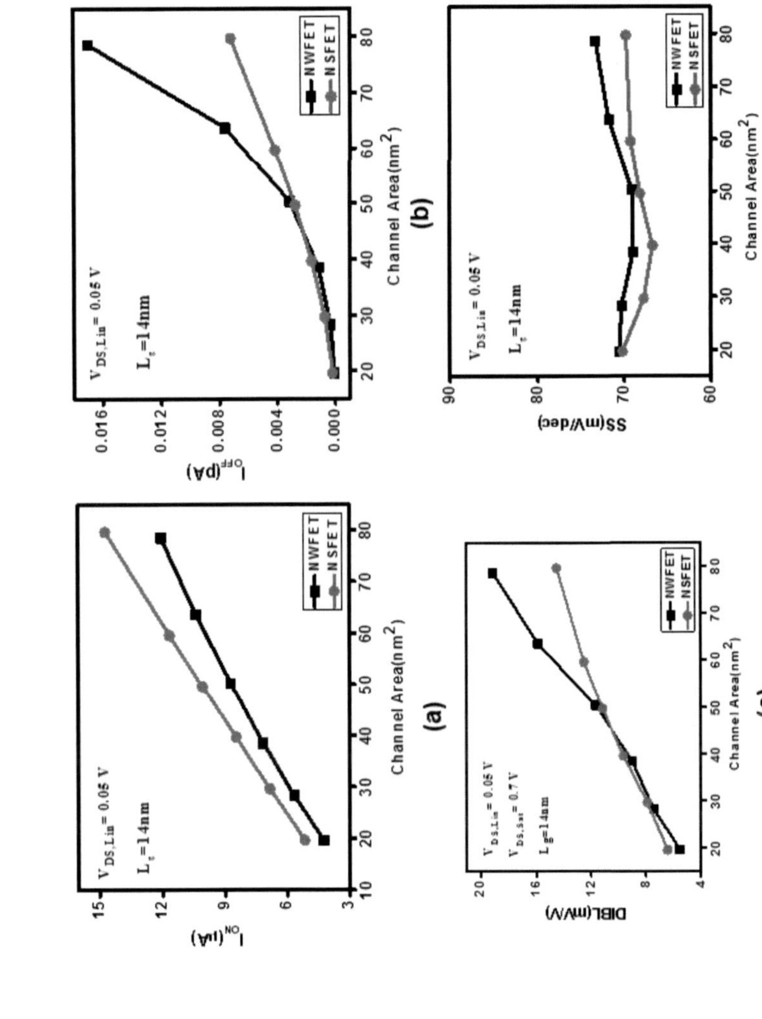

Figure 5.4 (a) I_{ON} (b) I_{OFF} (c) DIBL, and (d) SS as a function of channel area of the GAA NWFET and GAA NSFET at V_D = 50 mV.

5.2 Design and Simulation of Nanosheet Transistors

GAA nanosheet transistors are now considered to be the most suitable and feasible alternative structures for ongoing rapid scaling even below 7 nm technology nodes. The width of nanosheets in NSFETs is not constrained by fin pitch or fin quantization, allowing for higher effective width in circuit design [30]. Current drivability, electrostatics, and parasitic components are all affected by the width. Process differences such as line edge roughness and metal gate granularity are likely to occur in advanced technologies. Manufacturing variability is a random departure that alters the structure of the transistor and, as a result, the electrical properties of a circuit. It is thus critical to assess device design alternatives for NSFET design optimization.

5.3 Effects of Sheet Width Variation

We compare and analyze device figures of merit, viz., the threshold voltage (V_{TH}), I_{ON}, I_{OFF}, subthreshold slope (SS), and drain-induced barrier lowering. The transfer characteristic (I_D–V_{GS}) plot is analyzed by varying W_{NS} from 3 nm to 12 nm with fixed H_{NS} at 5 nm, as shown in Figure 5.5(a). It is observed that NSFET provides an improvement in the drive current for higher nanosheet width. I_D significantly increases by 183.4%, with an increase in nanosheet width from 3 nm to 12 nm. Figure 5.5(b) shows I_{ON} as a function of the $V_{TH,sat}$ for NSFETs with W_{NS} (3–12 nm) at L_g (8–14 nm). By reducing the gate length to 8 nm, the reduction of V_{TH} results in an increase of I_{ON}. The nanosheet width reduction also induces a higher $V_{TH,sat}$ due to the stronger confinement potential. It is also observed that the increase in I_{ON} is due to enhanced effective width. The distributions of electron current density contours for the simulated NSFET at on-state and off-state are illustrated in Figure 5.6.

Figure 5.7 (a and b) shows the impact of an increase in nanosheet width on SS and DIBL for different gate lengths, respectively. It is observed that the SS degradation at a gate length of 14 nm is marginal. However, 11 mV/dec SS degradation is observed when the gate length reduces to 8 nm. NSFETs with thinner width have smaller SS degradation due to enhanced gate control over the channel. Figure 5.7 (b) shows DIBL that has a similar trend with SS, and wider nanosheet provides higher DIBL. The impact of nanosheet width on the I_{OFF} and I_{ON}/I_{OFF} ratio is illustrated in Figure 5.7 (c) and Figure 5.7 (d), respectively. It is observed that

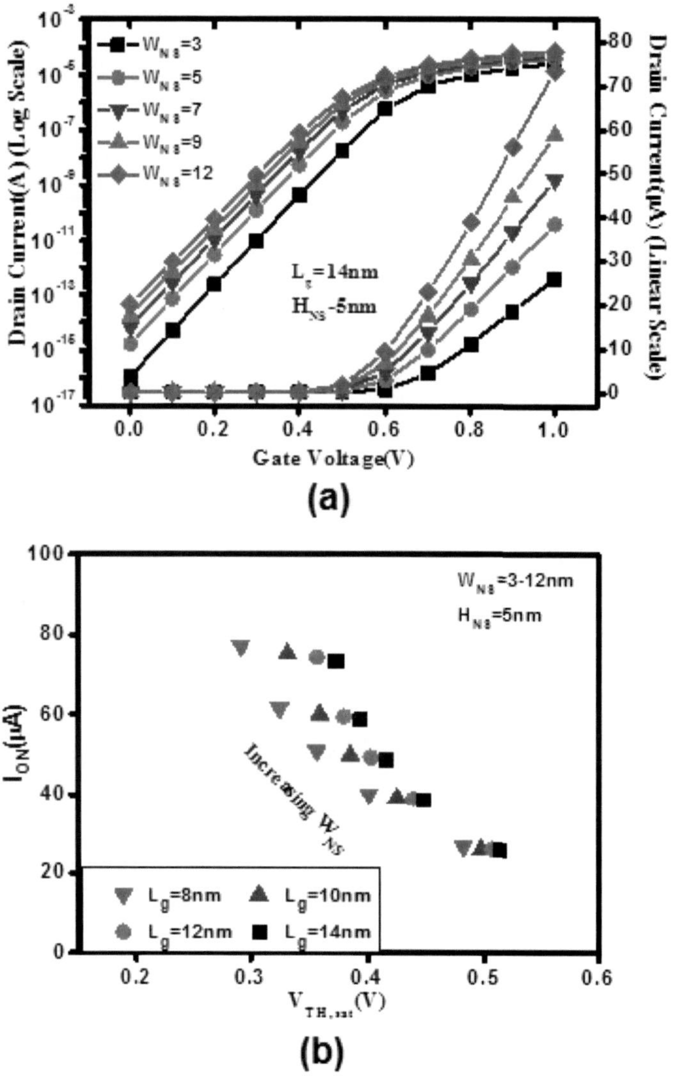

Figure 5.5 (a) I_D–V_{GS} of NSFETs having different W_{NS} at V_{DS} = 0.7 V and (b) I_{ON} as a function of $V_{TH,sat}$ for NSFETs with W_{NS} = 3–12 nm at L_g = 8–14 nm.

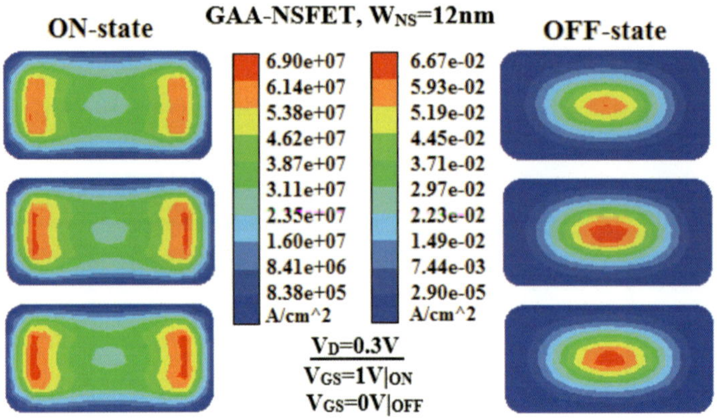

Figure 5.6 Electron current density contour at on-state and off-state for NSFET across the channel with W_{NS} = 12 nm and H_{NS} = 5 nm.

smaller nanosheet width provides a better I_{ON}/I_{OFF} ratio, due to larger overdrive voltage. However, a wider nanosheet provides larger I_{ON} and smaller I_{OFF}, so I_{ON}/I_{OFF} ratio gets better due to enhanced effective width. Figure 5.7 (d) shows I_{OFF} increases with an increase in the nanosheet width, which can be mitigated by using a thinner nanosheet. Finally, optimized NSFET is presented considering maximum performance improvement with W_{NS} and H_{NS} of 12 nm and 6 nm, respectively. The I_{ON}, DIBL, and SS of presented NSFET are 73.4 μA, 23.41 mV V^{-1}, and 67 mV/dec, respectively.

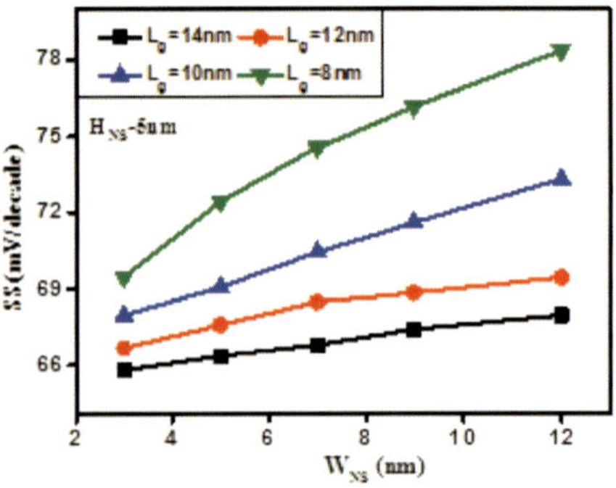

Figure 5.7 (*Continued*)

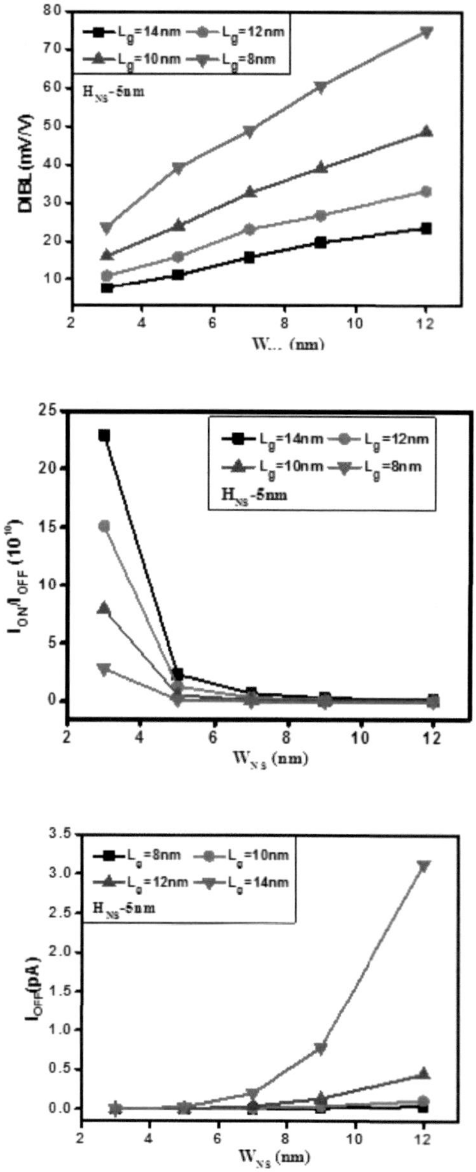

Figure 5.7 (a) SS, (b) DIBL, (c) I_{ON}/I_{OFF}, and (d) I_{OFF} as a function of W_{NS} for NSFETs having different L_g at V_{DS} = 0.7 V.

5.4 Effects of Sheet Height Variation

Figure 5.8 (a and b) shows I_{OFF} and I_{ON} as a function of W_{NS} for different H_{NS} at V_{DS} = 0.7 V. I_{OFF} increases with increasing H_{NS} due to the reduction in potential barrier height, and it can be reduced by using a thinner nanosheet. The variations of DIBL as a function of sheet width, which is shown in Figure 5.8 (c), indicate the trade-off between sheet width, which dominates the intrinsic performance of the devices and the immunity to the short channel effect.

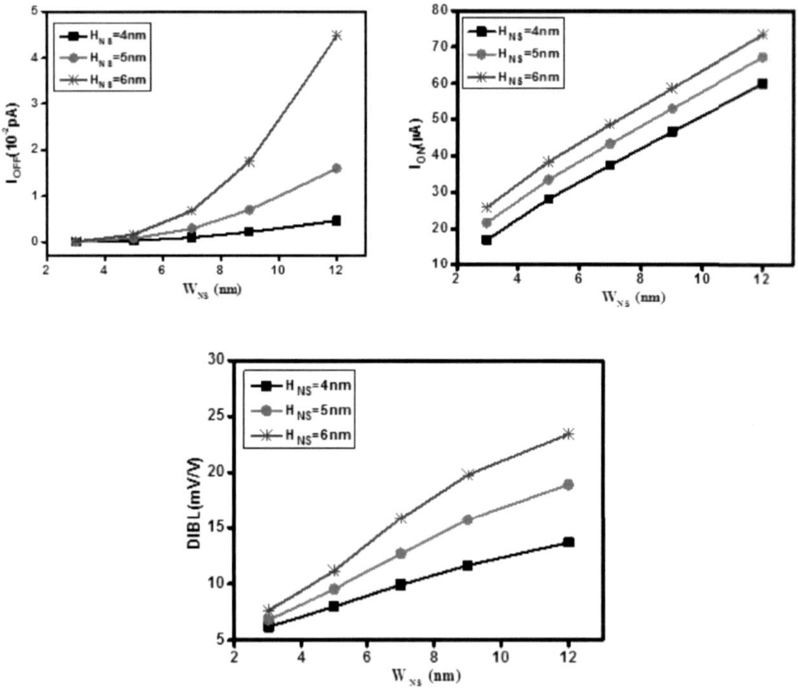

Figure 5.8 (a) I_{OFF}, (b) I_{ON}, and (c) DIBL of NSFETs with different W_{NS} and H_{NS} at L_g = 14 nm.

5.5 Vertically Stacked Lateral Nanosheet Transistors

The device structure for NSFET having a single channel (1 NS), two channels (2 NS), and three channels (3 NS) have been shown in

Figure 5.9. The output characteristics I_D–V_G curves of stacked (1–3) NSFETs at V_G = 0.7 V are shown in Figure 5.10. The high saturation current of the stacked SNSFET can be attributed to more number of electron-conducting layers, higher mobility, and lower S/D parasitic resistance compared to the single SNSFET. The I_D value becomes almost double and thrice in 2NS and 3NS compared to 1NS for saturated drain voltages in the transfer characteristics (see Figure 5.10). These enhancements can be justified with the help of the density of charges. As in multiple sheets, the density of holes gets doubled and thrice as compared to a single sheet. Besides, the density of charges and more number of conductive channels, low S/D parasitic resistance, and higher mobility also contribute to the increment in the current. This can be attributed to the decreased S/D parasitic resistance (RSD) of the stacked multiple NS in comparison with that of the single SNSFET. The RSD value of the stacked NS is nearly 8.7 times lower compared to the FinFET due to the larger cross-sectional area of stacked NSFET for conducting current at S/D.

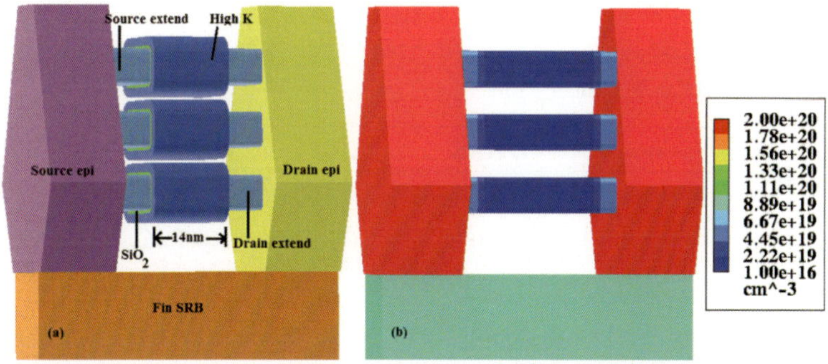

Figure 5.9 (a) A schematic of NSFETs (L_g =14 nm) and (b) Net doping concentration of NSFET is shown.

Other performance metrics, such as I_{ON}, I_{OFF}, $V_{TH,}$ DIBL, and SS are extracted from the transfer and output characteristics. The device parameters such as V_{TH}, SS, I_{ON}, and I_{OFF} have been found out and are shown in Table 5.2. It can be observed that the V_{TH} of 1NS is –259.5 mV and is decreasing with more sheets. It makes the multi-sheet devices more suitable for low-power applications. The SS values also reduce from 170 mV/dec to 61 mV/dec from 1 NS to 3 NS. The I_{ON}/I_{OFF} ratio maintains an order of 10^2 irrespective of the number of sheets. The overall subthreshold performances of NSFETs indicate that 3NS NSFET is a feasible and very promising device for future technology nodes.

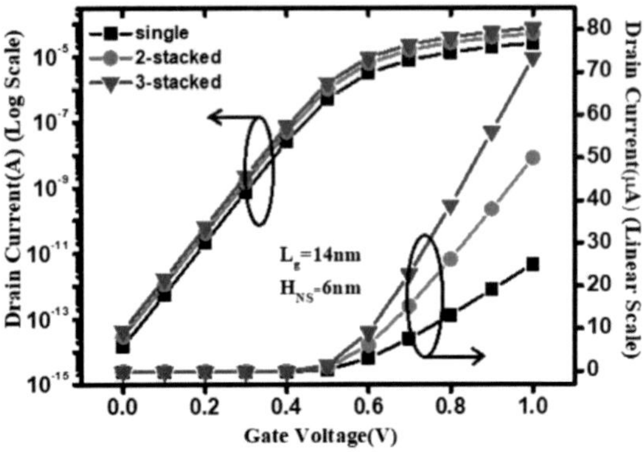

Figure 5.10 I_D–V_G plot of stacked (1–3 nanosheets) nanosheet transistor at V_D = 0.7 V.

Table 5.2 Comparison of stacked nanosheet device parameters such as V_{TH}, SS, I_{ON}, and I_{OFF}

Parameter	1NS	2NS	3NS
V_T_lin (mV)	0.42182	0.4001	0.38848
V_T_sat (mV)	405.5	385.22	373.26
I_{ON}_lin (μA)	12.07646	23.77895	34.54673
I_{ON}_sat (μA)	25.07941	49.86237	73.37389
I_{OFF}_lin (pA)	0.00816	0.01622	0.02486
I_{OFF}_sat (pA)	0.01454	0.02872	0.04477
SS_lin(mV/dec)	71.32928	69.13429	67.9715
SS_sat (mV/dec)	71.70713	68.63716	66.94848
DIBL (mV/V)	25.10395	22.88933	23.40727

5.6 Strain-Engineered NSFETs

Channel strain engineering continues to play a prominent role for future CMOS technology; however, to keep the same level of channel strain by embedded S/D SiGe is challenging as transistor dimension shrinks. This is because the volume of the stressor (volume of SiGe in the S/D region) reduces as the transistor dimension shrinks. It has also been reported that SiGe with epitaxial S/D on the strain-relaxed buffer (SRB) will continue to be effective boosters [31]. In this study, the effect of S/D $Si_{1Cx}Ge_x$ material is studied. While the channel is fixed to pure silicon, the S/D $Si_{1-x}Ge_x$ is varied from 0% to 50%. The compressive stress induced in the Si channel enhances the carrier mobility. The stress/strain engineering has been implemented in the simulation environment using the VSP tool [32] for analysis of the stress profile in the device. A 3D stress profile of the sheet (14 nm long) has been shown in Figure 5.11.

The longitudinal stress distribution in the channel regions of the device with a 30% Ge mole fraction in the S/D SiGe stressor is shown in Figure 5.11(a). Effective stress values are extracted and plotted in Figure 5.11(b). This figure also shows the stress distribution in the channel of NSFET for different Ge contents in the stressor. With the increase in the Ge mole fraction, the carrier effective mass decreases, and the hole mobility increases due to the S/D stressor. This results in strain on the channel which enhances the hole mobility of the Si channel. The longitudinal stress for different Ge mole fraction contents of 10%, 30%, and 50% are 0.512, 1.49, and 2.41GPa, respectively.

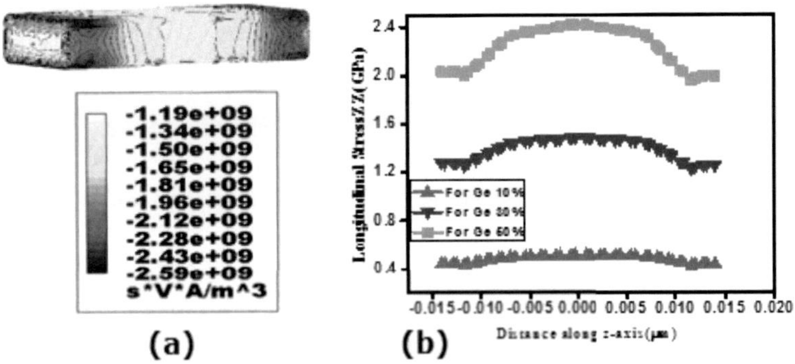

Figure 5.11 (a) The 3D contour of stress distribution in nanosheet with 30% Ge mole fraction. (b) 1D longitudinal stress profiles in the channel, due to diamond-shaped SiGe S/D stressor for different Ge-mole fractions.

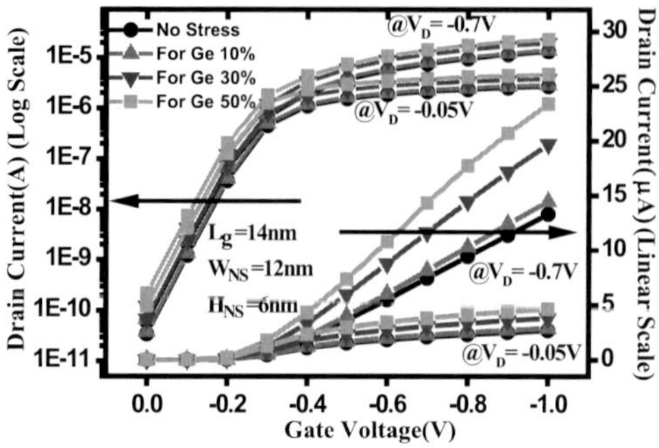

Figure 5.12 I_D–V_G characteristics of NSFET devices with SiGe S/D stressors having a gate length of 14 nm.

The impact of the Ge mole fraction on the I_D–V_G characteristics for the proposed NSFET design is shown in Figure 5.12. For the proposed device, drain current increases with raising the Ge mole fraction in the S/D region at V_D = 0.05 V (linear) and V_D = 0.7 V (saturation). This is due to the fact of higher stress in the SiGe S/D region indicates a higher hole mobility enhancement due to the larger induced stress with an increase in the Ge mole fraction. Superior I_D–V_G characteristics are demonstrated by the NSFET with 50% Ge mole fraction for both linear and saturation regimes. The enhancement in higher saturation driving current due to 50% Ge mole fraction is 75% higher than the Si channel device with no stress effect in the S/D region.

The device output characteristics I_D–V_D is shown in Figure 5.13. An enhancement in drain current is observed as strain in the sheet increases as Ge-content increases from 10% to 50%. The enhancement of I_{DSAT} is around 9%, 55%, and 89%, as compared to control Si NSFET without any stress by increasing the Ge content to 10%, 30%, and 50%, respectively.

Figure 5.14 shows the variation of V_{TH}, SS, and DIBL for NSFETs with different Ge mole fractions. The threshold voltage decreases significantly with an increase in the Ge mole fraction from 0% to 50% as shown in Figure 5.13. In the no-stress condition, the value of V_{TH} is maximum, which is according to the requirements of ITRS [33]. This decreases V_{TH} with the increase in the Ge mole fraction, due to the reduction in the energy bandgap, which results in a decrease in flat-band voltage, and hence an early onset of inversion [34]. The DIBL is extracted at $V_{DS} = 0.05$ V and $V_{DS} = 0.7$ V by constant current method. The DIBL decreases with an increase in the Ge mole fraction as shown in Figure 5.14. As mole fraction increases, the reduction of the bandgap at source and drain regime happens and carrier mobility increases that resulting in the decrease in DIBL. The SS value decreases with an increase in the Ge mole fraction in $Si_{1-x}Ge_x$ due to higher carrier mobility.

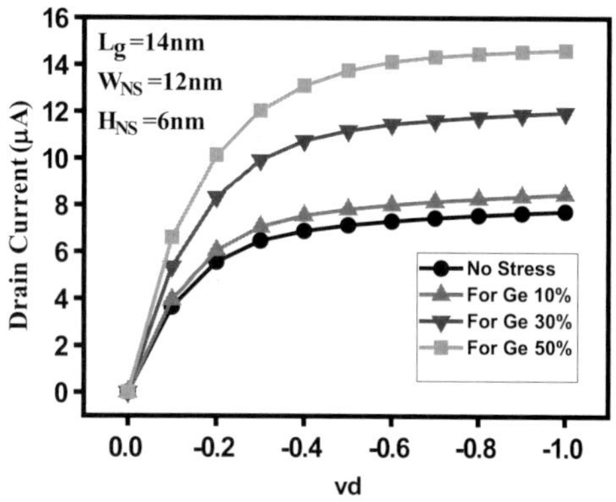

Figure 5.13 I_D–V_D characteristics of NSFET devices with SiGe S/D stressors having a gate length of 14 nm.

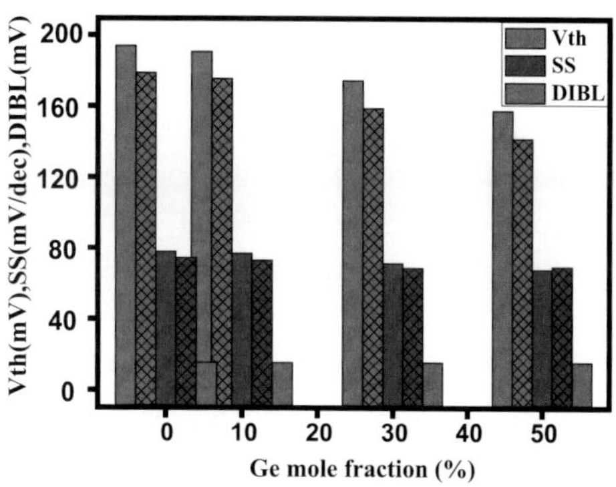

Figure 5.14 Impact of Ge mole fraction on V_{TH}, SS, and DIBL of $Si_{1-x}Ge_x$ S/D stressors in NSFET. The lines with stripped marks represent the saturation region ($V_D = 0.7$ V) and the plain lines represent the linear region of operation ($V_D = 0.05$ V).

The effect of variation of Ge-mole fraction on I_{ON} and I_{OFF} is shown in Figure 5.15. It is observed that the values of I_{ON} and I_{OFF} increase with rising in the Ge mole fraction. With the increase in Ge mole fraction, the bandgap decreases, which results in enhancement of drain current both in linear and saturation regions. The short channel effect and subthreshold performance are serious concerns for devices at this small dimension. The critical comparison of all electrical parameters (I_{ON}, I_{OFF}, V_{TH}, and SS) is also studied and plotted in Figure 5.16.

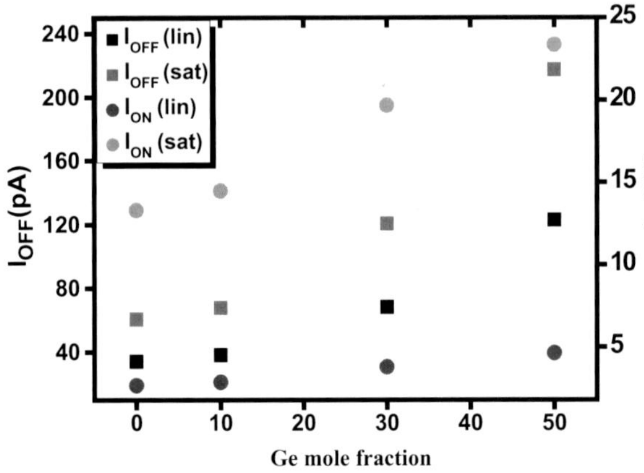

Figure 5.15 Impact of Ge-mole fraction on I_{ON} and I_{OFF} of $Si_{1-x}Ge_x$ S/D stressors in NSFET.

As shown in Figure 5.16, the extracted normalized electrical parameters are in saturation. The variations of the five electrical parameters with different Ge mole fractions are normalized concerning the control Si channel NSFET without S/D stressors. Furthermore, to boost the performance of NSFET uniaxial strained SiGe channel may be employed. The stress can be introduced in the channel by changing the material of the source-drain extension region and the channel region. It has been found that the uniaxial strained-SiGe NSFET device (with 30% Ge mole fraction resulting compressively strained) is having the highest drain current than the other devices. A further enhancement is also possible with 3-stacked NSFET having 30% Ge concentration to increase the lateral compressive stress resulting in improvement in the device drain current by the hole mobility enhancement.

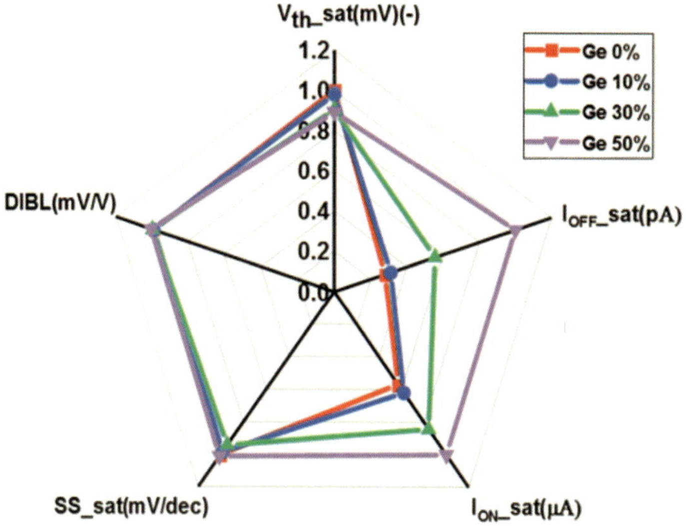

Figure 5.16 Normalized five electrical parameters (performance metrics) as a function of Ge mole fraction in the saturation regime. Red, blue, green, and purple color corresponds to 0%, 10%, 30%, and 50% Ge mole fraction values, respectively.

5.7 Variability in NSFETs

An increase in the complexity of the stacked nanosheet structures can lead to significant device variability. Using numerical simulation, we study the characteristics fluctuations induced by the random discrete dopants and metal grain granularity in NS GAA transistors. We use 3D quantum-mechanically corrected transport models in the simulation. Metal gate granularity is an issue mainly connected with the "gate first" process technology, whereby the gate metal is deposited before any high-temperature annealing procedure [35]. During high-temperature processing, the nominally amorphous metal gate material becomes poly-crystalline. This results in the formation of grains with different crystallographic orientations and metal work functions. Within the case of polysilicon gates, the interfaces between grains cause Fermi level pinning and doping non-uniformity as a consequence of rapid diffusion along grain boundaries [36, 37]. Both effects introduce significant variability in the performance of the devices. The effect of MGG on device performance is closely correlated with the metal grain size, concerning the overall gate size. In the case of a metal gate, a large grain size concerning overall gate size has a bimodal effect on transistor performance, whereby differing device instances are controlled by different grain work functions. Interestingly, small grain size, relative to the overall gate area, causes self-averaging of the grain work functions and has a smaller influence on overall device performance. In the gate last processes, where the gate is deposited post-annealing (and other high-temperature processes), the gate granularity could be significantly reduced.

The MGG induced random work function variation (WFV) is another important source of random variability for the technology nodes that use a high-k/metal gate stack [38]. Depending on their crystal orientations, the work function of each metal grain varies randomly. Therefore, instead of a single work function value, the transistor gate contains multiple grains with different work function values that cause random fluctuation of device performance parameters [39, 40]. With the use of a high-k dielectric material (e.g., HfO_2), the ever-increasing gate leakage current can be decreased and the gate-to-channel capacitive coupling can be improved. In addition, the thinner gate oxide layer can provide better channel control capability, which can effectively suppress the short channel effect and maintain excellent compatibility with traditional process technology [36, 37].

Recent studies have demonstrated that high-k materials are incompatible with polysilicon gates because of two insurmountable technical problems-Fermi-level pinning effect and phonon scattering, which result in shifting of the V_{TH} to higher values and reduction of the mobility of electrons, respectively [35, 39]. To overcome these limitations, metals are used in the gate electrode instead of polysilicon [40, 41]. However, as the work function of each metal grain depends on its crystalline orientation, the use of a metal gate introduces a new random fluctuation source, i.e., WFV, which can seriously affect V_{TH} and other device characteristics. Further, with the scaling down of the device size, the impact of WFV becomes increasingly serious, which significantly affects the performances of ICs [39, 42, 43].

5.8 Modeling of WFV

Assuming the gate length and width of the metal-gate are L and W, respectively. The calculation of the number of grains on the gate surface (N_{eff}) is based on the binomial distribution model. Therefore, N_{eff} distributed on the metal-gate surface is written as follows:

$$N_{\text{eff}} = \frac{\text{Area}}{\pi \times \left(\dfrac{\text{Grain Size}}{2} \right)^2}. \tag{5.1}$$

The grain probability is mapped with work functions for easy distribution of grains with a condition that the total probability should be one. The work function of the gate metal will be a probabilistic approach rather than a deterministic value. Assuming $X_1, X_2, X_3 \ldots \ldots + X_n$ to be the random variables (where $N_{\text{eff}} = X_1 + X_2 + X_3 \ldots \ldots + X_n$) that denote the number of grains having work function values of ϕ_1, ϕ_2, $\phi_3 \ldots \ldots \phi_n$, respectively. The probabilistic distribution of work function in a given metal can be realized as [44]:

$$\phi_{\text{Metal}} = \left(\frac{X_1}{N_{\text{eff}}} \right) \cdot \phi_1 + \left(\frac{X_2}{N_{\text{eff}}} \right) \cdot \phi_2 + \left(\frac{X_3}{N_{\text{eff}}} \right) \cdot \phi_3 + \ldots\ldots.. + \left(\frac{X_n}{N_{\text{eff}}} \right) \phi_n. \tag{5.2}$$

The first term $\dfrac{X_1}{N_{\text{eff}}}$ represents the percentage of the gate area covered with grains having work function of ϕ_1. The probability of getting the above-mentioned work function can be calculated from the probability density function (PDF) ($P_X(k)$):

$$P_X(k) = \binom{N_{\text{eff}}}{k} \times P_1^{\,k} \times (1 - P_1)^{N_{\text{eff}} - k}. \tag{5.3}$$

In this case, two random variables X_1 and X_2 (where $N_{\text{eff}} = X_1 + X_2$) that denote the number of grains having work function values of ϕ_1 and ϕ_2 in the binomial distribution model with probabilities of P_1 and P_2 (Assume $\phi_1 = 4.6$ eV and $\phi_2 = 4.35$ eV, $P_1 = 0.6$ and $P_2 = 1 - P_1 = 0.4$), respectively.

These two random variables are not independent ($X_2 = N_{\text{eff}} - X_1$). The probability of getting exactly $X_1 = k$ of such grains can be calculated by a binomial distribution where the PDF is given by:

$$P_{X_1}(k)=\frac{N_{\text{eff}}!}{k!(N_{\text{eff}}-k)!}0.6^k 0.4^{N_{\text{eff}}-k} \quad \text{where} \quad \binom{N_{\text{eff}}}{k}=\frac{N_{\text{eff}}!}{k!(N_{\text{eff}}-k)!} \tag{5.4}$$

and

$$\phi_{\text{Metal}}=\left(\frac{X_1}{N_{\text{eff}}}\right)(4.6\text{ eV})+\left(\frac{X_2}{N_{\text{eff}}}\right)(4.35\text{ eV})$$
$$=\left(\frac{X_1}{N_{\text{eff}}}\right)(4.6\text{ eV})+\left(\frac{N_{\text{eff}}-X_1}{N_{\text{eff}}}\right)(4.35\text{ eV}).$$

The Voronoi diagram has been used to investigate cells, metal grains, and solid-state physics [45–48]. With the assumption that the grain growth rate of each grain is the same for describing metal grain, the grain boundary will locate at the perpendicular bisector of neighboring grain seeds in the Voronoi diagram. For constructing the Voronoi diagram, we place the grain seeds randomly and the number of grain seeds is determined by the size of the grain, which is expressed in Eq. (5.1). We connect each seed by a straight dash line. Figure 5.17(b) shows that we can construct perpendicular bisectors of dash lines, and the perpendicular bisector of each dashed line constructs the grain boundary. Finally, we can construct the Voronoi pattern after the grain seeds and dash lines are removed. From Figure 5.17(b), we can see that the irregular shape of grain can physically reflect the grain patterns. The boundaries of the Voronoi pattern are used for efficient TCAD simulation.

In the following study, we employ TiN as the gate material for NSFETs. For the TiN metal gate, two possible grain orientations (<200> and <111>) with distinct work functions and probability should be considered. Table 5.3 summarizes the two-orientation characteristics of the TiN metal gate. To generate various metal-gate grain patterns for macroscopically identical devices, four factors regarding the grain should be considered: (i) seed position, (ii) shape, (iii) size, and (iv) orientation. Figure 5.17(a) shows the simulation flow of our proposed Voronoi approach to simulate WFV. First, the assigned average grain size is used to estimate the number of grains (seeds) for constructing the random Voronoi grain pattern. The determined seeds are randomly placed in the metal gate region (black solid points in Figure 5.17(b)) and the distributions of seed position are different for each device. With the seed location shown in Figure 5.17(b), the Voronoi pattern is constructed by connecting the solid lines that are the perpendicular bisector of each dashed line. Figure 5.17(b) shows the generated Voronoi pattern for the simulation of WFV in the metal gate region. Notice that in addition to the shape of metal grain, the effective work function for each grain varies with orientation.

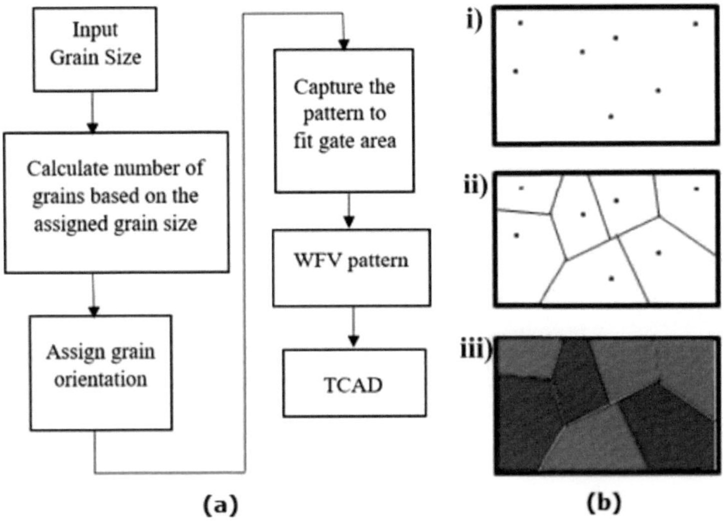

Figure 5.17 (a) Flow chart for determining the WFV pattern (N_{grain} = (Area)/($\pi\times$(grain size/2)^2)). (b) Formation of the Voronoi pattern in the TCAD simulation [49]. The boundary of grain is defined as a solid line which is the perpendicular bisectors of the connecting line of each grain seed.

Table 5.3 Grain orientations with corresponding probability and work function for a TiN metal gate

TiN metal gate		
Orientation	**Probability**	**Work Function [eV]**
<200>	60%	4.6
<111>	40%	4.35

The MGG induced variations due to different surface orientations of metal grains lead to variability in the effective work function using the Voronoi approach. To generate various metal-gate grain patterns, the grain depends on various factors, e.g., grain position, shape, size, and orientation. TiN is used as the metal gate material. The different work function values of TiN that depend on grain orientation further lead to variation in the electrical parameters of the device. The TiN has different work function values of 4.6 eV and 4.4 eV with probability of occurrence of 60% and 40% along <200> and <111>, respectively [49]. In the proposed model, it is assumed that all the grains with different orientations have a fixed grain size. Figure 5.17(b) shows two different work function patterns generated with three different average grain sizes of 1, 3, and 5 nm. Poisson–Voronoi approach creates the appropriate metal grain area distribution, which is a similar distribution observed in experimental results [50]. I_D–V_G plots of 14 nm gate length NSFET for MGG variations are shown in Figure 5.17(b). The σV_{TH} distribution due to MGG can be related to grain size as:

$$\sigma V_{\mathrm{TH}} \propto \frac{\text{Grain size}}{\sqrt{W_{\mathrm{eff}} \times L_{\mathrm{eff}}}}. \tag{5.5}$$

The profile for WFV in the metal gate with 1 nm grain size for NSFET is shown in Figure 5.18. I_D–V_G characteristics of 100 samples for NSFET for grain size 1 nm are represented in Figure 5.19.

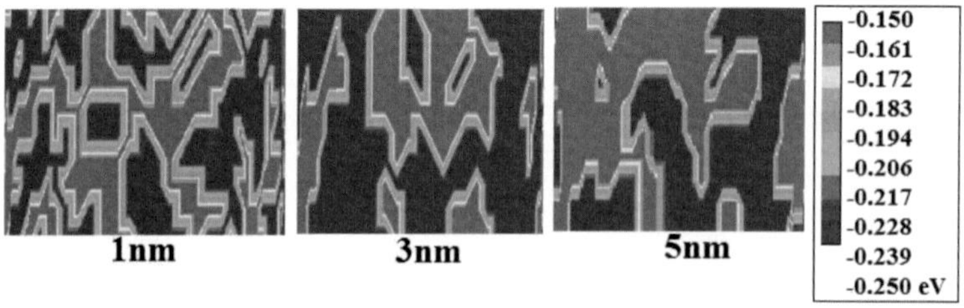

Figure 5.18 MGG profile shows the generated Voronoi pattern for grain sizes of 1, 3, and 5 nm of the 14 nm gate length NSFET.

The V_{TH} distribution, due to MGG, spreads more toward the head and tail regions with an increase in grain size is shown in Figure 5.20. As nano-sized grains are considered, these distribution curves are very close [51]. This results in more deviation of the distribution shape from its normal distribution. As the grain size increases, the V_{TH} distribution due to MGG for both $V_{\mathrm{TH,Lin}}$, and $V_{\mathrm{TH,Sat}}$ becomes

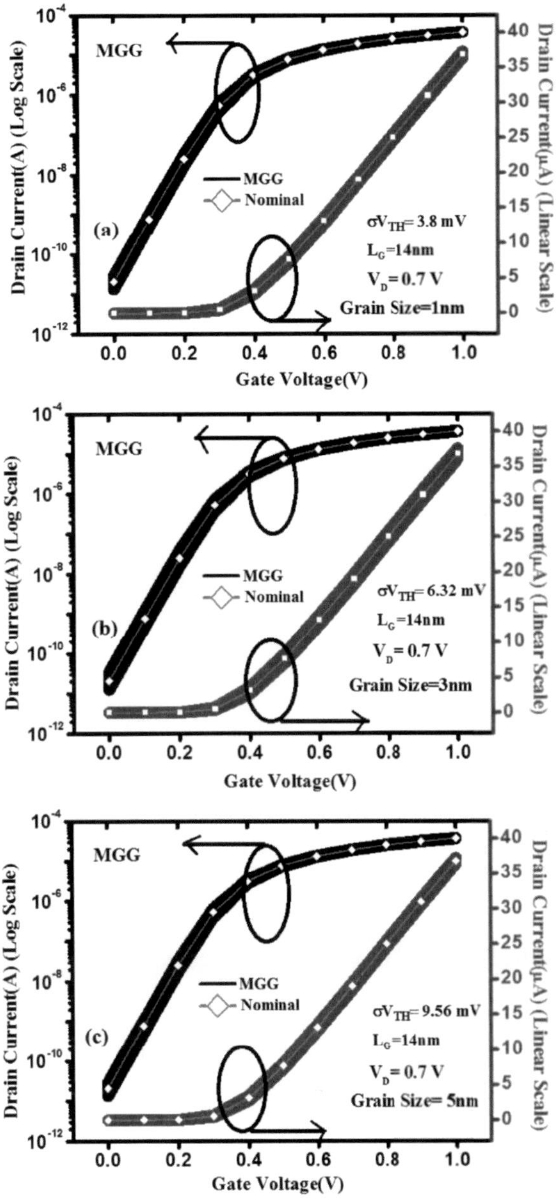

Figure 5.19 Transfer characteristics (I_D–V_G) of 14 nm gate length NSFET with MGG variations at an average grain size of 1, 3, and 5 nm with 200 different probability configurations. Grain size: 1 nm (black color), 3 nm (red color), and 5 nm (blue color).

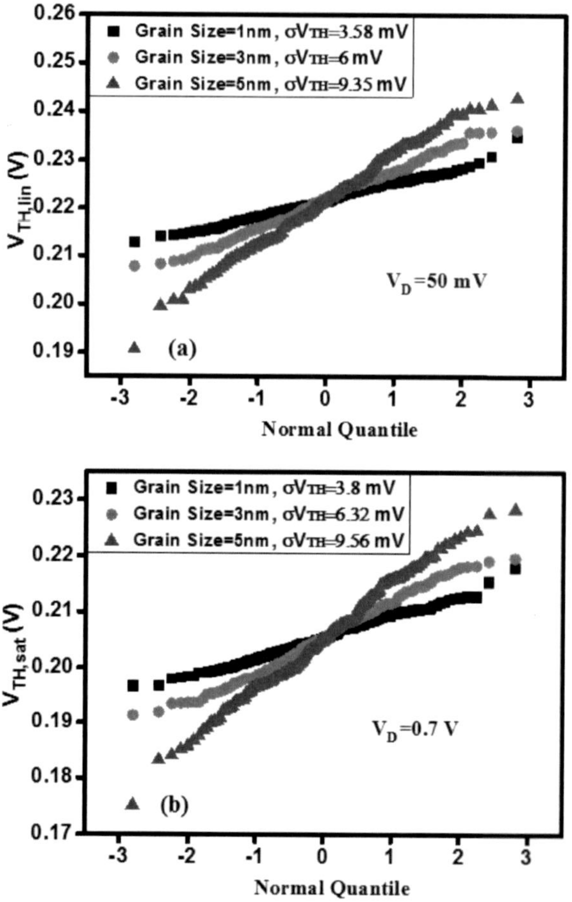

Figure 5.20 *Q–Q* plot of V_{TH} distribution due to MGG for different metal gate grain sizes in (a) 50 mV and (b) 0.7 V of drain bias, respectively.

more and more constrained at both the upper and lower tails alike to the behavior reported for bulk MOSFETs [52]. Due to the increase in the grain size, the total number of grains present in the gate surface decreases. As a result, the work function probability distribution diverges from the normal distribution and becomes discrete. Therefore, the increase in potential fluctuations will increase V_{TH} variability. With the increase in grain size from 1 nm to 5 nm, σV_{TH} increases by 161.2% and 151.2% in linear and saturation regions, respectively.

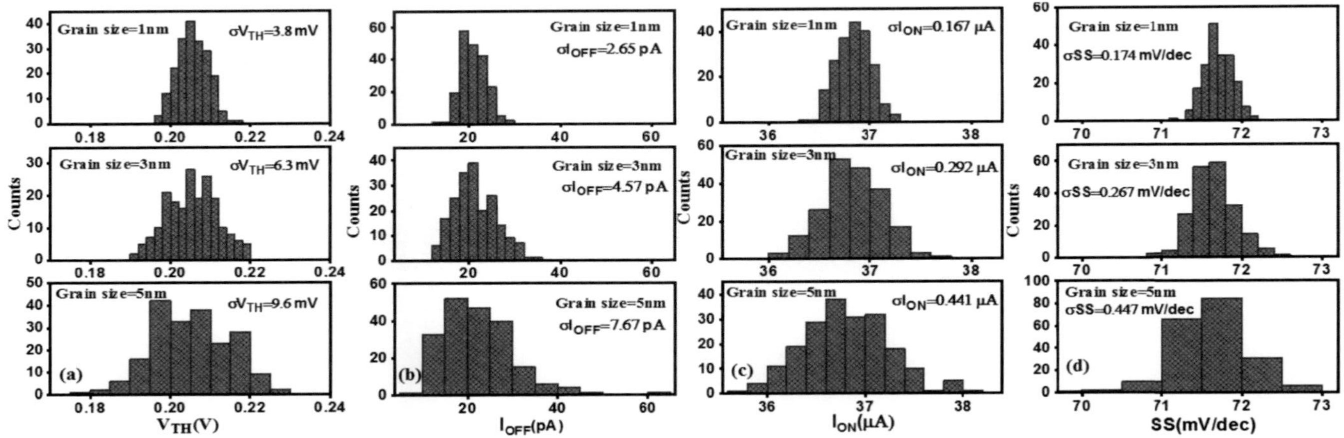

Figure 5.21 Normalized distribution of V_{TH}, I_{OFF}, I_{ON}, and SS due to grain size variations for the 14 nm NSFET at $V_D = 0.7$ V.

Figure 5.21 shows that the σ value of V_{TH}, I_{OFF}, I_{ON}, and SS variations increases with larger grain size at both the linear and saturation mode in NSFETs. It is observed that when the grain size decreases, the σ is reduced. This reduction is almost linear, dependent on the grain size for σV_{TH}, σI_{ON}, σI_{OFF} and σSS. In Figure 5.21, as the grain size is increased from 1 nm to 5 nm, the σV_{TH} is increased from 3.8 mV to 9.6 mV (a factor of 2). The standard deviation (σ) of the subthreshold characteristics is summarized in Table 5.4.

Table 5.4 The footprint of grain size on σ of V_{TH}, I_{OFF}, I_{ON}, and SS at V_D = 0.7 V

I. MGG II. (Grain Size)	σV_{TH} (mV)	σI_{OFF} (pA)	σI_{ON} (μA)	σSS (mV/dec)
1 nm	3.8	2.65	0.167	0.174
3 nm	6.3	4.57	0.292	0.267
5 nm	9.6	7.67	0.441	0.447

The effect of MGG on V_{TH} with different grain sizes, i.e., 1, 3, and 5 nm, is shown in Figure 5.22. The σV_{TH} in saturation mode is found to be 3.8 mV, 6.3 mV, and 9.6 mV for 1 nm, 3 nm, and 5 nm grain sizes, respectively. The effect of MGG on the V_{TH} is more noticeable in 5 nm grain size as the grain size is linearly dependent on the σV_{TH}. The V_{TH} variation is more severe when the grain size increases, as shown in Figure 5.22. This is because of the large number of grains in the metal gate area, suppressing the WFV-induced σV_{TH}.

The DIBL variability as a function of V_{TH} for different grain sizes at low and high drain biases is shown in Figure 5.22. The DIBL shows a minimal correlation with V_{TH}, mainly for grain sizes of 3 nm and 5 nm. The linear dependence between these variables is evaluated by the Pearson correlation coefficient (CC). Even the grain size reduces from 5 nm to 1 nm, the CC of the DIBL is larger for $V_{TH,Lin}$ than $V_{TH,Sat}$. A similar trend has also been reported by Seoane *et al.* [53]. The significant separation between the linear and saturation drain biases for the 1 nm grain size is pronounced due to the DIBL effect.

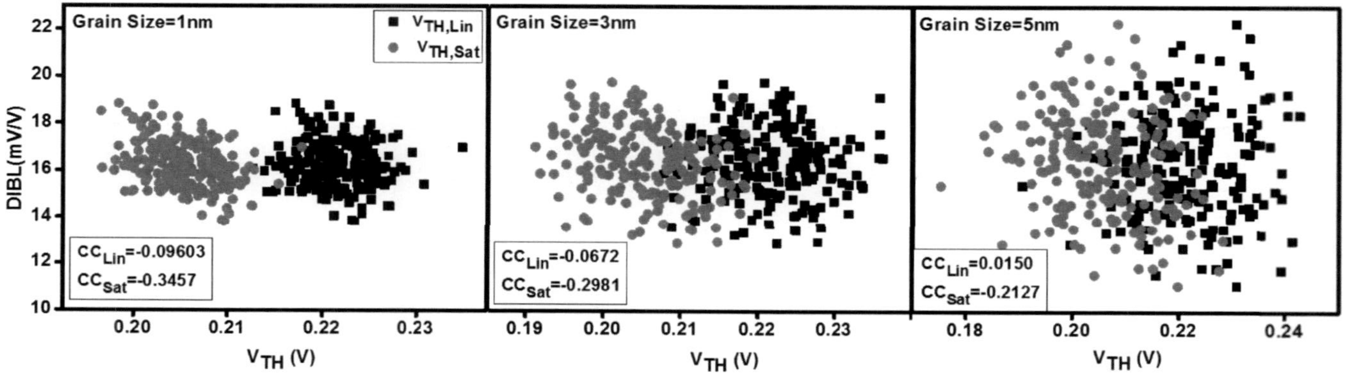

Figure 5.22 Scatter plots due to MGG variation in the 14 nm NSFET for three different grain sizes: 1 nm, 3 nm, and 5 nm for the DIBL as a function of V_{TH}.

5.9 Modeling of Random Dopant Fluctuation

The random and discrete nature of dopant charges changes the device characteristics in the channel region can be observed in highly-scaled MOSFETs and multigate FETs. The impact of RDD has been discussed in earlier research of both multi-gate MOSFET and traditional MOSFET [54], both experimentally [55, 56] and in simulation analysis [57–59]. The existing literature addresses the issues arising from the use of fine mesh for individual charge resolution in atomistic DD simulation [60]. Here, the electrostatic potential, derived from the output of the Poisson equation, determines the electron concentration when using Fermi-Dirac or Boltzmann statistics in the traditional DD model. The mobile charge trapping in the deep Coulomb potential wells, resolved by the fine mesh occurs. In quantum mechanics terms, due to the high ground electron state in the Coulomb well, this trapping is non-physical. Due to the greater resistance in the S/D regions, the transfer characteristics and the threshold voltage are modified.

Two of the approaches addressing the above issues in atomistic simulation scenarios include the application of screening [61] criteria to divide Coulomb potential into short- and long-range elements [60], as well as charge smearing [61]. The latter method, which is entirely empirical, can arbitrarily reduce the RDD impact particularly, in smaller devices. In the former method, limitations include the risk of double-counting of mobile charge screening, as well as the random selection of cut-off points. Confinement effects associated with the Coulomb potential of discrete dopants are accurately represented through the application of DG quantum corrections, which also takes into account the confinement effects within the channel. DG corrections in DD simulations can be calibrated in respect of the more accurate results obtained from a coupled Poisson/Schrödinger solution.

The RDF can be simulated by using a randomly non-uniform concentration of the dopant atoms in the active silicon region. To study the impact of random dopant fluctuations, the number of dopant atoms and the positions of the given dopant atoms in the S/D region are taken into consideration. The selection of the number of dopants in each device can be determined from Poisson's distribution of the average number of dopants. The distribution of the V_{TH} can be mathematically evaluated through the discrete convolution of the Poisson distribution with the mean value of the statistically significant region (i.e., S/D extension region). The deviation in the transfer characteristics is due to the variation in location and number of dopants. The equivalent charge density corresponding to the number of dopants suggested by Sano *et al.* [60] is given as:

$$n(r) = N_f \cdot \frac{k_c^3}{2\pi^2} \frac{\sin(k_c r) - (k_c r)\cos(k_c r)}{(k_c r)^3}, \tag{5.6}$$

where, N_f is the normalization factor, r is the distance from the discrete dopant, and k_c is the screening factor determined from $k_c \approx 2(N_{D/A})^{1/3}$ where $N_{D/A}$ denotes donor/acceptor concentration. The σV_{TH} distribution due to RDD is modeled following reference [56]:

$$\sigma_{V_{TH}} = \sqrt[4]{4q^3 \varepsilon_{Si} N_a \varphi_F} \frac{T_{ox}}{\varepsilon_{ox}} \frac{1}{\sqrt{2W * L}}. \tag{5.7}$$

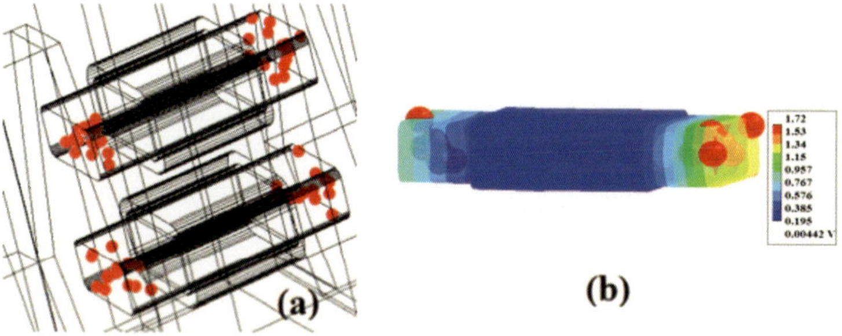

Figure 5.23 (a) NSFET featuring discrete dopant fluctuations in the S/D extension region and (b) potential distribution in the stacked-NSFET device at $V_G = 0$ V with dopant fluctuations in the S/D extension region.

The characteristic fluctuations curves of the studied device induced by RDD in the S/D-extension have been shown in Figure 5.23(a). The fluctuation of dopants can be used to analyze the impact of RDD in terms of their number and position. Figure 5.23(a) shows that the number of dopants follows Poisson distribution, and the location of dopants follows a uniform distribution. Based on the doping concentration, the dopant position is determined randomly. Here the dopant discretization is only executed in the source and drain extension regions [57, 62, 63]. Both position and dopant number variations affect the transfer characteristics and other device parameters. The discrete dopants are positioned in the device randomly, and the electrostatics associated with these random dopants are calculated by using the Sano method [60]. Potential distribution in the stacked-NSFET device at $V_G = 0$ V with dopant fluctuations in the S/D extension region is shown in Figure 5.23(b). The number density of the discrete dopant is given by:

$$n(r) = N_f \frac{k_c^3}{2\pi^2} \frac{\sin(k_c r) - (k_c r)\cos(k_c r)}{(k_c r)^3}, \quad (5.8)$$

where N_f is the normalization factor, k_c is the inverse of screening length, and r is the distance from the discrete dopant. The screening length can be calculated by using the Conwell-Weisskopf model [64, 65] and defined as $k_c \approx 2(N_{D/A})^{1/3}$ where $N_{D/A}$ represent donor/acceptor concentration.

I_D–V_G plots of 14 nm gate length NSFET are shown in Figure 5.24, including discrete dopants fluctuations in the S/D extension regions at a drain bias of 50 mV and 0.7 V. The random dopant fluctuations in the source and drain extension induce the change in the V_{TH}, channel electrostatic potential, and source-channel potential barrier, which further results in I_{ON} variation in these regions. The variation is caused by a change both in the number of dopants and in their placement in the source and the drain extension region.

5.10 Modeling of Random Discrete Dopant

An analytical model for V_{TH}, due to statistical dopant fluctuations, has been reported for a 2D cross-section in reference [65]. As slices of a 2D sheet can be integrated over the depth to give a 3D sheet, the use of the reported analytical model can be extended to compute V_{TH} for the 3D NSFET. The variance of dopant fluctuations in a sheet is the total number of dopants in the sheet. The number of dopants fluctuations per unit depth of cross-section of the sheet of dopant is considered as $N_{sheet} = N_A W_{NS} H_{NS}$. Since $dN_{sheet}(y)$ is the random variable for the area of $W_{NS}H_{NS}$ at a depth y of the sheet, $dV_{TH}(y)$ becomes the random function in this sheet of depth dy and can be expressed as [66]:

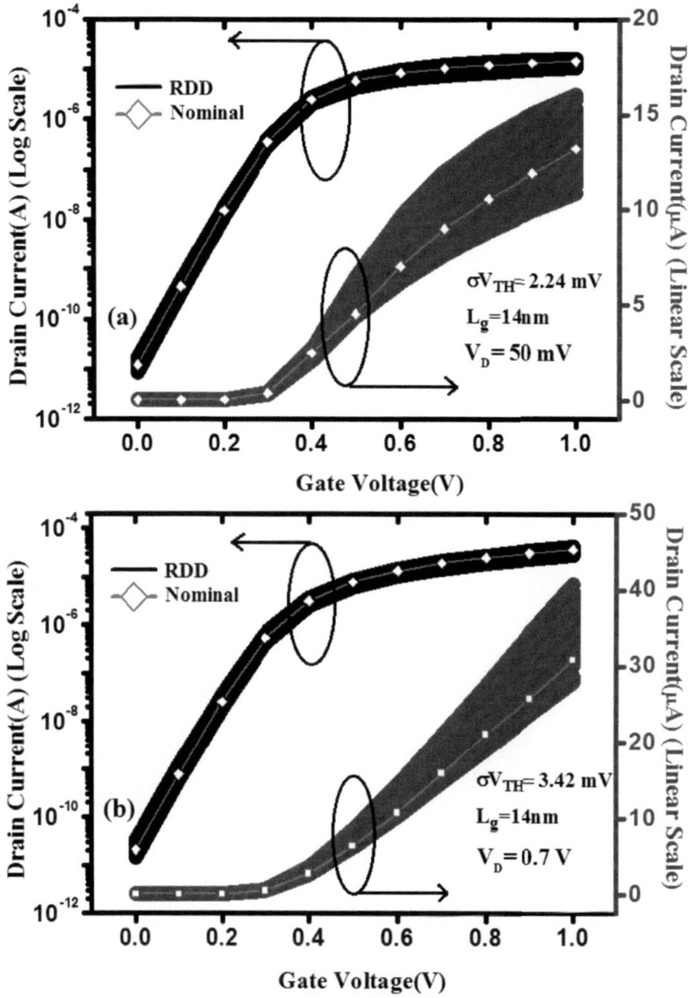

Figure 5.24 Transfer characteristics (I_D–V_G) of 14 nm gate length NSFET in linear and log scale including random discrete dopants in the S/D extension at a drain bias of (a) 50 mV and (b) 0.7 V.

$$dV_{\mathrm{TH}}(y) = \frac{q}{\varepsilon_{\mathrm{Si}}} \frac{dN_{\mathrm{sheet}}(y)}{W_{\mathrm{NS}} \times H_{\mathrm{NS}}} y, \tag{5.9}$$

where, q is the electronic charge and $\varepsilon_{\mathrm{Si}}$ is the relative permittivity of Si. $dV_{\mathrm{TH}}(y)$ is solved for a particular value of dz along the longitudinal axis (z-axis). The solution of each cross-section of the sheet further can be integrated over the entire length (L_{eff}) to consider the entire sheet. The V_{TH} of the NSFET as a function of individual nanosheet can then be expressed as [67]:

$$V_{TH} = -\beta\eta \ln\left(\sum_{i=n} e^{\frac{-V_{TH,i}}{\beta\eta}}\right), \tag{5.10}$$

where, $\beta = \frac{kT}{q} = 26\text{ mV}$ at room temperature, η = (SS(mV/dec)/60), and $V_{TH,i}$ is the threshold voltage of each sheet. Further, the impact of RDD induced fluctuations on the σ of V_{TH} can be expressed as:

$$\sigma V_{TH,RDD} \propto \frac{\sqrt[4]{N_A}}{\sqrt{W_{NS} \times H_{NS}}}, \tag{5.11}$$

where, N_A denotes the average substrate dopant concentration.

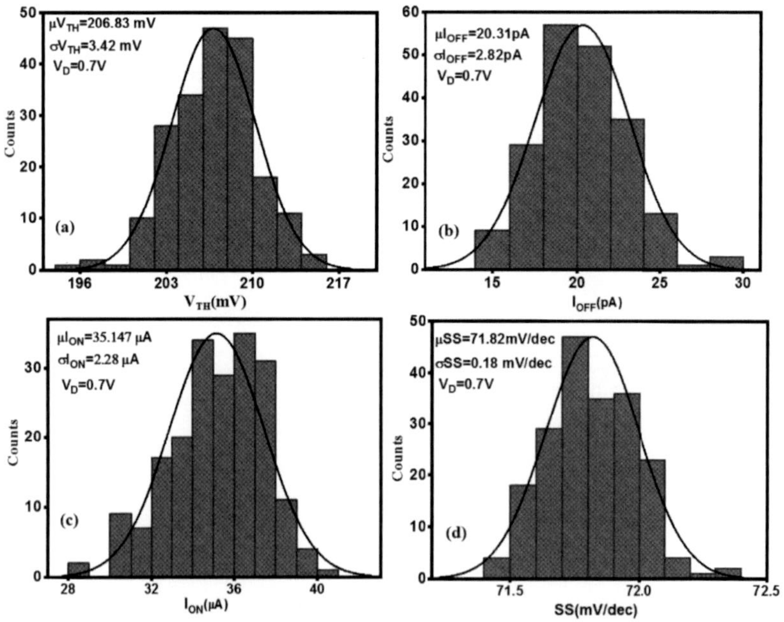

Figure 5.25 Distributions of the impact of RDD-induced fluctuations in the S/D extension on (a) V_{TH}, (b) I_{OFF}, (c) I_{ON}, and (d) SS for 200 different configurations.

From Figure 5.25, it is observed that the figure of merits: V_{TH}, I_{OFF}, I_{ON}, and SS distributions are subject to random dopants, i.e., RDD in the S/D extension. The mean (μ) and σ of V_{TH}, I_{OFF}, I_{ON}, and SS are evaluated.

5.11 Combined Variability (RDD and MGG) in NSFETs

In the following, we examine the combined effects of RDD and MGG on V_T and SS. Figure 5.26 illustrates histograms and quantile–quantile plots of V_{TH} and SS distributions arising based on the individual and combined RDD and MGG effects. Table 5.5 shows the footprint of grain size on σ of V_{TH} and SS at V_D = 0.7 V. Combined variability is found to have a more significant impact than RDD and MGG individually. In this study, if RDD and MGG are independent, the approximate σ of the relevant transistor figure of merit corresponding to the combined effect of the two parameters is given in Eq. (5.12):

$$\sigma_{\text{total}} = \sqrt{(\sigma_{\text{RDD}})^2 + (\sigma_{\text{MGG}})^2}. \tag{5.12}$$

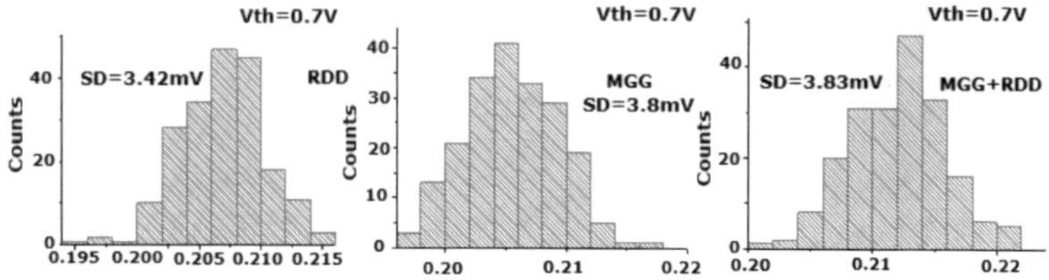

Figure 5.26 Normalized distribution of V_{TH} due to grain size variations for the 14 nm NSFET at V_D = 0.7 V.

Table 5.5 The footprint of grain size on σ of V_{TH}, and SS at V_D = 0.7 V.

	V_{TH} (mV)		SS (mV/dec)	
NSFETs	**Mean**	σV_{TH}	**Mean**	σSS
RDD	206.83	3.4	71.83	0.18
MGG	205.53	3.8	71.71	0.17
RDD+MGG	212.03	3.83	72.02	0.28

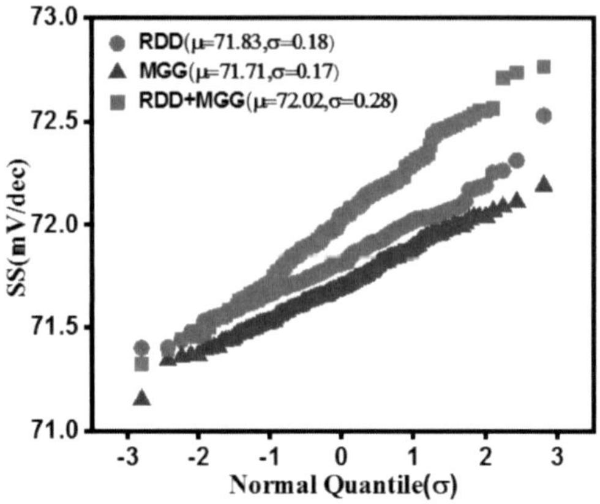

Figure 5.27 Quantile plot of SS distribution due to MGG and RDF variability sources of GAA NSFET at high drain bias.

Figure 5.27 shows the normal quantile plot of SS distribution, due to both the variability sources MGG, RDD, and combined variability effect close to a Gaussian distribution. RDD+MGG produces a skewed SS distribution with a prolonged left and right tail compared to RDD and MGG variability. An increase in the σSS is due to combined variability sources and increases by 55.6% and 58.8% compared to MGG and RDD alone, respectively. The spread due to the combined statistical variability increases and departs further from the normal distribution. As nano-sized grains are considered, these distribution curves are very close [51]. This results in more deviation of the distribution shape from its normal distribution. As the grain size increases, the V_{TH} distribution due to MGG becomes more and more constrained at both the upper and lower tails alike to the behavior reported for bulk MOSFETs [52].

The σSS, due to combined variability sources, increases by 65% concerning MGG variation, as shown in Figure 5.28. As discussed before, (I_D–V_G of MGG), WFV gives rise to V_{TH} variability, and in Figure 5.28 (a, b, and c), the correlation plot between SS and V_{TH} has been presented. The SS shows negative correlations with $V_{\mathrm{TH},}$ and a higher correlation (ρ = −0.46) is also observed in the combined variability effect. The higher the correlation coefficient (ρ) value, the less sensitive the variability is to a change in the drain bias. The linear regressions indicate the sensitivity of SS as a function of V_{TH} at a high drain bias.

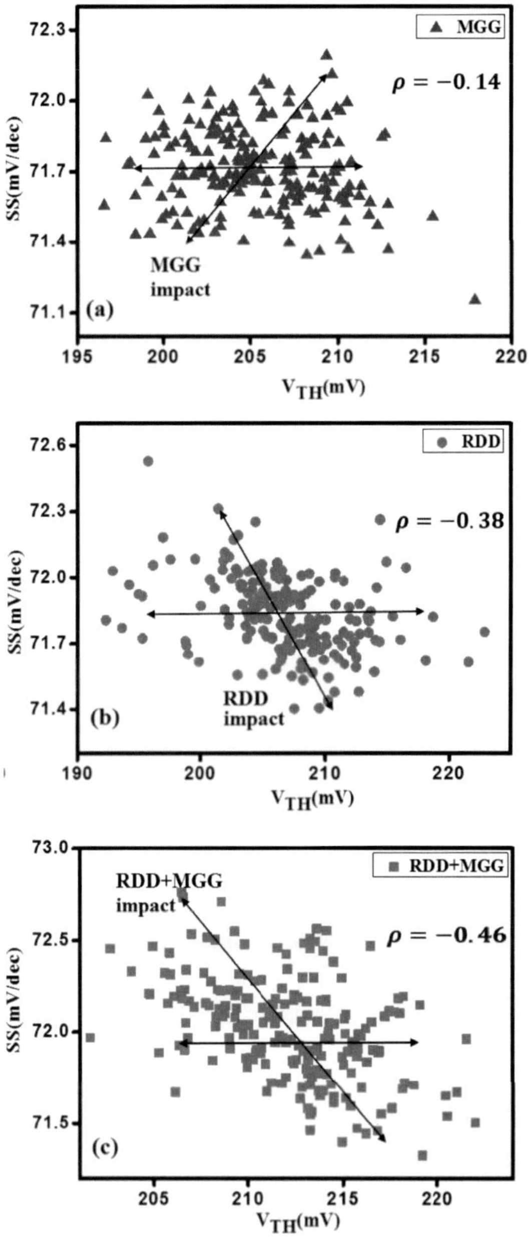

Figure 5.28 Scatter plots showing the correlation between SS and V_{TH} considering the impact of RDD and MGG variability in NSFET at $V_D = 0.7$ V.

5.12 Summary

Vertically stacked horizontal NS GAA transistors seem to be one of the viable solutions toward scaling down below sub-7 nm technology nodes. In this chapter, TCAD calibration with reported experimental data of nanosheet transistor parameters has been performed first. The NSFET has been designed with important design parameters at a 5 nm technology node. Then the performance of NSFET has been analyzed by comparing it with NWFET. We compare electrical performance, including variability studies of several horizontal nanosheet transistors toward transistor structure optimization. We explore the impacts of nanosheet width and thickness on electrical performance and outline important design guidelines necessary for vertically stacked NSFETs.

In this chapter, the possibility of further enhancement of performance of NSFET has been attempted by the introduction of different strain engineering techniques, which lead to the enhancement in the mobility of carriers. The applied mechanical stress has been varied from 0 GPa to 3 GPa with a step of 1 GPa. With the increasing applied stress, the drain current is found to improve. In addition, the SiGe source-drain stressor has been used to generate compressive stress in the channel, which enhances the carrier mobility. The linear elasticity theory has been implemented in simulation to describe the evolution of stress. The stress transfer efficiency is shown for different process conditions with various Ge mole fractions.

Using physics-based 3D TCAD simulations and device optimization beyond the 7 nm technology node, we can show possible improvements in critical design parameters for the vertically stacked NSFETs to surpass NWFET technology. It has been found out that when the grain size approaches the gate length, the distribution of V_{TH} deviates from the normal distribution. We have studied the effects of stacking multiple nanowires for possible performance improvement of lateral nanowire transistors using predictive simulation. We have shown that stacked nanowires offer versatile design options for the performance and power management essential for More-than-Moore applications beyond 7 nm technology node.

It is observed that the σV_{TH} due to MGG variability is 12% higher than RDD variability while the RDD variability strongly influences the I_{ON}. The statistical simulation results predict that the presence of combined variability due to RDD and MGG strongly influences the threshold voltage variation (σV_{TH}) in nanoscale devices. This approach may be applied to the different types of variability, the geometry of the device including the vertical and lateral dimensions of the transistor, and the biasing conditions. Finally, it is concluded that NSFETs with larger widths induce higher on-state drive current and smaller metal grains to improve device statistical variability immunity for future sub-7 nm node devices.

References

1. C. Auth *et al.*, "A 22nm high performance and low-power CMOS technology featuring fully-depleted tri-gate transistors, self-aligned contacts, and high density MIM capacitors," in *Digest of Technical Papers - Symposium on VLSI Technology*, 2012, pp. 131–132. doi: 10.1109/VLSIT.2012.6242496.
2. A. Razavieh, P. Zeitzoff, and E. J. Nowak, "Challenges and limitations of CMOS scaling for FinFET and beyond architectures," *IEEE Trans. Nanotechnol.*, vol. 18, pp. 999–1004, 2019. doi: 10.1109/TNANO.2019.2942456.
3. S. C. Song *et al.*, "Holistic technology optimization and key enablers for 7 nm mobile SoC," in *2015 Symposium on VLSI Circuits (VLSI Circuits)*, 2015, pp. T198–T199. doi: 10.1109/VLSIC.2015.7231373.
4. L. Liebmann, J. Zeng, X. Zhu, L. Yuan, G. Bouche, and J. Kye, "Overcoming scaling barriers through design technology cooptimization," in *2016 IEEE Symposium on VLSI Technology*, 2016, pp. 1–2. doi: 10.1109/VLSIT.2016.7573398.
5. M. G. Bardon *et al.*, "Extreme scaling enabled by 5 tracks cells: Holistic design-device co-optimization for FinFETs and lateral nanowires," in *2016 IEEE International Electron Devices Meeting (IEDM)*, Jan. 2016, pp. 28.2.1–28.2.4. doi: 10.1109/IEDM.2016.7838497.
6. H. Mertens *et al.*, "Vertically stacked gate-all-around Si nanowire transistors: Key process optimizations and ring oscillator demonstration," in *2017 IEEE International Electron Devices Meeting (IEDM)*, Jan. 2017, pp. 37.4.1–37.4.4. doi: 10.1109/IEDM.2017.8268511.
7. J. Appenzeller, J. Knoch, M. T. Björk, H. Riel, H. Schmid, and W. Riess, "Toward nanowire electronics," *IEEE Trans. Electron Devices*, vol. 55, no. 11, p. 2827, 2008. doi: 10.1109/TED.2008.2008011.

8. A. Asenov *et al.*, "Nanowire transistor solutions for 5nm and beyond," in *2016 17th International Symposium on Quality Electronic Design (ISQED)*, May 2016, vol. 2016-May, pp. 269–274. doi: 10.1109/ISQED.2016.7479212.
9. A. Veloso *et al.*, "Nanowire & nanosheet FETs for ultra-scaled, high-density logic and memory applications," *Solid. State. Electron.*, vol. 168, p. 107736, Jun. 2020. doi: 10.1016/j.sse.2019.107736.
10. M. Bohr, "The evolution of scaling from the homogeneous era to the heterogeneous era," 2011. doi: 10.1109/IEDM.2011.6131469.
11. M. De Marchi *et al.*, "Polarity control in double-gate, gate-all-around vertically stacked silicon nanowire FETs," 2012. doi: 10.1109/IEDM.2012.6479004.
12. S. Bangsaruntip *et al.*, "Density scaling with gate-all-around silicon nanowire MOSFETs for the 10 nm node and beyond," in *2013 IEEE International Electron Devices Meeting*, 2013, pp. 20.2.1–20.2.4. doi: 10.1109/IEDM.2013.6724667.
13. S. G. Hur *et al.*, "A practical Si nanowire technology with nanowire-on-insulator structure for beyond 10 nm logic technologies," 2013. doi: 10.1109/IEDM.2013.6724698.
14. A. Veloso *et al.*, "Gate-all-around NWFETs vs. triple-gate FinFETs: Junctionless vs. extensionless and conventional junction devices with controlled EWF modulation for multi-VT CMOS," in *2015 Symposium on VLSI Technology (VLSI Technology)*, 2015, pp. T138–T139. doi: 10.1109/VLSIT.2015.7223652.
15. I. Lauer *et al.*, "Si nanowire CMOS fabricated with minimal deviation from RMG FinFET technology showing record performance," *VLSI Technol. Symp.*, pp. T140–T141, 2015. doi: 10.1109/VLSIT.2015.7223653.
16. H. Wu, W. Wu, M. Si, and P. D. Ye, "First demonstration of Ge nanowire CMOS circuits: Lowest SS of 64 mV/dec, highest gmax of 1057 μS/μm in Ge nFETs and highest maximum voltage gain of 54 V/V in Ge CMOS inverters," in *2015 IEEE International Electron Devices Meeting (IEDM)*, Feb. 2015, vol. 2016-Febru, pp. 2.1.1–2.1.4. doi: 10.1109/IEDM.2015.7409610.
17. A. Veloso *et al.*, "Junctionless gate-all-around lateral and vertical nanowire FETs with simplified processing for advanced logic and analog/RF applications and scaled SRAM cells," in *2016 IEEE Symposium on VLSI Technology*, 2016, pp. 1–2.
18. Loubet *et al.*, "Stacked nanosheet gate-all-around transistor to enable scaling beyond FinFET," in *2017 Symposium on VLSI Technology*, Jul. 2017, pp. T230–T231. doi: 10.23919/VLSIT.2017.7998183.

19. S. Barraud *et al.*, "Performance and design considerations for gate-all-around stacked-nanowires FETs," in *2017 IEEE International Electron Devices Meeting (IEDM)*, Jan. 2017, pp. 29.2.1–29.2.4. doi: 10.1109/IEDM.2017.8268473.

20. Y. M. Lee *et al.*, "Accurate performance evaluation for the horizontal nanosheet standard-cell design space beyond 7nm technology," in *2017 IEEE International Electron Devices Meeting (IEDM)*, Jan. 2017, pp. 29.3.1–29.3.4. doi: 10.1109/IEDM.2017.8268474.

21. D. Jang *et al.*, "Device exploration of nanosheet transistors for sub-7-nm technology node," *IEEE Trans. Electron Devices*, vol. 64, no. 6, pp. 2707–2713, Jun. 2017. doi: 10.1109/TED.2017.2695455.

22. H. J. Cho *et al.*, "Si FinFET based 10nm technology with multi Vt gate stack for low power and high performance applications," in *Digest of Technical Papers - Symposium on VLSI Technology*, Sep. 2016, vol. 2016-September. doi: 10.1109/VLSIT.2016.7573359.

23. S. Y. Wu *et al.*, "A 7nm CMOS platform technology featuring 4th generation FinFET transistors with a 0.027um2 high density 6-T SRAM cell for mobile SoC applications," in *2016 IEEE International Electron Devices Meeting (IEDM)*, Jan. 2016, pp. 2.6.1–2.6.4. doi: 10.1109/IEDM.2016.7838333.

24. R. Xie *et al.*, "A 7nm FinFET technology featuring EUV patterning and dual strained high mobility channels," in *2016 IEEE International Electron Devices Meeting (IEDM)*, 2016, pp. 2.7.1–2.7.4. doi: 10.1109/IEDM.2016.7838334.

25. S. Yang *et al.*, "10nm high performance mobile SoC design and technology co-developed for performance, power, and area scaling," in *Digest of Technical Papers - Symposium on VLSI Technology*, Jul. 2017, pp. T70–T71. doi: 10.23919/VLSIT.2017.7998203.

26. D. Nagy, G. Indalecio, A. J. Garc, A-Loureiro, M. A. Elmessary, K. Kalna, and N. Seoane, "FinFET versus gate-all-around nanowire FET: Performance, scaling, and variability," *IEEE J. Electron Devices Soc.*, vol. 6, pp. 332–340, 2018. doi: 10.1109/JEDS.2018.2804383.

27. IEEE, International Roadmap for Devices and Systems, 2018.

28. T. Faraz, "Expanding the toolbox of atomic scale processing," PhD Thesis, Technische Universiteit Eindhoven, 2021.

29. C. K. Maiti, *Introducing Technology Computer-Aided Design (TCAD)*. Jenny Stanford Publishing, 2017.

30. S. Barraud *et al.*, "Top-down fabrication and electrical characterization of Si and SiGe nanowires for advanced CMOS technologies," *Semicond. Sci. Technol.*, vol. 34, no. 7, 2019. doi: 10.1088/1361-6641/ab1e5b.

31. G. V. Angelov, D. N. Nikolov, and M. H. Hristov, *Technology and Modeling of Nonclassical Transistor Devices*, vol. 2019, Hindawi Limited, 2019.

32. GTS Framework, VSP User Manual, 2020.

33. ITRS, ITRS Roadmap 2015 Edition, More Moore, 2015.

34. V. Venkataraman, S. Nawal, and M. J. Kumar, "Compact analytical threshold-voltage model of nanoscale fully depleted Strained-silicon on silicon-germanium-on-insulator (SGOI) MOSFETs," *IEEE Trans. Electron Devices*, vol. 54, no. 3, pp. 554–562, Mar. 2007. doi: 10.1109/TED.2006.890369.

35. A. R. Brown, N. M. Idris, J. R. Watling, and A. Asenov, "Impact of metal gate granularity on threshold voltage variability: A full-scale three-dimensional statistical simulation study," *IEEE Electron Device Lett.*, vol. 31, no. 11, pp. 1199–1201, 2010. doi: 10.1109/LED.2010.2069080.

36. C. C. Hobbs *et al.*, "Fermi-level pinning at the polysilicon/metal oxide interface-Part I," *IEEE Trans. Electron Devices*, vol. 51, no. 6, pp. 971–977, 2004. doi: 10.1109/TED.2004.829513.

37. C. C. Hobbs *et al.*, "Fermi-level pinning at the polysilicon/metal-oxide interface - Part II," *IEEE Trans. Electron Devices*, vol. 51, no. 6, pp. 978–984, Jun. 2004. doi: 10.1109/TED.2004.829510.

38. X. Wang, G. Roy, O. Saxod, A. Bajolet, A. Juge, and A. Asenov, "Simulation study of dominant statistical variability sources in 32-nm high-κ/metal gate CMOS," *IEEE Electron Device Lett.*, vol. 33, no. 5, pp. 643–645, 2012. doi: 10.1109/LED.2012.2188268.

39. H. Dadgour *et al.*, "Statistical modeling of metalgate work-function variability in emerging device technologies and implications for circuit design," *Proc. IEEE/ACM Int. Conf. Comput. Des.*, pp. 270–277, 2008. doi: 10.1109/ICCAD.2008.4681585.

40. H. F. Dadgour, K. Endo, V. K. De, and K. Banerjee, "Grain-orientation induced work function variation in nanoscale metal-gate transistors - Part I: Modeling, analysis, and experimental validation," *IEEE Trans. Electron Devices*, vol. 57, no. 10, pp. 2504–2514, Oct. 2010. doi: 10.1109/TED.2010.2063191.

41. S. Migita *et al.*, "Design and demonstration of very high-k (k~50) HfO_2 for ultra-scaled Si CMOS," in *Digest of Technical Papers - Symposium on VLSI Technology*, 2008, pp. 152–153. doi: 10.1109/VLSIT.2008.4588599.

42. H. Nam and C. Shin, "Study of high-k/metal-gate work function variation in FinFET: The modified RGG concept," *IEEE Electron Device Lett.*, vol. 34, no. 12, pp. 1560–1562, Dec. 2013. doi: 10.1109/LED.2013.2287283.

43. H. Nam, Y. Lee, J. Park, and C. Shin, "Study of work-function variation in high-κ /metal-gate gate-all-around nanowire MOSFET," *IEEE Trans. Electron Devices*, vol. 63, no. 8, pp. 3338–3341, 2016. doi: 10.1109/TED.2016.2574328.

44. H. F. Dadgour, K. Endo, V. K. De, and K. Banerjee, "Grain-orientation induced work function variation in nanoscale metal-gate transistors-part II: Implications for process, device, and circuit design," *IEEE Trans. Electron Devices*, vol. 57, no. 10, pp. 2515–2525, 2010. doi: 10.1109/TED.2010.2063270.

45. A. Okabe *et al.*, *Spatial tessellations: Concepts and applications of Voronoi diagrams*, J. Wiley and Sons, Chichester, 1992.

46. H. V. Swygenhoven *et al.*, "Grain-boundary structures in polycrystalline metals at the nanoscale," *Phys. Rev. B*, vol. 62, no. 2, pp. 831–838, 2000. doi: 10.1103/PhysRevB.62.831.

47. H. D. Espinosa and P. D. Zavattieri, "A grain level model for the study of failure initiation and evolution in polycrystalline brittle materials. Part I: Theory and numerical implementation," in *Mechanics of Materials*, Mar. 2003, vol. 35, no. 3–6, pp. 333–364. doi: 10.1016/S0167-6636(02)00285-5.

48. A. Leonardi *et al.*, "Realistic nano-polycrystalline microstructures: Beyond the classical Voronoi tessellation," *Phil. Mag.*, vol. 92, no. 8, pp. 986–1005, 2012. doi: 10.1080/14786435.2011.637984.

49. R. Wang *et al.*, "Investigation on variability in metal-gate Si nanowire MOSFETs: Analysis of variation sources and experimental characterization," *IEEE Trans. Electron Devices*, vol. 58, no. 8, pp. 2317–2325, 2011. doi: 10.1109/TED.2011.2115246.

50. K. Ohmori *et al.*, "Impact of additional factors in threshold voltage variability of metal/high-k gate stacks and its reduction by controlling crystalline structure and grain size in the metal gates," 2008. doi: 10.1109/IEDM.2008.4796707.

51. S. H. Chou, M. L. Fan, and P. Su, "Investigation and comparison of work function variation for FinFET and UTB SOI devices using a voronoi approach," *IEEE Trans. Electron Devices*, vol. 60, no. 4, pp. 1485–1489, 2013. doi: 10.1109/TED.2013.2248087.

52. X. Wang, A. R. Brown, N. Idris, S. Markov, G. Roy, and A. Asenov, "Statistical threshold-voltage variability in scaled decananometer bulk HKMG MOSFETs: A full-scale 3-D simulation scaling study," *IEEE Trans. Electron Devices*, vol. 58, no. 8, pp. 2293–2301, 2011. doi: 10.1109/TED.2011.2149531.

53. N. Seoane *et al.*, "Comparison of fin-edge roughness and metal grain work function variability in InGaAs and Si FinFETs," *IEEE Trans. Electron Devices*, vol. 63, no. 3, pp. 1209–1216, Mar. 2016. doi: 10.1109/TED.2016.2516921.

54. B. Hoeneisen and C. A. Mead, "Fundamental limitations in microelectronics-I. MOS technology," *Solid State Electron.*, vol. 15, no. 7, pp. 819–829, 1972. doi: 10.1016/0038-1101(72)90103-7.

55. J. T. Horstmann, U. Hilleringmann, and K. F. Goser, "Matching analysis of deposition defined 50-nm mosfet's," *IEEE Trans. Electron Devices*, vol. 45, no. 1, pp. 299–306, 1998. doi: 10.1109/16.658845.

56. T. Mizuno, J. ichi Okamura, and A. Toriumi, "Experimental study of threshold voltage fluctuation due to statistical variation of channel dopant number in MOSFET's," *IEEE Trans. Electron Devices*, vol. 41, no. 11, pp. 2216–2221, 1994. doi: 10.1109/16.333844.

57. A. Asenov, "Random dopant induced threshold voltage lowering and fluctuations in sub-0.1 /spl mu/m MOSFET's: A 3-D 'atomistic' simulation study," *IEEE Trans. Electron Devices*, vol. 45, no. 12, pp. 2505–2513, 1998. doi: 10.1109/16.735728.

58. P. A. Stolk, F. P. Widdershoven, and D. B. M. Klaassen, "Device modeling of statistical dopant fluctuations in MOS transistors," in *SISPAD '97. 1997 International Conference on Simulation of Semiconductor Processes and Devices, Technical Digest*, 1997, pp. 153–156. doi: 10.1109/SISPAD.1997.621360.

59. H. S. Wong and Y. Taur, "Three-dimensional 'atomistic' simulation of discrete random dopant distribution effects in sub-0.1 μm MOSFET's," in *Technical Digest - International Electron Devices Meeting*, 1993, pp. 705–708. doi: 10.1109/iedm.1993.347215.

60. N. Sano, K. Matsuzawa, M. Mukai, and N. Nakayama, "On discrete random dopant modeling in drift-diffusion simulations: Physical meaning of 'atomistic' dopants," *Microelectron. Reliab.*, vol. 42, no. 2, pp. 189–199, 2002. doi: 10.1016/S0026-2714(01)00138-X.

61. A. Asenov *et al.*, "Integrated atomistic process and device simulation of decananometre MOSFETs," in *International Conference on Simulation of Semiconductor Processes and Devices*, 2002, pp. 87–90. doi: 10.1109/SISPAD.2002.1034523.

62. D. Reid *et al.*, "Prediction of random dopant induced threshold voltage fluctuations in NanoCMOS transistors," in *2008 International Conference on Simulation of Semiconductor Processes and Devices*, 2008, pp. 21–24. doi: 10.1109/SISPAD.2008.4648227.

63. C. Shin, X. Sun, and T. J. K. Liu, "Study of random-dopant-fluctuation (RDF) effects for the trigate bulk MOSFET," *IEEE Trans. Electron Devices*, vol. 56, no. 7, pp. 1538–1542, 2009. doi: 10.1109/TED.2009.2020321.

64. N. Sano *et al.*, "Role of long-range and short-range Coulomb potentials in threshold characteristics under discrete dopants in sub-0.1μm Si-MOSFETs," *IEEE Int. Electron devices Meet.*, pp. 275–278, 2000. doi: 10.1109/IEDM.2000.904310.

65. S. K. Saha, "Modeling statistical dopant fluctuations effect on threshold voltage of scaled JFET devices," *IEEE Access*, vol. 4, pp. 507–513, 2016. doi: 10.1109/ACCESS.2016.2519039.

66. P. Harsha Vardhan *et al.*, "Threshold voltage variability in nanosheet GAA transistors," *IEEE Trans. Electron Devices*, vol. 66, no. 10, pp. 4433–4438, 2019. doi: 10.1109/TED.2019.2933061.

67. P. H. Vardhan, S. Mittal, S. Ganguly, and U. Ganguly, "Analytical estimation of threshold voltage variability by metal gate granularity in FinFET," *IEEE Trans. Electron Devices*, vol. 64, no. 8, pp. 3071–3076, 2017. doi: 10.1109/TED.2017.2712763.

Chapter 6

III-Nitride Flexible Electronic Devices

Chinmay K. Maiti
SOA University, Bhubaneswar 751030, Odisha, India
ckmaiti@outlook.com

Electronics have become pervasive in our daily lives as a result of the growth of information and communication technologies. However, integrating new electronic functions remains a challenge. For many emerging applications, the flexibility properties of the components are indispensable for use in confined areas or under heavy strain. Indeed, incorporating an electronic circuit in a cheap and little object is a major constraint in terms of cost and integration of relevant technologies. Because of the shape of the objects, often complex and non-rigid, and the constant need to integrate into minimum congestion, the development of flexible, compact, low-cost electronics has become essential [1]. At present, manufacturing and industrialization of this type of electronics are being developed and we are still far away from flexible products manufactured in mass production.

Fabless Semiconductor Manufacturing: In the Era of Internet of Things
Edited by Chinmay K. Maiti

ISBN 978-981-4968-27-0 (Hardcover), 978-1-003-31497-4 (eBook)
www.jennystanford.com

Silicon, a material at the base of conventional electronics, cannot meet the flexibility criteria because of its rigidity. Historically, the emergence of amorphous silicon, however, enables silicon to meet the challenges of flexibility, conformability, a derivative that has allowed a big breakthrough in the field of flexible electronics [2]. Followed by the implementation of organic semiconductor materials during 1970–1980 has paved the way for the development of flexible electronics [3]. Nevertheless, the relatively low mobility of these materials severely limits the potential applications. To meet these new expectations, new materials need to be explored such as semiconductor oxides, III-V materials, carbon nanotubes, or graphene.

The birth of the flexible electronics industry has attracted the attention of researchers for several years. It is a promising way to address these new applications [4]. Within this framework, the main objective of this chapter lies in the manufacture of flexible components in heterogeneous technology for high-frequency applications. An overview of the context of flexible electronics and their practical applications is presented. Existing flexible transistors based on different channel materials are introduced.

The first flexible transistor dates back to 1968 when Brody and his colleagues made a thin-film transistor (TFT) from Tellurium on a strip of paper and proposed to use them for display applications. In the years that followed, Brody's group manufactured TFTs on different types of flexible substrates, including mylar, polyethylene, and foils of anodized aluminum [5]. In the 1980s, the success of liquid crystal display (LCD) and the demonstration of flexible solar cells have stimulated research on flexible electronic circuits.

In flexible electronics, several materials have been used, namely silicon, organic materials, inorganic, and hybrid organic/inorganic for CMOS technology. In 2013, Samsung has integrated its OLED technology into its flexible display devices and LG has unveiled the first plastic electronic paper [6]. The semiconductor industry has strongly developed flexible display devices. However, research in the flexible electronics sector is aimed at a wide variety of domains, far beyond mere display. The development of these requires components having operating frequencies beyond gigahertz (GHz). Currently, the production of flexible components with frequencies comparable to rigid components remains a major challenge for the electronics industry. Surveillance, detection, or even diagnosis of diseases are being attempted from the skin. Indeed the treatment of physiological signals will provide this information, which until now is done by bulky equipment. Flexible electronics are a good candidate for the detection and signal processing functions directly on the skin via smart sensors [7–10]. Figure 6.1 shows a few examples of next-generation flexible electronic systems.

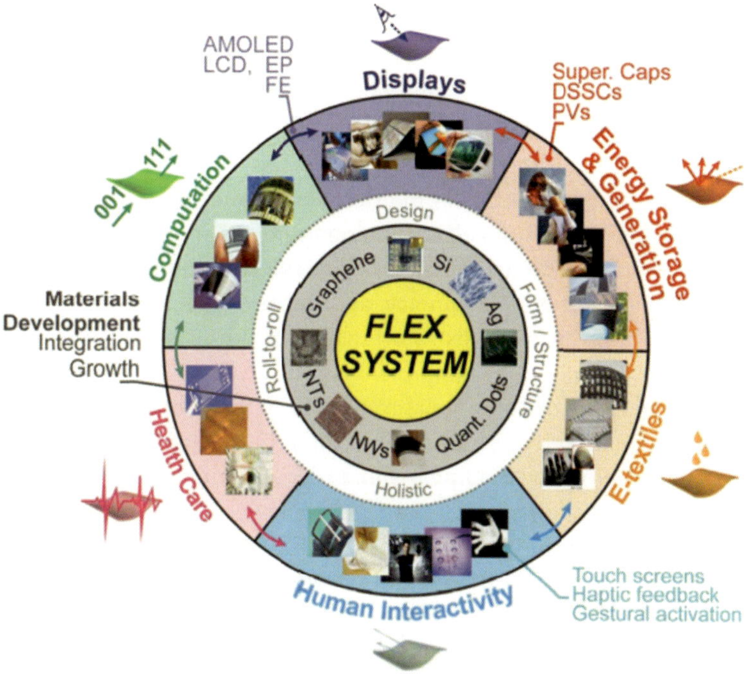

Figure 6.1 Next-generation flexible electronic systems and possible areas of applications [11].

Inorganic materials such as semiconductor oxides, silicon, and III-V materials also contribute to the evolution of flexible electronics, benefiting either from the maturity of the technology or from the nature of the materials. These materials represent good candidates compared to organic materials since they have significantly higher mobilities and they are compatible with the matured silicon technology allowing the miniaturization of components. Performances in the gigahertz range have been achieved with silicon nano-membranes [12, 13]. An f_T and $f_{\max}$ of 3.8 GHz and 12 GHz, respectively, have been reported [14] using a two-finger grid topology and reducing the access resistance. The measurements were carried out under torsional stress (radius R = 29 mm) and compression (R = 78 mm). Seo et al. have reported f_T and $f_{\max}$ frequencies of 5 GHz and 38 GHz, respectively, by measuring transistors based on nano-membranes on polyethylene terephthalate (PET) [15]. This development shows that silicon remains an interesting material for flexible electronics.

III-V materials are also characterized by remarkable physical properties. The more important is undoubtedly their bandgap, which reaches 3.4 eV for GaN and 6.2 eV for the AlN. Heterostructures can be made to increase mobility by allowing the creation of a 2D electron gas. In addition to their transport properties, III-V semiconductors have shown real potential for multiple communication systems such as 4/5G and Wi-Fi. High electron mobility transistors (HEMTs) and heterojunction bipolar transistors (HBTs) represent two mature technologies for high-frequency applications [16]. Silicon seems to be the appropriate substrate for the growth of heterostructure HEMTs especially for components dedicated to soft substrate transfer because of its low cost compared to SiC. Johnson et al. [17] have demonstrated the first record of microwave power density 12 W/mm combined with a linear gain of 15.3 dB and an added power output of 52.7% for a 700 nm gate length device at a drain bias of 50 V at a measurement frequency of 2.14 GHz. Several research groups have reported state-of-the-art HEMT components on Si [18–21].

As the HEMTs are attractive for flexible electronics applications, the development of new manufacturing techniques based on exploiting the maturity of existing technology has become important. However, so far the majority of the components were manufactured on a rigid substrate and transferred to flexible. Chen et al. succeeded in 2008 in transferring HEMTs based on InP on polyethylene naphthalate (PEN) [22, 23]. About AlGaN/GaN HEMTs, particular attention is paid to the optimization of the process of transferring components from rigid to flexible substrates. Defrance et al. [24] obtained frequencies f_T and $f_{\max}$ of 32 GHz and 52 GHz, respectively, on a topology HEMT AlGaN/GaN transistor. Chang et al. [25] have reported AlGaN/GaN HEMTs on PET and managed to have f_T and $f_{\max}$ frequencies, respectively, of 60 GHz and 115 GHz. An output power density (Pout) of 420 mW/mm associated with a linear power gain of 15.8 dB and power-added efficiency (PAE) of 29.6% has been reported [26].

In the carryover approach, components are manufactured by standard methods on rigid substrates such as silicon or glass. They are subsequently transferred to flexible substrates. This technique is used mainly in the context of silicon technologies and III-V materials [27]. The advantage is to provide high-performance devices with flexible supports. As for the disadvantages, they reside in the fragility of epitaxies to transfer, the risks of mechanical degradation of the components, in the small surface transferred, and also the high cost of the process.

The transferred components will usually be integrated into circuits already manufactured. Direct manufacturing requires: (i) relying on semiconductors that can be grown on a flexible substrate (which is not the case for gallium nitride), (ii) develop new process techniques, (iii) dispose of new materials and, (iv) to find a compromise between device performance and low temperatures tolerated by flexible substrates. The flexible substrates used must be light, inexpensive, and resistant to the different solvents used as part of the manufacturing process. Currently, several types of flexible substrates are in use such as paper, polyimides, textiles, etc. In Table 6.1, we present a set of polymers usually used in the field of flexible electronics and their characteristics in terms of the glass transition temperature, thermal conductivity, and coefficient of thermal expansion. Several specifications for flexible printing techniques are shown in Table 6.2.

Table 6.1 Comparison of different polymer parameters

Polymer	Transition temp. (°C)	Thermal conductivity (W/(m.K))	Coefficient of thermal expansion (ppm/°C)
PEN	155–160	0.15–0.24	13–21
PET	105–170	0.15–0.4 (23 °C)	20–80
PI (polyimide, Kapton)	250–320	0.10–0.35 (23 °C)	30–60
PDMS (polydimethylsiloxane)	127	0.17	310

Table 6.2 Flexible printing technique specifications

Print technique	Resolution (μm)	Feature
Flexography	30–75	Ecological technique, requires a low viscosity ink
Gravure Printing	10–75	Simple process, requires to use of smooth substrates
Jet Printing Ink	20–50	No contact with the substrate
Silkscreen	50–100	Allows the use of ink viscous and thick
3D Printing	15–35	Allows fast production and controllable

Inkjet printing techniques have been growing recently, due to their unique features, such as simplicity of manufacture, compatibility with different substrates, feasibility of non-contact, and mask-free patterns as well as low temperature and low cost [28, 29]. It is used in most laboratories and industries to print the patterns using nanoparticle inks on different supports including glass, semiconductor materials, paper, and textiles. Two categories of the printer are commonly used: (i) continuous jet (continuous inkjet [CIJ]) and (ii) drop-on-demand (DOD) printing. Screen printing is a printing technique that uses stencils. It is characterized by its simplicity and respect for the environment due to the use of ceramic inks and polymers. In electronics, it is used to make conductive interconnections between electronic components on printed circuits. In 1997, Bao et al. [30] fabricated the first FET by screen printing. They used a stainless steel screen with 400 counts/in to drop a polymer insulating layer and the source and drain electrodes with conductive ink. Similarly, Knobloch et al. [31] reported high-performance transistors fabricated by screen printing techniques organic FETs. 3D printing is a cost-effective technique for fast production. It is characterized by mold-free fabrication or the components are created layer by layer [32]. Noble metal conductors, organic conductors, semiconductors, and insulators can be used for printing. Organic TFTs [33, 34], engraving masks, and inorganic TFTs [35–38] were manufactured thanks to this printing technique.

In recent years, flexible electronics are the subject of many developments and offer new opportunities for different applications such as flexible RFID tags, screens flexible, etc. For these applications, flexible organic components are currently used but their operating frequencies are limited by their very low mobility and by the resolution of manufacturing techniques. Moreover, for inorganic semiconductors, the mobilities of TFTs based on silicon and semiconductor oxides are not allowed to reach high frequencies. As a result, interest in III-V materials such as GaN has occurred. Currently, components such as HEMTs and light-emitting diodes (LEDs) in the GaN sector have high-performance thanks to the improvement in material growth and process technology. New applications can be developed and offered to the industry by providing flexibility to these components. The objective is to design and realization of flexible HEMTs and LED components in the GaN industry. The goal is to get on the one hand flexible transistors delivering power of 2 W/mm at 10 GHz and secondly and flexible LEDs.

Gallium nitride is a semiconductor material made up of column III elements of the periodic table associated with the nitrogen element (N) of column V. These materials, commonly referred to as "III-N materials" (GaN, AlN, InN, and their alloys), crystallize according to two types of structure: the structure cubic blende zinc (ZnS) and the hexagonal structure wurtzite. GaN wurtzite is noted h-GaN, the cubic GaN is denoted c-GaN. The high ionicity of III-N bonds that the wurtzite phase is thermodynamically stable. Let's focus exclusively on this phase. The hexagonal wurtzite structure consists of two sub-hexagonal compact lattices, one containing the gallium atoms (Ga) and the other the nitrogen atoms (N). The crystal structure is related to the growing conditions as well as to the surface symmetry of the substrate. The zinc blende structure is obtained by growth under non-equilibrium conditions thermodynamics (low temperature) from the (001) planes of cubic substrates. The wurtzite structure is the most commonly used because it is the most stable thermodynamically. The ionic nature of the III-N bond provides great stability to the structure, due to the high electronegativity of nitrogen.

Figure 6.2 represents the elementary lattice of a wurtzite structure defined by the three lattice parameters: (i) length of the base hexagon (a), (ii) height of the elementary cell (c), which corresponds to the distance between two atoms of the same nature, and (iii) bond length between atoms III and N (u). The wurtzite structure consists of two hexagonal sub-networks, one occupied by element III and the other by nitrogen (N), interpenetrated and shifted along axis [0001] by 5/8 of the cell unit. In an ideal wurtzite crystal, the relation between these three parameters should be u = 3/8 and c/a = 1.633, which results in regular tetrahedra (each atom is linked to four neighbor atoms, three of which happen to be in the same crystallographic plane perpendicular to the axis). However, the III-N wurtzite structure deviates slightly from the ideal wurtzite structure. So, the element III atoms are in a somewhat distorted tetrahedral environment. Therefore, the sum of the dipole moments is non-zero considering that the atom of element III has a positive charge, and the more electronegative nitrogen atom presents a negative charge.

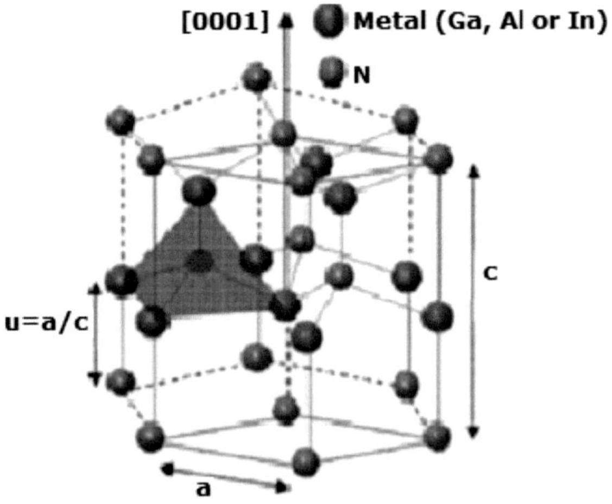

Figure 6.2 Wurtzite crystal structure of III-N. Atomic structure of GaN as wurtzite crystal. Ga atoms are represented in blue color and are relatively larger than N atoms represented in gray. The bonds are presented in two colors. Lattice parameters along with bond length are presented in blue arrows [10].

Theoretically, with a gap of 3.4 eV, the intrinsic carrier density in GaN is extremely weak (~ 10^{-6} cm^{-3}). In practice, the unintentionally doped material is always found of type n, with electronic densities evolving from 10^{16} to 10^{19} cm^{-3} (following growth patterns and conditions), defects occurring during growth (incorporation of impurities, gallium, or nitrogen vacancies). On the other hand, p-type doping is much more difficult to achieve. The breakdown field (V/cm) is the value of the maximum electric field that the material can support. It can be assumed that the high value of the bandgap must generate a relatively high maximum field. Indeed, in theory, it is estimated at 2 MV.cm^{-1} and from the measurements, it lies between 1 and 3 MV cm^{-1}. Note that these values are comparable to those obtained in SiC.

There is a non-zero macroscopic polarization in Wurtzite GaN. This polarization is called spontaneous polarization. It results from the non-superposition of centroids of positive and negative charges [39]. The value of the spontaneous polarization of GaN in gallium polarization is estimated at −0.029 cm^{-2}, which corresponds to an electric field of the order 3 MV.cm^{-1} and a surface charge density of 1.8×10^{13} cm^{-2}. GaN is often produced on sapphire, SiC, or silicon substrate. The disagreement mesh between GaN and its substrate imposes a constraint (extension or compression) in the plane basic GaN layers. Deformation of GaN mesh results in a displacement of centroids and thus a change in spontaneous polarization. This effect can be considered as the appearance of a so-called "piezoelectric" polarization, which reinforces or weakens spontaneous polarization [39].

The problem encountered for the epitaxy of gallium nitride is that there is a mismatch with the substrate. The most widely used substrates at present are sapphire (Al_2O_3), silicon carbide (6H-SiC), and silicon (Si). Their characteristics relating to epitaxy are shown in Table 6.3.

Table 6.3 Characteristics of substrates most used for GaN growth

Material	Lattice parameter (A)	Coefficient of thermal expansion (10^{-6} K^{-1})	Thermal conductivity (W·K^{-1}·cm^{-1})
Al_2O_3	a = 4.788	7.5	0.5
	c = 12.99	8.5	
6H-SiC	a = 3.08	4.2	4.9
	c = 12.12	4.68	
Si	5.4301	3.59	1.5
GaN	a = 3.189	a = 5.59	1.3
	c = 5.18	c = 3.17	

Silicon has a significant advantage because of its much lower cost than other materials, its availability in the large wafer (up to 12"), and good thermal conductivity that make it a prime candidate for power components. In addition, the growth of GaN on silicon could allow the integration of nitride-based devices from group III with the numerous technologies of the silicon industry. Finally, the possibility of chemically etching the silicon substrate allows the realization of the backside (electrical connections, heat sink), or obtaining layers self-supporting GaN that can be transferred to other substrates taking advantage of the transparency of the GaN layers. However, silicon has a large mismatch and a significant thermal expansion coefficient difference with GaN. The extensive stress generated during cooling remains a critical issue for the growth of uncracked layers of thickness greater than 3 μm. The choice of such a substrate is the compromise between the device's performance and its cost. The problem of lattice mismatch can otherwise be reduced by buffer layers.

6.1 $Al_xGa_{1-x}N$ Alloy

To calculate the effects of polarization in $Al_xGa_{1-x}N$ alloy, we need values of lattice parameters, spontaneous polarization, piezoelectric constants, and elastic constants of $Al_xGa_{1-x}N$. These values are estimated from Vegard's law (linear interpolation between the values of GaN and AlN), which is written in the form [39]:

$$a_{\mathrm{Al}_x\mathrm{Ga}_{1-x}\mathrm{N}(x)} = x\, a_{\mathrm{AlN}} + (1-x)\, x\, a_{\mathrm{GaN}}. \tag{6.1}$$

The $Al_xGa_{1-x}N$ alloy is a direct gap semiconductor regardless of the aluminum content. Against this, the variation of the bandgap of the $Al_xGa_{1-x}N$ alloy as a function of x is not quite linear. It is expressed by [39]:

$$E^g_{\mathrm{Al}_x\mathrm{Ga}_{1-x}\mathrm{N}} = xE^g_{\mathrm{AlN}} + (1-x)\,E^g_{\mathrm{GaN}} - bx(1-x), \tag{6.2}$$

where $$E^g_{\mathrm{Al}_x\mathrm{Ga}_{1-x}\mathrm{N}},\ E^g_{\mathrm{AlN}},\ \text{and}\ E^g_{\mathrm{GaN}} \tag{6.3}$$

are respectively the bandgap of the alloy $Al_xGa_{1-x}N$, AlN, and GaN, and b is the nonlinearity term representing the deviation compared to the Vegard linear law. The diversity of values of b reported in the literature can be explained by a wide dispersion in the quality of the epitaxial layers and the difficulty in separate effects, due to stresses and fluctuations in the alloy composition. By analyzing theoretical and experimental values, Vurgaftman et al. [40] proposed $b = 0.7$ eV for two types of AlGaN wurtzite and zinc-blende. The energy of the forbidden band of III-N alloys according to the lattice parameter, a, as shown in Figure 6.3.

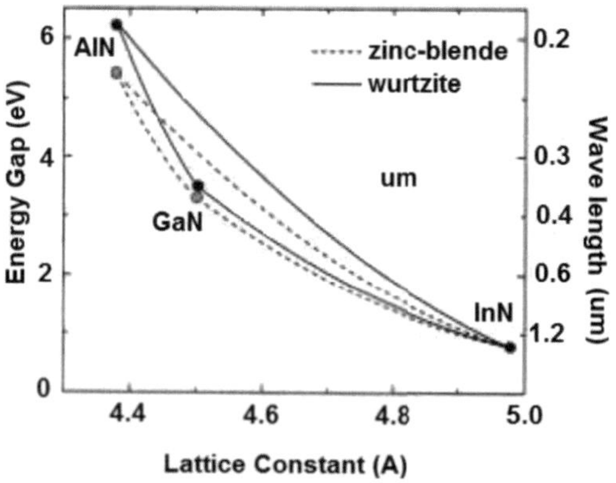

Figure 6.3 The bandgap energy of III-N alloys as a function of the lattice parameter (a) [10].

Consider an AlGaN layer on a GaN layer without constraints. They are two gallium polarities. The lattice parameters of the AlGaN layer are lower than those of the GaN layer. The AlGaN layer is therefore voltage-stressed. Vectors of polarization in these two layers result in a 2D electron gas. In this case, a positive fixed charge density is obtained at the interface such that:

$$\sigma^+ = |P_{\text{spAlGaN}}| - |P_{\text{spGaN}}| + |P_{\text{pzAlGaN}}|. \tag{6.4}$$

This positive charge is compensated by the presence of electrons at the interface forming a gas confined to this interface called 2D electron gas. The AlGaN layer is called the "barrier", the GaN layer where find the electrons at the interface is called the "channel". Note that there is no intentional doping of the AlGaN barrier. Electrons confined to the interface can come from several sources: intrinsic defects, impurities in the GaN buffer layer, AlGaN barrier, or surface conditions [41–43]. Electrons are confined to the interface by the strong electric field of polarization. Therefore, the density of electrons in 2D electron gas depends on the stress state of the layers, the aluminum content, and the thickness of the barrier.

6.2 GaN HEMT Manufacturing Technology

In recent years, we have witnessed the rise of the new electronics industry based on flexible media. Many applications are difficult to achieve by conventional electronics, such as RFID tags, mobile sensors, flexible screens, etc. Flexible electronics are essentially based on organic materials in which the low mobility ($<1\ cm^2/V.s$) [44] considerably limits the microwave performance of components. In this context, the heterogeneous integration of GaN components on a flexible substrate appears to be a promising solution for microwave power applications where conformability on a non-planar surface is desired. Indeed, the HEMTs of the GaN industry have been the subject of numerous studies over the last 10 years for microwave power applications [19, 45]. Thanks to the properties of the GaN material, remarkable performances have been demonstrated [46–48]. The properties of gallium nitride are very interesting for manufacturing flexible components for various sectors. In this section, we present the process technology for the fabrication of flexible AlGaN/GaN HEMTs. Fabrication begins with the realization of components on the rigid substrate followed by a specific method of transfer on a flexible substrate.

The flexible transistor manufacturing process is based on three main steps. The first step is to grow the layers using the technique of molecular beam epitaxy (MBE) on the silicon substrate (111). The heterostructure consists of a 44 nm nucleation layer of AlN, a 250 nm Al0.15Ga0.85N/130 nm AlN transition layer to make the transition between the substrate, and the GaN buffer layer. The latter, also called a buffer, is made up of 1.73 μm of Al0.05Ga0.95N allows the confinement of the carriers transferred in the 2D electron gas present in a 10 nm thick channel layer of GaN. The latter is covered by an exclusion layer, also called a spacer, of 1 nm AIN and an AlGaN barrier with a thickness of 14 nm and an Al content of 29%. The AlN layer makes it possible to minimize the penetration of the electron gas into the surface. The structure is covered by a thin layer of GaN, which stabilizes the surface conditions avoiding the degradation of the electrical properties of the transistors during operation and also preventing oxidation of the aluminum present in the barrier layer. The epitaxial structure is shown in Figure 6.4.

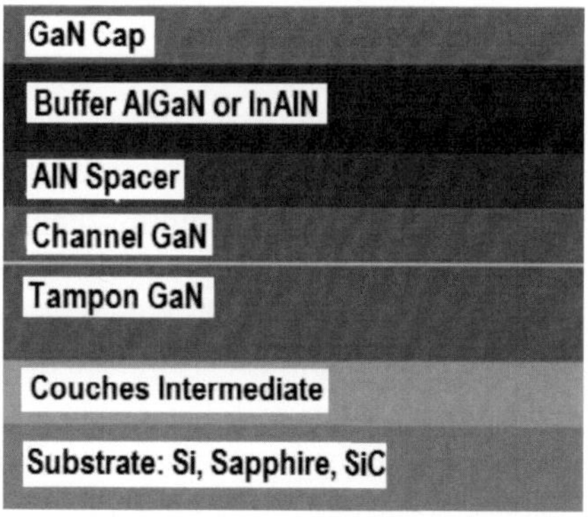

Figure 6.4 Structure of the HEMT transistor.

The second step of the process is the fabrication of the submicron gate transistors by e-beam lithography [21]. The ohmic contacts are deposited by evaporation following the metallic sequence Ti/Al/Ni/Au. This is followed by rapid annealing at 850 °C for 30 seconds. The components are then isolated by N+ implantation. Subsequently, the "T" gates of length L_g = 80 nm, are defined by e-beam lithography sequentially using a tri-layer of resin and metalized with a Ni/Au. Finally, the transistors are passivated with a 50 nm/100 nm SiN/SiO_2 dielectric bilayer deposited by plasma-assisted chemical vapor deposition (PECVD) at 340 °C after treatment with N_2O plasma. The contact pads are defined with a metallic Ti/Au. At this stage, the measurements were carried out using the Van der Pauw patterns to determine the Hall mobility (μ = 1782 cm^2/Vs) and a carrier density Ns of 8.7×10^{12} cm^{-2} as well as a sheet resistance value R of 403 Ω/sq. Finally, the manufactured components are transferred to the flexible substrate supplied by the 3M Company after the removal of the silicon substrate.

We present below the results obtained on a HEMT AlGaN/GaN component applied to a flexible substrate. From components produced on an AlGaN/GaN epitaxy on a silicon substrate, the transfer process is based on the removal of the growth substrate and the bonding to the flexible substrate [49]. The flexible substrate used is chosen such that it can withstand high temperatures, has a thermal conductivity of 0.8 W/mK, and has excellent electrical insulation properties.

For the transfer, the front face of the components is protected and bonded with a resin on a sapphire host substrate (Al_2O_3). This step allows thinning of the silicon on the rear face by mechanical polishing followed by selective dry etching with xenon di-fluoride (XeF_2) to remove the silicon remaining after thinning using the nucleation layer as a stop layer. Once the silicon is removed, the backside of the components is glued to the flexible substrate. The temporary sapphire support is detached from the front face of the components by dissolving the layer of photosensitive resin in acetone. The hose used is composed of a thermally reinforced copolymer backing, coated with an acrylic adhesive and having a thermal conductivity of 0.8 W/mK. All the steps of the component transfer process are shown in Figure 6.5.

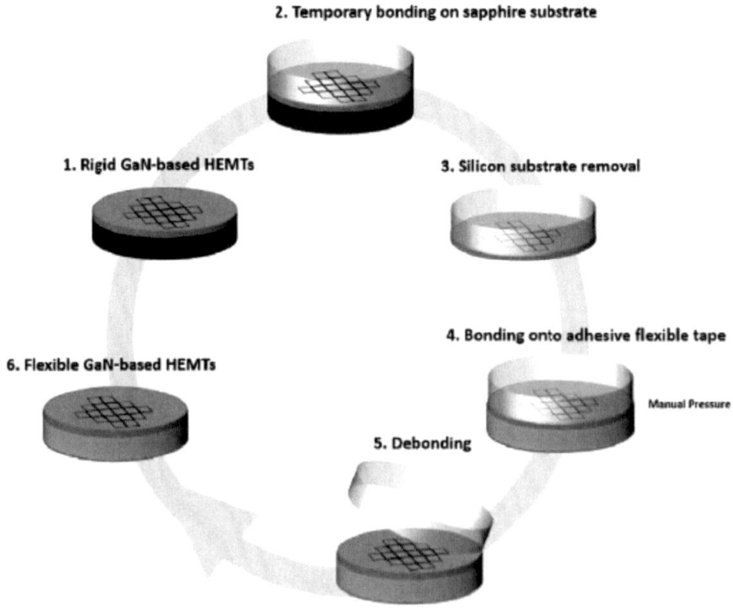

Figure 6.5 Main technology steps of the transfer process from silicon to flexible substrate [10].

A maximum I_{DS} current density of 270 mA/mm and an extrinsic G_{mext} transconductance of 170 ms/mm were obtained from a transistor of gate length $L_g = 80$ nm. The device has a cutoff frequency (f_T) and a maximum oscillation frequency (F_{max}) around 45 GHz.

Power measurements were performed at 10 GHz in continuous wave (CW) mode; at V_{DS} = 5 V an output power density of 110 mW/mm is obtained, associated with a PAE efficiency of 14% and a linear gain of 12 dB. These results show the potential of the GaN industry for flexible electronics.

6.3 Losses in Flexible Substrates

To make the coplanar lines, a sequential Ti/Au metal with a thickness of 50/400 nm was deposited through a mechanical mask made with a silicon substrate. Figure 6.6 shows the transmission parameters according to frequency. The loss measured is 0.3 dB.mm^{-1} at 10 GHz at ambient temperature. We notice that these values are of the same order of magnitude as the current losses obtained on silicon, thus showing the potential of flexible substrate for circuit design.

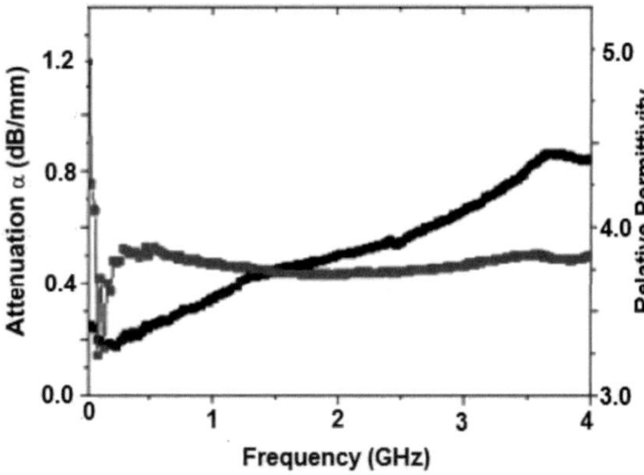

Figure 6.6 Transmission loss versus frequency [10].

6.4 Characterization Flexible GaN HEMTs

All the measures presented in this work are carried out on a HEMT AlGaN/GaN transistor with two gate fingers 50 μm wide and 80 nm long. The distances between the source and the drain and the source and the gate are 1.75 μm and 0.6 μm, respectively. In this case, the DC measurements are carried out using an Agilent HP4142B parameter analyzer. It may be noted that all static, frequency, and power measurements were made flat (without deformation of the flexible substrate). Figure 6.7 shows the I_{DS}–V_{DS} transfer characteristic for different values of V_{GS}. A current density maximum of 270 mA/mm is obtained at V_{GS} = 0 V (800 mA/mm on silicon substrate before transfer). A maximum transconductance value of 170 mA/mm was obtained at V_{DS} = 4 V for a V_{GS} = −2.5 V (290 mS/mm before transfer). The S parameters are measured using a coplanar tip device consisting of a Keysight vector network analyzer (E8361A) in the frequency band [20 MHz to 67 GHz] calibrated in using the line-reflect-reflect-match (LRRM) technique [50]. Open-short type de-embedding [51] is used to remove the contribution of accesses coplanar (effects of capacitive parasitic elements and inductive) to return to the intrinsic performance of the component. S parameters depend on the frequency. Figure 6.8 shows the modulus of the current gain |H21| and Mason's unilateral gain (U). At the point of polarization corresponding to the maximum of the transconductance, the component has a transition frequency (F_t) and a maximum F_{max} of 46 GHz as extrapolated by the slope to −20 dB/decade.

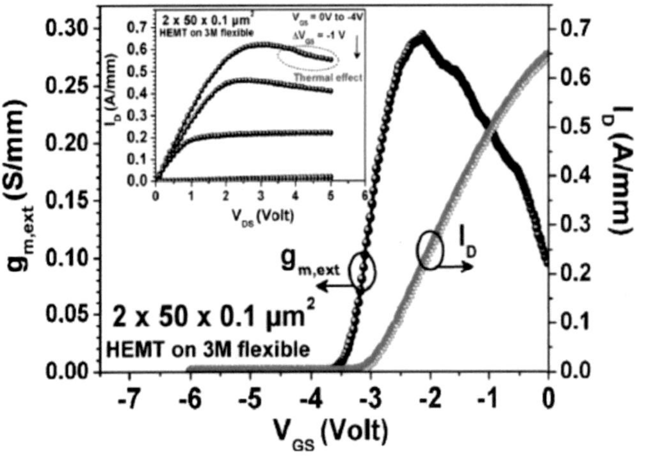

Figure 6.7 I_D–V_{DS} characteristics for a 2×50×0.1 um² flexible AlGaN/GaN HEMT (K_{th} = 1.6 WmK^{-1}) [10].

Large signal power measurements are carried out under peaks at 10 GHz using an active load configuration "load-pull" in CW conditions with a large signal network analyzer (LSNA) operating up to 50 GHz. A V_{DS} = 5 V and I_{DS} = 16 mA/mm, corresponding to operation in class AB, the selected components show very good performance: a power density of 110 mW/mm associated with a linear gain of 12 dB and a 14% PAE were obtained at V_{DS} = 5 V (see Figure 6.9).

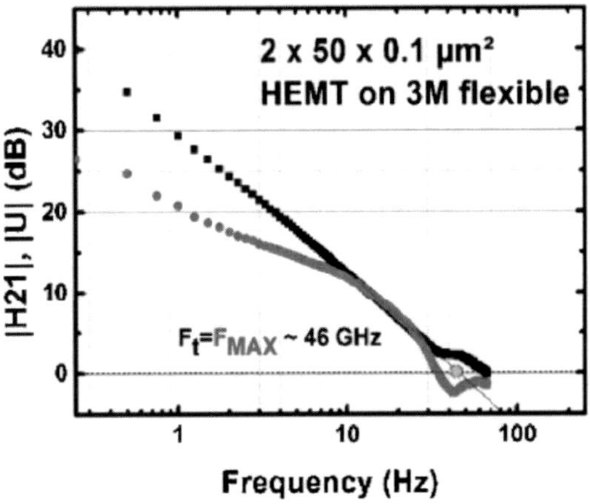

Figure 6.8 Modulus of the current gain $|H_{21}|$ and Mason's unilateral gain (U) [10].

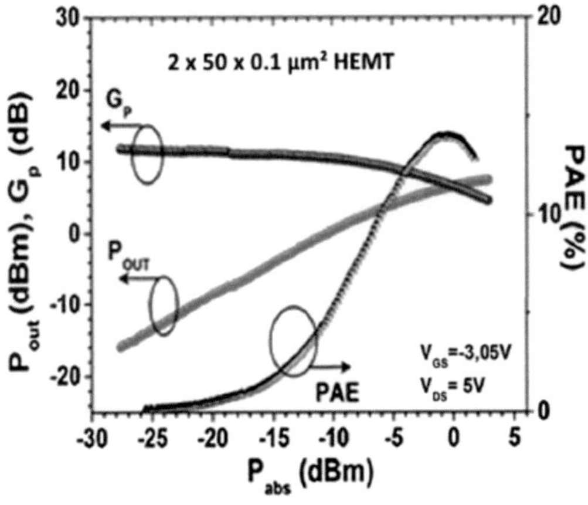

Figure 6.9 Output power, power gain, and PAE versus absorbed power at 10 GHz for a 2×50×0.1 um^2 AlGaN/GaN HEMT on 3M flexible at V_{DS} = 5 V (K_{th} = 0.8 WmK^{-1}). The use of a higher thermal conductivity flexible substrate will be an interesting solution to minimize thermal effects and improve the device's performance and reliability [10].

6.5 Stress/Strain Analysis in GaN HEMTs

Bending, stretching, or twisting of flexible devices causes a significant distortion. The flexibility of devices is frequently investigated using analyzes based on the neutral plane (or neutral line in a 2D cross-sectional presentation) along which no stress or compression occurs while bending. The neutral plane, on the other hand, exists in a restricted volume. As a result, it cannot be used in flexible devices with multi-component vertically stacked layered structures that entail various forms of bending deformations. The local stress/strain in the device multilayers must be known to overcome mechanical and electrical reliability challenges. The mechanical stress distribution in multi-layered flexible structures is rarely reported, even though the mechanical stress of multilayered systems on rigid substrates has been extensively investigated. Inorganic, organic, and printed electronics make up the majority of flexible electronic devices.

Computer simulations can be used to anticipate the electrical characteristics of a device design rapidly and cheaply instead of going through an expensive and time-consuming fabrication procedure. The main focus of this section is computational prototyping, which shows how the hierarchical TCAD can be effectively used for mechanical deformation-aware designing and modeling of flexible electrical devices. Figure 6.10 depicts a typical device schematic view (a) in the device's 2D cut plane (zero strain), (b) in compression, and (c) in tension. Flexible devices have a defined strain profile when flexed or bent, including a zero-strain neutral line.

For the analysis of flexible devices under mechanical deformation, the concept of a "neutral plane" is often utilized [52]. Even with modest bending radii, the devices operate reliably in the neutral plane [44, 53, 54]. Figure 6.11 shows that when the device's inner radius (b) is compressed, the device's outer radius (c) is in tension. The stack is the neutral plane (or neutral line in a 2D cross-sectional depiction) at some point in the layer, along which there is neither tension nor compression when bending. However, because the neutral plane occurs in a restricted space, it cannot be used in flexible systems that have multiple components stacked vertically and require diverse forms of bending deformations such as tensile, compressive, shearing, bending, and torsional.

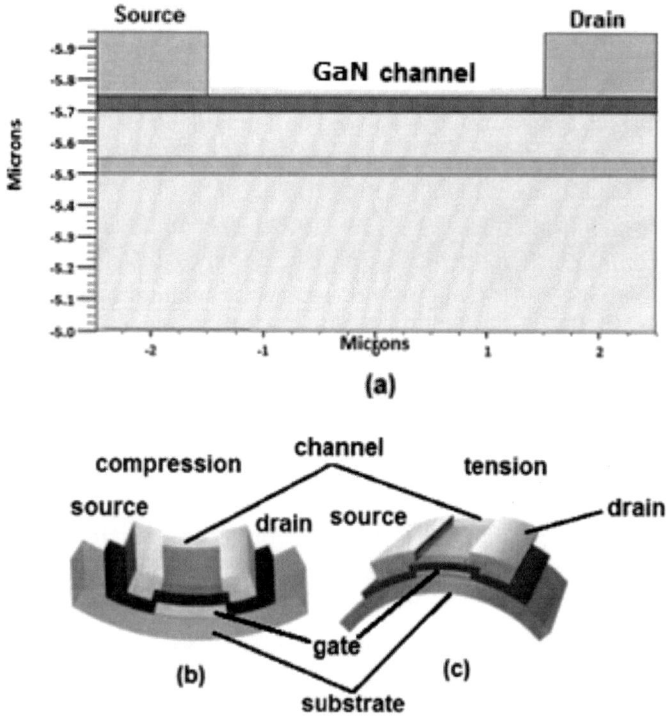

Figure 6.10 Schematic view of a GaN HEMT, considered in bending test: (a) device, (b) under compression, and (c) under tension.

Some inorganic layers may become brittle, and other films in the stack may buckle or break if the curvature of bending a flexible device is too high. Buckling or cracking of this nature impairs electrical performance and poses a major reliability risk. Flexible electronic components, as a result, require a mechanical support structure (a plastic sheet, a thin metal sheet, or a more complicated structure) to guide the device's movement while bending. As a result, the mechanical flexibility of each layer in response to applied physical force from bending while maintaining electrical stability is critical for the system's overall flexibility.

When flexible electronic devices are subjected to repetitive flexing, the bending of the film stack causes a variety of failure mechanisms. Because device degradation is expected, neutral plane management has been used to reduce bending stress by facilitating significant shear stresses between layers, which minimize shear coupling and increase buckling resistance [55]. It has been attempted to boost the reliability of flexible electronics by positioning the most vulnerable layers nearer to the neutral plane of the devices [56]. A specific cell structure design utilized in plastic LCDs has also been attempted to address the upper and bottom substrate alignment displacement under bending stress [57].

For practical applications, the flexible electronics assembly with electronic components installed must stretch to a desired radius of curvature without buckling or breaking while maintaining consistent electrical performance. The fluctuation of electrical properties is commonly observed following repetitive mechanical stress (bending test). It has been shown that when repeated width-axis tensile stress is applied, there is more degradation than when repeated width-axis compressive stress is applied. Mechanical bending has also been shown to affect flexible ferroelectric memory systems [58]. Despite substantial experimental characterization data on the effects of bending stress/strain, these submissions contain no information on the device's local stress/strain distribution. In general, stated electrical performance degradation is based on qualitative stress data [59]. It should be emphasized that limits can be unpredictable from the standpoint of applications, and their impact should be reduced as much as feasible. Unfortunately, at present time, there is no systematic way for analyzing the reliability of flexible electronics that we want to use. In the following, we use a predictive TCAD framework to (a) build multi-layered flexible electronic device architectures with substantial deformations, and (b) simulate and analyze stress/strain profiles using a predictive TCAD framework.

Multiple thin composite sheets of various materials, thicknesses, and even sizes are stacked to create flexible electronic devices. When attempting to answer the following issue, it is critical to consider the mechanical aspects: What effect will external mechanical stresses have on the electrical properties of the devices and circuits that are being studied? The term "flexible" for an electronic component frequently alludes to the system's deformability. Flexible substrates are classified into two categories: (a) substrates that only enable inextensible bending and (b) substrates that allow huge arbitrary bending. Depending on the applications, each flexible component must be able to tolerate large deformations [60, 61].

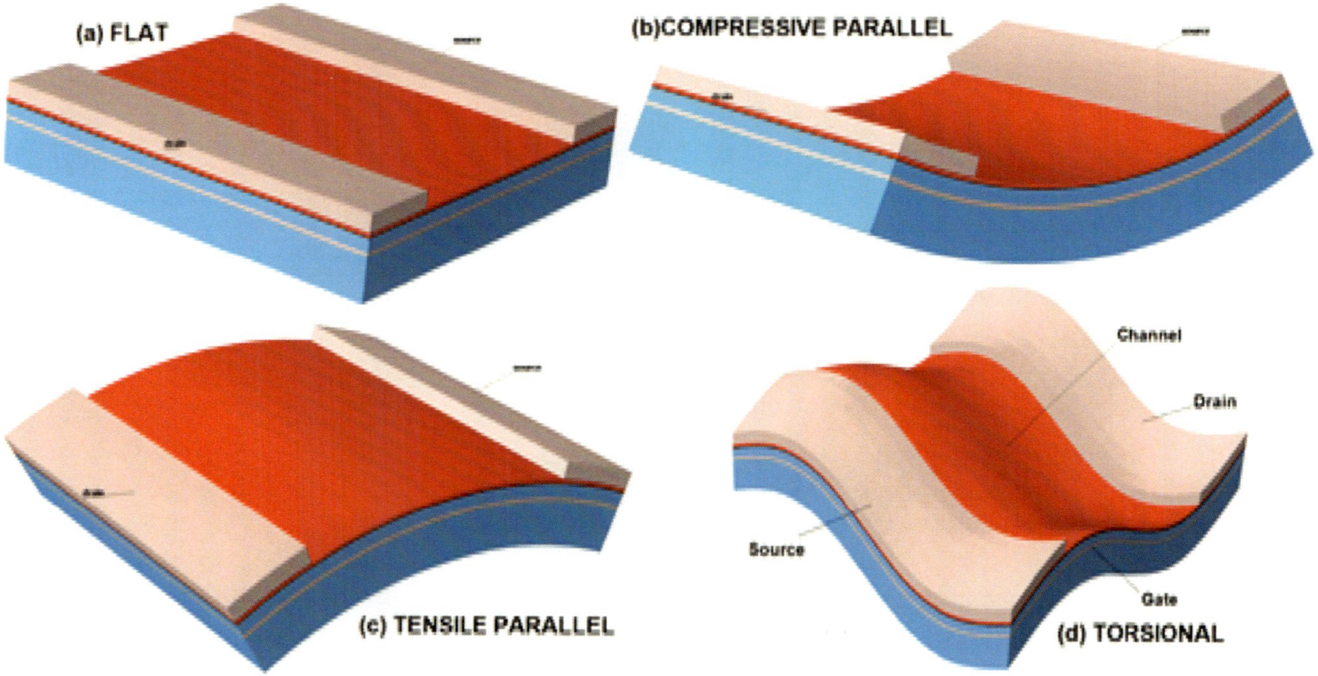

Figure 6.11 Different bending configurations generated from the planar structure: (a) flat (no stress), (b) compressive bending, (c) tensile bending, and (d) torsional (sine) bending.

The elasticity theory is used in a lot of mechanical analyzes in microelectronics. Its fundamental equation, on the other hand, is difficult to solve. For many geometries and conditions encountered by the engineering design, the solution is frequently unknown. The area of mechanics known as continuum mechanics studies the kinematics and mechanical behavior of materials in terms of strain and stress. The elastic equations can be computed using the finite element technique (FEM) for general geometries and circumstances for study. By discretizing its core equations, it gives a way to mimic elasticity problems. The body will be in static equilibrium, and tensorial therapy will be utilized. For deformations far smaller than the body dimensions, infinitesimal strain theory [62] defines solid behavior. The linearization of Lagrangian or Eulerian strain tensors of finite strain theory [63], a generalization of infinite strain theory, is used to analyze tiny deformations of materials. To model tiny displacements, it was originally intended to exclusively deal with bending difficulties [64–66]. Different mechanics and piezoresistivity ideas used in studies will be briefly reviewed in the following sections.

To anticipate the stress distribution, we shall use nonlinear deformation theory following the simulation methodology described in reference [67]. When the stress is not too high, it can be estimated using Young's modulus and Poisson's ratio, as stress is proportional to strain in linear elasticity theory. By applying bending stress to the ends of a beam, the stress in the film can be calculated. When the device is constructed by stacking numerous layers of different materials with arbitrary geometry (as in the case of GaN HEMT), nonlinear deformation theory must be used to estimate the stress.

The stress-strain relationship for a perfectly elastic material may be calculated using the strain energy density function, which is dependent on the deformation gradient matrix [68]. The FEM is commonly used to simulate the mechanical reaction of isotropic materials. Because flexible devices can encounter a variety of deformations (tensile, compressive, shearing, bending, and torsional), specified motions at boundaries can be used to deal with any combination of these deformation modes. A fixed origin is used to create the reference frame, which is made up of rectangular coordinate axes. The structure is bent, and the stresses induced by rotation and shifting the left and right sides of the structure are calculated. The motion of nodal points at the rigid surface follows the rigid body kinematics when rigid surfaces at boundaries are considered, and is given as:

$$x_i = x_{\text{ref}} + R(\theta_{\text{ref}})\,(X_i - X_{\text{ref}}), \tag{6.5}$$

where, x_{ref} (reference point translation) and θ_{ref} (reference point rotation) are simulation variables, and R is a 3×3 rotation matrix for the given rotation angle. Following reference [67], we consider the linear system of algebraic equations resulting from finite element discretization as:

$$Kd = f, \tag{6.6}$$

where, d is the unknown displacement vector, f is the given force vector, and K is the symmetric positive definite stiffness matrix. The second Piola-Kirchhoff, S stress is thus given by:

$$S = DE, \tag{6.7}$$

where, D is constant elastic modulus and the Green-Lagrange strain, E is given in terms of the deformation tensor as:

$$E = \frac{1}{2}(FF^T - I), \tag{6.8}$$

where, F is the deformation gradient. Cauchy stress, where both the normal and forces are in the deformed frame, is often used in engineering practices. The following transformation can be used to get the Cauchy stresses:

$$\sigma = \frac{1}{\det(F)} FSF^T. \tag{6.9}$$

A conjugate strain to the Cauchy stress in large deformations is Almansi strain, which is defined as:

$$e = \frac{1}{2}(I - F^{-T}F^{-1}). \tag{6.10}$$

In terms of the displacement field, it is written as:

$$e = \frac{1}{2}\left(\frac{\partial u^T}{\partial x} + \frac{\partial u}{\partial x} - \frac{\partial u^T}{\partial x}\frac{\partial u}{\partial x}\right). \tag{6.11}$$

The method described above is used to generate different structures of the multi-layered GaN HEMTs on flexible substrates to determine structural integrity, performance, and reliability, as well as predict the structural failures. Several structures are created for the deformation-induced stress analysis. The stress profile in bent and twisted structures has been attempted to be determined. The stress-induced by such deformation has been calculated for the flat (no stress) and torsional (sine) bending cases. Figure 6.11 shows the possible different bending configurations generated from the planar structure: (a) flat (no stress), (b) compressive bending (parallel to the current direction), (c) tensile bending (parallel to the current direction), (d) torsional (sine) bending. The radius of curvature is 2 μm.

The body deforms or changes size when external or internal forces are applied to it. The body reflects the continuity before and after deformation, even if it is distorted. As a result, all body point coordinates after deformation are continuous, particular functions of their corresponding beginning coordinates. The linearized elastic equations are solved using the elastic model. The displacement vectors in the device and crystal coordinates are provided. On the mesh elements, the strain tensor is computed from the displacement vectors.

The elastic model accounts for the strain caused by a lattice mismatch between materials. The material database for GaN is used to retrieve the lattice constants. In the simulation, all segments are subjected to an initial strain condition that forces their lattice constant to match that of the substrate segments. By solving the elastic equations, the system can relax and reduce its overall energy, resulting in the proper strain conditions at material interfaces. With the components C_{11}, C_{12}, and C_{44}, the elasticity tensor defines the connection between stress and strain. The default material database is used to retrieve the components and lattice constants.

VictoryStress [69] is used for the stress/strain calculations, and the command "rebalance stress" is used to allow for free movement of side walls in the structure. After the 'STRESS' statement is processed, the structure contains the displacement field describing the structure's deformation caused by the existing stresses, as described in the simulation environment. After the computation of stress/stress, the structure contains the displacement field describing the structure's deformation (shape) caused by the existing stresses. The displacement profiles are plotted. A cut plane is a plane (2D slice) that is drawn through a 3D structure (see Figure 6.12). The cut plane may be used so that one can examine the inside of a structure or to perform a 2D device simulation. We have calculated the internal stresses caused by the deformation of the flexible structures. Figures 6.13 and 6.14 show the 2D (XX) stress distribution in bent GaN HEMT structures for both flat (no bending) and torsional bending, respectively.

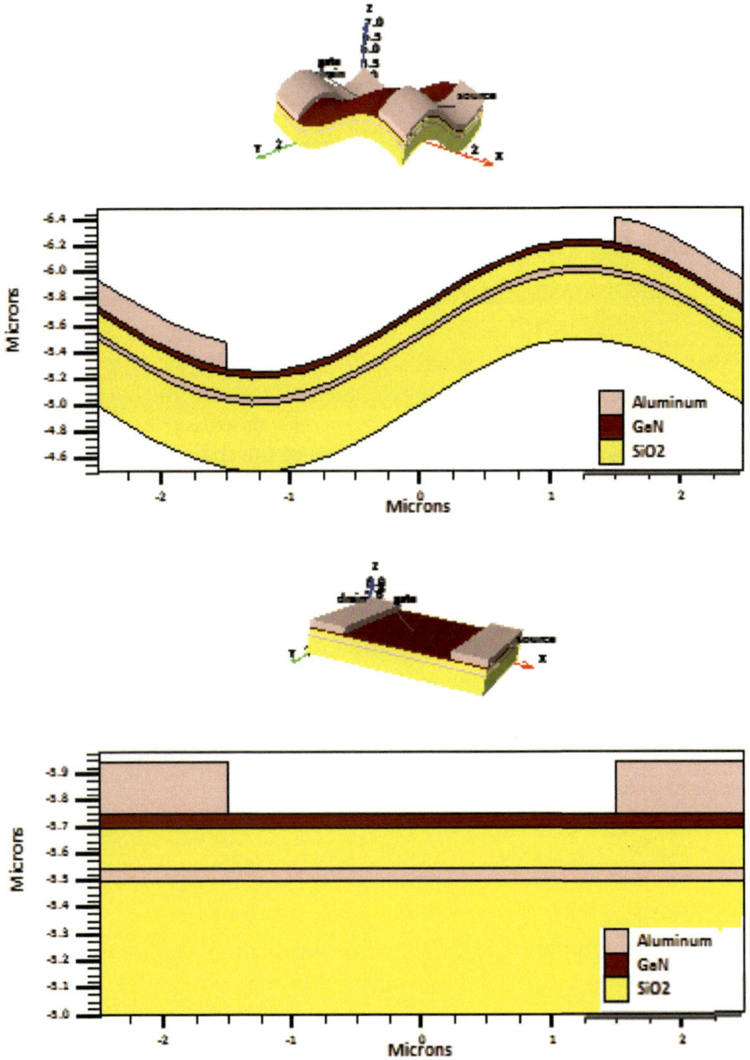

Figure 6.12 2D cut plane drawn through the 3D structure (XZ plane). The cut plane is used to examine the stress/strain profile inside a structure or to perform a 2D device simulation.

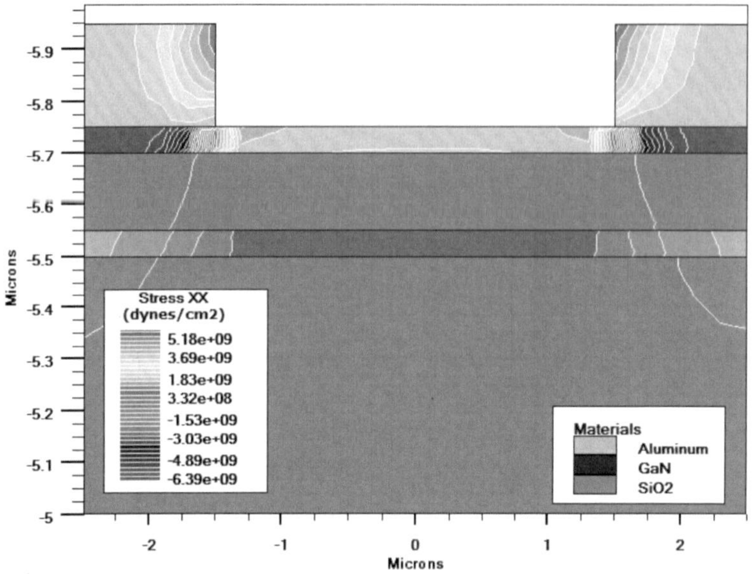

Figure 6.13 2D XX stress distribution in bent GaN HEMT structure for flat (no stress) condition. The simulated stress shows a lower stress (~5.18e+09 dynes/cm^2) for no bending case. These components of the strains and stresses are, for reasons of symmetry, only plotted against the thickness of the deformed flexible structure.

Because shear and tensile stresses contribute to the driving force for interfacial delamination, the von Mises stress along the interface is critical for the investigation of film dependability. The von Mises stress is a combined shear and tensile stress indicator. State stain energy density is equal to the area under the stress-strain curve in the case of uniaxial stress. The strain energy density, U_0, in the general 3D situation is represented as:

$$U_0 = \frac{1}{2}\sigma_x\varepsilon_x + \sigma_y\varepsilon_y + \sigma_z\varepsilon_z + \tau_{yz}\gamma_{yz} + \tau_{zx}\gamma_{zx} + \tau_{xy}\gamma_{xy}, \qquad (6.12)$$

where, σ_x, σ_y, and σ_z are normal stress components; ε_x, ε_y, and ε_z are strain components; τ_{yz}, τ_{zx}, and τ_{xy} are shear stress components; and γ_{yz}, γ_{zx}, and γ_{xy} are surface energy per unit area. For a coordinate system that is parallel to the principal stress directions, no shear components exist. Extending Eq. (6.12) to this stress states yields:

$$U_0 = \frac{1}{2}(\sigma_1\varepsilon_1 + \sigma_2\varepsilon_2 + \sigma_3\varepsilon_3), \qquad (6.13)$$

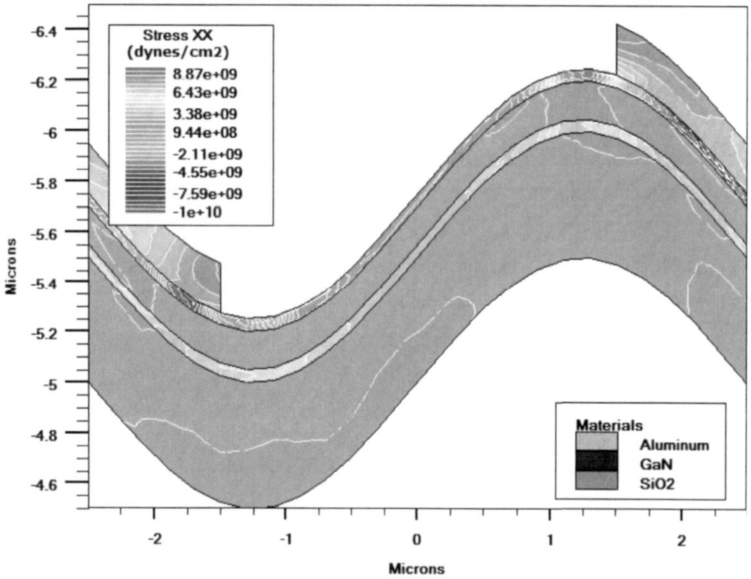

Figure 6.14 2D XX stress distribution in bent GaN HEMT structure for torsional bending. The simulated stress shows a higher stress (~8.87e+09 dynes/cm^2) for torsional bending. These components of the strains and stresses are, for reasons of symmetry, only plotted against the thickness of the deformed flexible structure.

where, σ_1, σ_2, and σ_3 are normal stress components and ε_1, ε_2, and ε_3 are strain components. As stresses and strains are related through the linear elastic relations, principal stresses and strains are given as:

$$\begin{cases} \varepsilon_1 = \dfrac{1}{E}(\sigma_1 - v\sigma_2 - v\sigma_3) \\ \varepsilon_2 = \dfrac{1}{E}(\sigma_2 - v\sigma_1 - v\sigma_3) \\ \varepsilon_3 = \dfrac{1}{E}(\sigma_3 - v\sigma_1 - v\sigma_2), \end{cases} \tag{6.14}$$

where, v is Poisson's ratio and E is Young's modulus. Substituting from Eq. (6.14) into Eq. (6.13), the strain energy density in terms of principal stresses is given as:

$$U_0 = \frac{1}{2E}\left[\sigma_1^2 + \sigma_2^2 + \sigma_3^2 - 2v(\sigma_1\sigma_2 + \sigma_2\sigma_3 + \sigma_1\sigma_3)\right]. \tag{6.15}$$

The von Mises stress σ_{VM} is defined in terms of principal stresses as:

$$\sigma_{VM} = \sqrt{\frac{(\sigma_1 - \sigma_2)^2 + (\sigma_2 - \sigma_3)^2 + (\sigma_3 - \sigma_1)^2}{2}}. \tag{6.16}$$

The von Mises stress can be rewritten in terms of stress components as:

$$\sigma_{VM} = \sqrt{\frac{(\sigma_{xx} - \sigma_{yy})^2 + (\sigma_{yy} - \sigma_{zz})^2 + (\sigma_{zz} - \sigma_{xx})^2 + 6(\tau_{xy}^2 + \tau_{yz}^2 + \tau_{zx}^2)}{2}}, \tag{6.17}$$

where, σ_{xx}, σ_{yy}, and σ_{zz} are normal stress components.

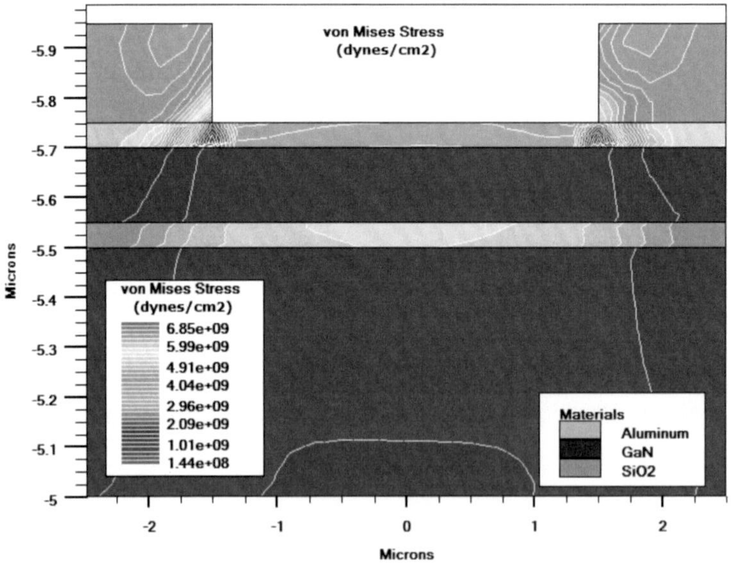

Figure 6.15 2D von Mises stress (XZ) distribution in flat GaN HEMT structure. The simulated stress shows a much lower stress (~6.85e+09 dynes/cm^2) in a flat structure (no bending condition).

Figures 6.15 and 6.16 show the 2D von Mises stress distribution in bent GaN HEMT structures for both flat (no bending) and torsional bending, respectively. It is observed that torsional bending induces considerable higher stress/strain in GaN HEMT devices. Also, we can see, most of the stress appears at the metal edge of the source/drain contact material nearby the bending point.

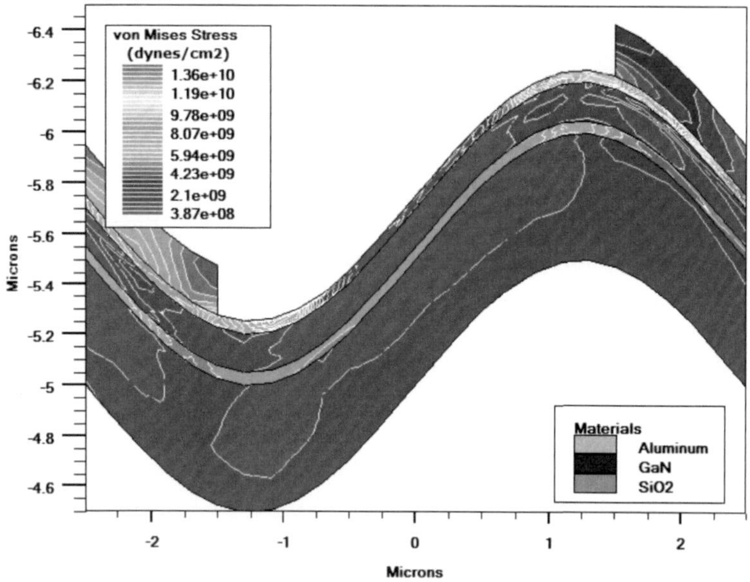

Figure 6.16 2D von Mises stress (XZ) distribution in GaN HEMT (torsional bending) structures. The simulated stress shows a much higher stress (~1.36e+10 dynes/cm^2) for torsional bending.

6.6 Summary

In this chapter, we have presented the properties of III-nitride (GaN) materials that can serve as a channel for flexible HEMTs. We have shown the prospects of III-nitride devices for the flexible electronics industry paving the way for new manufacturing techniques and the exploitation of new materials. Organic materials have a low process cost and are compatible with flexible substrates. Nevertheless, they are limited by their low mobility. The appearance of new materials including graphene opens a new avenue of research for the integration of the latter in flexible transistors. It should be noted, however, that silicon and III-V technologies have attracted much attention for applications requiring conformability with rigid substrates. However, an improvement of the manufacturing processes, as well as the reliability of the devices, remains a major challenge for the development of the flexible sector.

The introduction of flexible electronics requires the exploitation of new materials and new manufacturing and integration techniques. To bring conformability to the electronic components, two techniques are considered: (i) the first is to manufacture the components on a rigid substrate and to transfer to a flexible substrate layer, and (ii) the second is to make the components directly on the flexible substrate. This technique is used mainly in the context of silicon technologies and III-V materials [27]. The advantage is to provide high-performance devices with flexible supports. As for the disadvantages, they reside in the fragility of epitaxies to transfer, risks of mechanical degradation of the components, small surface transferred, and high cost of the process.

Fabrication of AlGaN/GaN HEMT transferred to the flexible substrate has been reported. The device provides a maximum current density of 270 mA/mm and an extrinsic transconductance of 170 ms/mm. The device also has an f_t and a maximum F_{max}, around 40 GHz. Measurements performed in load-pull carried out at 10 GHz have shown a power density output of 110 mW/mm, a PAE efficiency of 14%, and a linear gain of 12 dB at V_{DS} = 5 V. This constitutes the first step in flexible component development for RF power applications.

This study also outlines a TCAD application methodology for designing strain-engineered simulation-based GaN HEMTs. To compute stress/strain from the deformation gradient, we first explained the elasticity concepts and then proposed a method based on nonlinear FEM. To cope with deformation-induced stress determination in flexible electronic devices, a geometric nonlinear finite element methodology was employed to handle the ever-increasing demand for stress/strain mapping in GaN HEMTs. It is also worth noting that, in terms of commercial applications, limitations can be random, and their impact should be reduced as much as feasible. To test structural integrity, performance, and dependability, as well as predict structural failures, a 3D TCAD framework was used.

We have shown how to use TCAD to design stress/strain engineered GaN HEMTs and how to reduce the effect of deformations on electrical properties. This method will aid in the visualization of stress/strain distribution in flexible devices, which is difficult to measure in small volumes. The effects of several types of bending deformations on the electrical characteristics of strained GaN HEMT are investigated, including flat (no stress) and torsional (sine) bending.

References

1. A. M. Hussain and M. M. Hussain, "CMOS-technology-enabled flexible and stretchable electronics for internet of everything applications," *Adv. Mater.*, vol. 28, no. 22, pp. 4219–4249, Jun. 2016. doi: 10.1002/adma.201504236.
2. R. Shringarpure, L. T. Clark, S. M. Venugopal, D. R. Allee, and S. G. Uppili, "Amorphous silicon logic circuits on flexible substrates," in *2008 IEEE Custom Integrated Circuits Conference*, 2008, pp. 181–184. doi: 10.1109/CICC.2008.4672053.
3. K.-H. Liu *et al.*, "14-2: Towards commercial organic electronics and comprehensive comparison of device performance and reliability of organic and a-Si:H thin-film transistor technologies," *SID Symp. Dig. Tech. Pap.*, vol. 48, no. 1, pp. 173–175, 2017. doi: 10.1002/sdtp.11614.
4. R. Courtland, "Plotting a Moore's law for flexible electronics [News]," *IEEE Spectr.*, vol. 54, no. 7, pp. 7–8, 2017. doi: 10.1109/MSPEC.2017.7951706.
5. T. Peter Brody, "The thin film transistor—A late flowering bloom," *IEEE Trans. Electron Devices*, vol. 31, no. 11, pp. 1614–1628, 1984. doi: 10.1109/T-ED.1984.21762.
6. A. F. Harper, P. J. Diemer, and O. D. Jurchescu, "Contact patterning by laser printing for flexible electronics on paper," *npj Flex. Electron.*, vol. 3, no. 1, Dec. 2019. doi: 10.1038/s41528-019-0055-3.
7. Y. Y. Hsu *et al.*, "Epidermal electronics: Skin sweat patch," in *Proceedings of Technical Papers - International Microsystems, Packaging, Assembly, and Circuits Technology Conference, IMPACT*, 2012, pp. 228–231. doi: 10.1109/IMPACT.2012.6420313.
8. S. Wang *et al.*, "Mechanics of epidermal electronics," *J. Appl. Mech. Trans. ASME*, vol. 79, no. 3, 2012. doi: 10.1115/1.4005963.
9. T. Q. Trung and N.-E. Lee, "Flexible and stretchable physical sensor integrated platforms for wearable human-activity monitoringand personal healthcare," *Adv. Mater.*, vol. 28, no. 22, pp. 4338–4372, Jun. 2016. doi: 10.1002/adma.201504244.
10. S. Mhedhbi, "Develpoment de composants flexibes en technologies heterogene (gan et graphene) pour des applications hautes frequences," PhD Thesis, Université de Lille, 2017.

11. J. Njeim, "Physical and electrical characterization of CVD graphene devices: Towards large scale and flexible graphene electronics," PhD Thesis, Sorbonne Université, 2018.
12. H. C. Yuan and Z. Ma, "Microwave thin-film transistors using Si nanomembranes on flexible polymer substrate," *Appl. Phys. Lett.*, vol. 89, no. 21, p. 212105, Nov. 2006. doi: 10.1063/1.2397038.
13. H. C. Yuan, G. K. Celler, and Z. Ma, "7.8-GHz flexible thin-film transistors on a low-temperature plastic substrate," *J. Appl. Phys.*, vol. 102, no. 3, p. 034501, Aug. 2007. doi: 10.1063/1.2761782.
14. L. Sun *et al.*, "12-GHz thin-film transistors on transferrable silicon nanomembranes for high-performance flexible electronics," *Small*, vol. 6, no. 22, pp. 2553–2557, Nov. 2010. doi: 10.1002/smll.201000522.
15. J. H. Seo *et al.*, "Fast flexible transistors with a nanotrench structure," *Sci. Rep.*, vol. 6, no. 1, pp. 1–9, Apr. 2016. doi: 10.1038/srep24771.
16. D. Streit, R. Lai, A. Oki, and A. Gutierrez-Aitken, "InP HEMT and HBT applications beyond 200 GHz," *Conf. Proceedings-International Conf. Indium Phosphide Relat. Mater.*, pp. 11–14, 2002. doi: 10.1109/ICIPRM.2002.1014077.
17. J. W. Johnson *et al.*, "12 W/mm AlGaN-GaN HFETs on silicon substrates," *IEEE Electron Device Lett.*, vol. 25, no. 7, pp. 459–461, Jul. 2004. doi: 10.1109/LED.2004.831190.
18. D. C. Dumka, T. M. Chou, F. Faili, D. Francis, and F. Ejeckam, "AlGaN/GaN HEMTs on diamond substrate with over 7W/mm output power density at 10 GHz," *Electron. Lett.*, vol. 49, no. 20, pp. 1298–1299, 2013. doi: 10.1049/el.2013.1973.
19. S. Tirelli *et al.*, "AlInN/GaN HEMTs on SiC and on silicon with regrown ohmic contacts by selective ammonia MBE," in *Digest of Papers - 2015 International Conference on Compound Semiconductor MaNufacturing TECHnology*, 2015, pp. 261–264.
20. F. Medjdoub, M. Zegaoui, B. Grimbert, D. Ducatteau, N. Rolland, and P. A. Rolland, "First demonstration of high-power GaN-on-silicon transistors at 40 GHz," *IEEE Electron Device Lett.*, vol. 33, no. 8, pp. 1168–1170, 2012. doi: 10.1109/LED.2012.2198192.
21. P. Altuntas *et al.*, "Power performance at 40 GHz of AlGaN/GaN high-electron mobility transistors grown by molecular beam epitaxy on Si(111) substrate," *IEEE Electron Device Lett.*, vol. 36, no. 4, pp. 303–305, Apr. 2015. doi: 10.1109/LED.2015.2404358.
22. W. Chen, T. L. Alford, T. F. Kuech, and S. S. Lau, "High electron mobility transistors on plastic flexible substrates," *Appl. Phys. Lett.*, vol. 98, no. 20, May 2011. doi: 10.1063/1.3593006.

23. W. Chen, P. Chen, J. E. Pulsifer, T. L. Alford, T. F. Kuech, and S. S. Lau, "Integration of thin layers of single-crystalline InP with flexible substrates," *Appl. Phys. Lett.*, vol. 92, no. 21, p. 212109, May 2008. doi: 10.1063/1.2937409.

24. N. Defrance *et al.*, "Fabrication, characterization, and physical analysis of AlGaN/GaN HEMTs on flexible substrates," *IEEE Trans. Electron Devices*, vol. 60, no. 3, pp. 1054–1059, 2013. doi: 10.1109/TED.2013.2238943.

25. Tzu-Hsuan Chang *et al.*, "High power fast flexible electronics: Transparent RF AlGaN/GaN HEMTs on plastic substrates," in *2015 IEEE MTT-S International Microwave Symposium*, 2015, pp. 1–4. doi: 10.1109/MWSYM.2015.7167085.

26. S. Mhedhbi *et al.*, "First power performance demonstration of flexible AlGaN/GaN high electron mobility transistor," *IEEE Electron Device Lett.*, vol. 37, no. 5, pp. 553–555, May 2016. doi: 10.1109/LED.2016.2542921.

27. Y. Sun, W. M. Choi, H. Jiang, Y. Y. Huang, and J. A. Rogers, "Controlled buckling of semiconductor nanoribbons for stretchable electronics," *Nat. Nanotechnol.*, vol. 1, no. 3, pp. 201–207, Dec. 2006. doi: 10.1038/nnano.2006.131.

28. I. I. Smolyaninov, C. C. Davis, and A. V. Zayats, "Light-controlled photon tunneling," *Appl. Phys. Lett.*, vol. 81, no. 18, pp. 3314–3316, Oct. 2002. doi: 10.1063/1.1518778.

29. H. Usta, A. Facchetti, and T. J. Marks, "N-channel semiconductor materials design for organic complementary circuits," *Acc. Chem. Res.*, vol. 44, no. 7, pp. 501–510, Jul. 2011. doi: 10.1021/ar200006r.

30. Z. Bao, Y. Feng, A. Dodabalapur, V. R. Raju, and A. J. Lovinger, "High-performance plastic transistors fabricated by printing techniques," *Chem. Mater.*, vol. 9, no. 6, pp. 1299–1301, 1997. doi: 10.1021/cm9701163.

31. A. Knobloch, A. Bernds, and W. Clemens, "Printed polymer transistors," in *First International IEEE Conference on Polymers and Adhesives in Microelectronics and Photonics: Incoperating POLY, PEP and Adhasives in Electronics*, 2001, pp. 84–90. doi: 10.1109/polytr.2001.973262.

32. S. Wagner, H. Gleskova, J. C. Sturm, and Z. Suo, "Novel processing technology for macroelectronics," Springer, Berlin, Heidelberg, 2000, pp. 222–251.

33. H. Sirringhaus *et al.*, "High-resolution inkjet printing of all-polymer transistor circuits," *Science (80-.)*., vol. 290, no. 5499, pp. 2123–2126, Dec. 2000. doi: 10.1126/science.290.5499.2123.

34. F. Garnier, R. Hajlaoui, A. Yassar, and P. Srivastava, "All-polymer field-effect transistor realized by printing techniques," *Science (80-.).*, vol. 265, no. 5179, pp. 1684–1686, 1994. doi: 10.1126/science.265.5179.1684.

35. B. A. Ridley, B. Nivi, and J. M. Jacobson, "All-inorganic field effect transistors fabricated by printing," *Science (80-.).*, vol. 286, no. 5440, pp. 746–749, Oct. 1999. doi: 10.1126/science.286.5440.746.

36. T. Shimoda *et al.*, "Solution-processed silicon films and transistors," *Nature*, vol. 440, no. 7085, pp. 783–786, Apr. 2006. doi: 10.1038/nature04613.

37. D. S. Shen, "Electrophotographic patterning of thin-film silicon on glass foil," *IEEE Electron Device Lett.*, vol. 16, no. 10, pp. 418–420, 1995. doi: 10.1109/55.464803.

38. W. S. Wong *et al.*, "Amorphous silicon thin-film transistors and arrays fabricated by jet printing," *Appl. Phys. Lett.*, vol. 80, no. 4, pp. 610–612, Jan. 2002. doi: 10.1063/1.1436273.

39. O. Ambacher *et al.*, "Two-dimensional electron gases induced by spontaneous and piezoelectric polarization charges in N- and Ga-face AIGaN/GaN heterostructures," *J. Appl. Phys.*, vol. 85, no. 6, pp. 3222–3233, 1999. doi: 10.1063/1.369664.

40. I. Vurgaftman and J. R. Meyer, "Band parameters for nitrogen-containing semiconductors," *Journal of Applied Physics*, vol. 94, no. 6. American Institute of PhysicsAIP, pp. 3675–3696, 2003. doi: 10.1063/1.1600519.

41. J. P. Ibbetson, P. T. Fini, K. D. Ness, S. P. DenBaars, J. S. Speck, and U. K. Mishra, "Polarization effects, surface states, and the source of electrons in AlGaN/GaN heterostructure field effect transistors," *Appl. Phys. Lett.*, vol. 77, no. 2, pp. 250–252, 2000. doi: 10.1063/1.126940.

42. G. Koley and M. G. Spencer, "On the origin of the two-dimensional electron gas at the AlGaN/GaN heterostructure interface," *Appl. Phys. Lett.*, vol. 86, no. 4, 2005. doi: 10.1063/1.1850600.

43. L. Hsu and W. Walukiewicz, "Effect of polarization fields on transport properties in AlGaN/GaN heterostructures," *J. Appl. Phys.*, vol. 89, no. 3, pp. 1783–1789, Feb. 2001. doi: 10.1063/1.1339858.

44. T. Sekitani *et al.*, "Flexible organic transistors and circuits with extreme bending stability," *Nat. Mater.*, vol. 9, no. 12, pp. 1015–1022, 2010. doi: 10.1038/nmat2896.

45. T. Palacios *et al.*, "High-power AlGaN/GaN HEMTs for Ka-band applications," *IEEE Electron Device Lett.*, vol. 26, no. 11, pp. 781–783, Nov. 2005. doi: 10.1109/LED.2005.857701.

46. P. N. Chyurlia *et al.*, "Monolithic integration of AlGaN/GaN HFET with MOS on silicon ⟨111⟩ substrates," *Electron. Lett.*, vol. 46, no. 3, pp. 240–242, 2010. doi: 10.1049/el.2010.3167.

47. D. C. Dumka, C. Lee, H. Q. Tserng, P. Saunier, and M. Kumar, "AlGaN/GaN HEMTs on Si substrate with 7 W/mm output power density at 10 GHz," *Electron. Lett.*, vol. 40, no. 16, pp. 1023–1024, 2004. doi: 10.1049/el:20045292.

48. D. C. Dumka and P. Saunier, "GaN on Si HEMT with 65 power added efficiency at 10GHz," *Electron. Lett.*, vol. 46, no. 13, pp. 946–947, Jun. 2010. doi: 10.1049/el.2010.1284.

49. M. Lesecq *et al.*, "High performance of AlGaN/GaN HEMTs reported on adhesive flexible tape," *IEEE Electron Device Lett.*, vol. 32, no. 2, pp. 143–145, Feb. 2011. doi: 10.1109/LED.2010.2091251.

50. A. Davidson, K. Jones, and E. Strid, "LRM and LRRM calibrations with automatic determination of load inductance," in *36th ARFTG Conference Digest - Fall 1990*, 1990, pp. 57–63. doi: 10.1109/ARFTG.1990.323996.

51. L. F. Tiemeijer and R. J. Havens, "A calibrated lumped-element de-embedding technique for on-wafer RF characterization of high-quality inductors and high-speed transistors," *IEEE Trans. Electron Devices*, vol. 50, no. 3, pp. 822–829, Mar. 2003. doi: 10.1109/TED.2003.811396.

52. D. K. Lee, K. Park, J. H. Ahn, N. E. Lee, and Y. J. Kim, "A numerical study on the mechanical characteristics of zinc oxide-based transparent thin film transistors," in *Journal of Nanoscience and Nanotechnology*, Jul. 2011, vol. 11, no. 7, pp. 5870–5875. doi: 10.1166/jnn.2011.4470.

53. Y. H. Kim, E. Lee, J. G. Um, M. Mativenga, and J. Jang, "Highly robust neutral plane oxide TFTs withstanding 0.25mm bending radius for stretchable electronics," *Sci. Rep.*, vol. 6, May 2016. doi: 10.1038/srep25734.

54. M. J. Park *et al.*, "Improvements in the bending performance and bias stability of flexible InGaZnO thin film transistors and optimum barrier structures for plastic poly(ethylene naphthalate) substrates," *J. Mater. Chem. C*, vol. 3, no. 18, pp. 4779–4786, May 2015. doi: 10.1039/c5tc00048c.

55. F. Salmon, A. Everaerts, C. Campbell, B. Pennington, B. Erdogan-Haug, and G. Caldwell, "64-1: Modeling the mechanical performance of a foldable display panel bonded by 3M optically clear adhesives," *SID Symp. Dig. Tech. Pap.*, vol. 48, no. 1, pp. 938–941, 2017. doi: 10.1002/sdtp.11796.

56. Y.-H. Lin, M.-T. Lee, and A. T. Huang, "5-1: Invited paper: Booming flexible applications enabled by AMOLED technologies," *SID Symp. Dig. Tech. Pap.*, vol. 48, no. 1, pp. 29–32, 2017. doi: 10.1002/sdtp.11570.
57. T.-H. Huang *et al.*, "P-124: Bending shift free plastic liquid crystal display," *SID Symp. Dig. Tech. Pap.*, vol. 48, no. 1, pp. 1732–1734, 2017. doi: 10.1002/sdtp.11997.
58. L. Petti *et al.*, "Influence of mechanical bending on flexible InGaZnO-based ferroelectric memory TFTs," *IEEE Trans. Electron Devices*, vol. 61, no. 4, pp. 1085–1092, 2014. doi: 10.1109/TED.2014.2304307.
59. B. Chen *et al.*, "Effects of repetitive mechanical bending strain on various dimensions of foldable low temperature polysilicon TFTs fabricated on polyimide," *IEEE Electron Device Lett.*, vol. 37, no. 8, pp. 1010–1013, 2016. doi: 10.1109/LED.2016.2584138.
60. S. Wagner and S. Bauer, "Materials for stretchable electronics," *MRS Bull.*, vol. 37, no. 3, pp. 207–213, Mar. 2012. doi: 10.1557/mrs.2012.37.
61. K. D. Harris, A. L. Elias, and H. J. Chung, "Flexible electronics under strain: A review of mechanical characterization and durability enhancement strategies," *Journal of Materials Science*, vol. 51, no. 6, Springer New York LLC, pp. 2771–2805, Mar. 01, 2016. doi: 10.1007/s10853-015-9643-3.
62. G. Q. Z. W. van Driel and X. Fan, *Mechanics of Microelectronics*, Springer, 2006.
63. de Souza Neto, D. Peric and D. R. J. Owen, *Computational Methods for Plasticity: Theory and Applications*, John Wiley & Sons, 2011.
64. S. M. Lee, D. Shin, and I. Yun, "Degradation mechanisms of amorphous InGaZnO thin-film transistors used in foldable displays by dynamic mechanical stress," *IEEE Trans. Electron Devices*, vol. 64, no. 1, pp. 170–175, 2017. doi: 10.1109/TED.2016.2631597.
65. M. M. Billah *et al.*, "Effect of tensile and compressive bending stress on electrical performance of flexible a-IGZO TFTs," *IEEE Electron Device Lett.*, vol. 38, no. 7, pp. 890–893, 2017. doi: 10.1109/LED.2017.2707279.
66. H. H. N. Wacker *et al.*, "Bending induced electrical response variations in ultra-thin flexible chips and device modeling," *Appl. Phys. Rev.*, vol. 4, no. 3, p. 31101, Sep. 2017. doi: 10.1063/1.4991532.
67. H. Lim, S. Kong, E. Guichard, and A. Hoessinger, "A general approach for deformation induced stress on flexible electronics," in *2018 International Conference on Simulation of Semiconductor Processes*

and Devices (SISPAD), Nov. 2018, vol. 2018-Septe, pp. 276–279. doi: 10.1109/SISPAD.2018.8551752.

68. P. G. Ciarlet, *Mathematical elasticity, vol. 1, Three-dimensional elasticity,* Elsevier Science Publishers B. V., 1988.

69. Silvaco International, VictoryStress User Manual, 2018.

Chapter 7

FDSOI RF Flexible Electronics

Chinmay K. Maiti
SOA University, Bhubaneswar 751030, Odisha, India
ckmaiti@outlook.com

Since the late 1960s, the semiconductor industry has grown steadily up to generating nearly $335 billion now. Thus, the constant evolution of technologies, characterized by improved performance and integration density in addition to the reduction in manufacturing costs, allowed the unprecedented diversification in the electronics market, ranging from automotive to telecommunications through IT. For more than half a century, research efforts allowed to meet miniaturization specifications related to "Moore's Law" [1]. Until the early 2000s, the continued progression of performance of CMOS technologies based on a miniaturization approach had enabled us to adhere to predictions of "Moore's Law" without encountering technological major obstacles. Figure 7.1(a) shows the historical evolution of the manufacturing cost of integrated circuits as their complexity increases, in addition to the concomitant increase of wafer diameter. Considered as a prediction at the infancy of the semiconductor industry, this trend became the driver of the world

Fabless Semiconductor Manufacturing: In the Era of Internet of Things
Edited by Chinmay K. Maiti

ISBN 978-981-4968-27-0 (Hardcover), 978-1-003-31497-4 (eBook)
www.jennystanford.com

semiconductor industry. This is illustrated in Figure 7.1(b) where it can be observed that the number of transistors per microprocessor was multiplied by a factor of one million in about half a century (reaching 3.1 billion in Intel® Itanium processor in 2011), with a concomitant decrease of the dimensions of the transistor (by a factor of almost one thousand in 50 years) and cost (from $1 per transistor in the late 1960s to less than a µ$ nowadays).

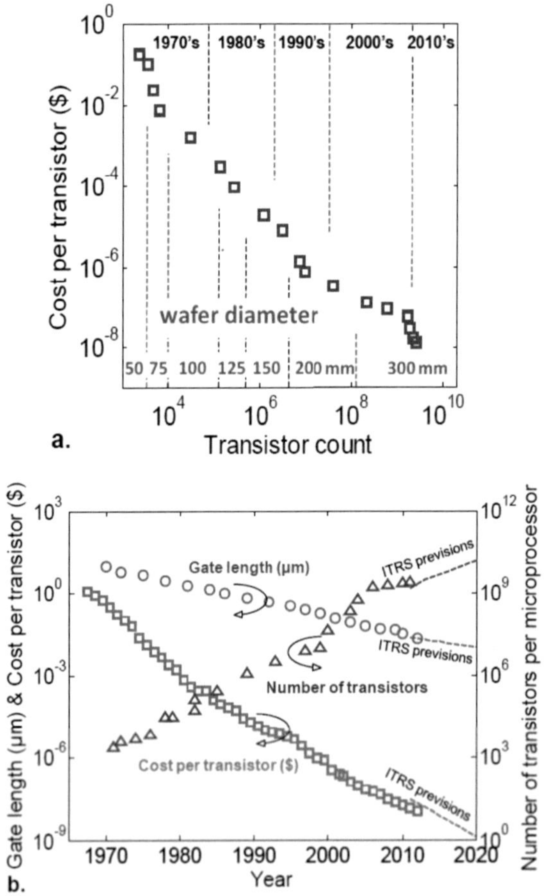

Figure 7.1 (a) Evolution of the manufacturing cost of integrated circuits as their complexity increases, and (b) Moore's law as it is commonly represented the number of transistors per microprocessor as a function of time, gate length, and cost per transistors, in addition to the International Technology Roadmap for Semiconductors previsions [2, 3].

In the continuity of this period known as happy scaling, the increase in high-frequency performance and low-power consumption have required the implementation of advanced technological options such as strain, metal gate, and high-k dielectric. The exploration of advanced device architectures (silicon-on-insulator (SOI), multi-gates) also emerged. These advanced technological options are often taken up in the literature under the term "technological boosters". Alternatively, according to a qualified approach of "More than Moore", many technological enhancements were also possible thanks to a diversification of functions not exclusively based on geometry scaling [4]. However, a physical limit of the current downscaling methodology of transistor dimensions can be foreseen as it is approaching the atomic scale. These observations lead to the emergence and growth of a new field, called "More-than-Moore", where the added value does not rely on downscaling and increasing intrinsic performance but is rather provided by the incorporation of novel functionalities on the same chip.

The "More-than-Moore" field has been growing in parallel to the conventional "More-Moore" efforts during the last decades. A large number of nano- or micro-electromechanical systems (NEMS or MEMS), active or passive radio-frequency (RF) and analog-mixed signal devices, high voltage power devices, various sensors, and actuators have been packaged together in embedded systems to provide better interaction with the environment. This integration of novel functionalities provides diversification from the baseline CMOS downscaling results in high-value systems and is driven by the International Technology Roadmap for Semiconductors [4]. Figure 7.2(a) shows a chart detailing that higher value systems can be fabricated by following: (i) the "More-Moore" downscaling of CMOS technology, memory, and logic, (ii) by providing functional diversification to connect the CMOS chips with the environment, and (iii) the ITRS recommendations are to combine both approaches in systems-on-chip (SoC) and systems-in-package (SiP) [4, 5]. The co-integration of a large variety of non-digital systems (Figure 7.2(b)) providing additional functionalities to the digital signal and data processing chip gave rise to the so-called SoC and SiP. The former describes the integration on the same chip of novel functionalities, whereas SiP corresponds to the bonding in the same packaging of several chips with different functionalities. An increasing number of references to the "More-than-Moore" development area, and especially to flexible electronics, can be seen in the latest IRDS reports [6]. This interest in the "More-than-Moore" trend has also been observed in other research areas [7, 8].

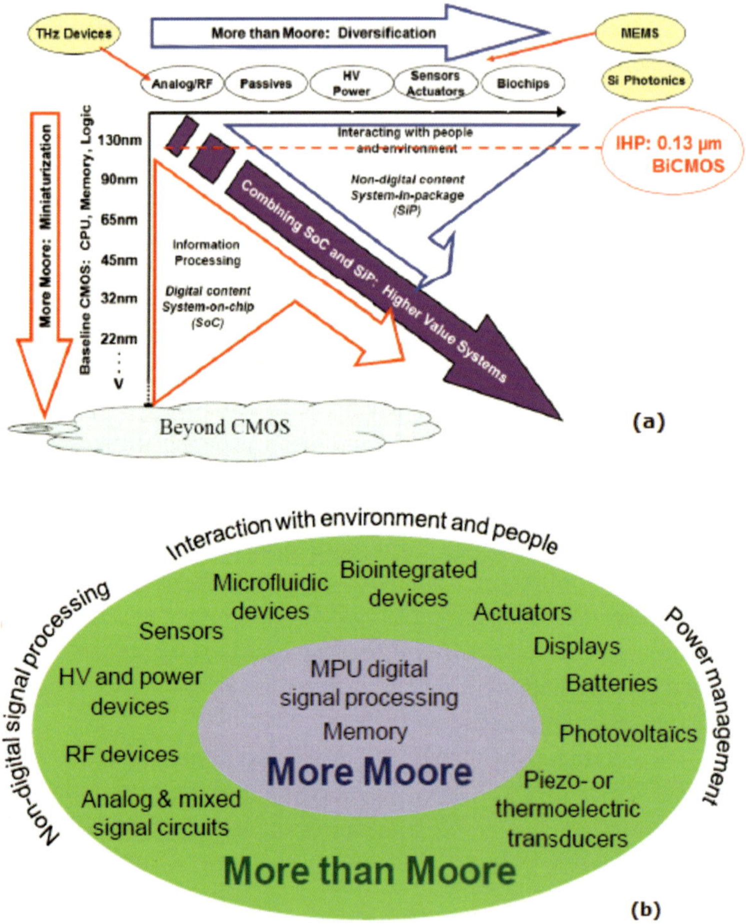

Figure 7.2 (a) Parallel developments of the "More-Moore" and "More-than-Moore" fields. (b) "More-than-Moore" components can be integrated for digital signal processing and non-digital processing capabilities and power management into the "More-Moore" core [5].

Macroelectronics, as first defined by Reuss et al. [9], describes the field of flexible and large-area electronics. Thanks to the use of organic materials, this is often synonym to thin, lightweight, transparent, bendable, and/or stretchable devices fabricated at high throughput and low cost on large areas. However, current organic semiconductors feature limitations and the co-integration of thin inorganic materials on organic films proved to open the way to novel and promising applications in this field. Indeed, smart and communicating systems, such as wireless sensor networks, biomedical, health infrastructure, or environmental monitoring systems require electrical performance that organic electronics cannot currently achieve. Figure 7.3(a) represents the inherent advantages and drawbacks of both nanoelectronics and macroelectronics. However, the realization of hybrid devices could benefit from several advantages of both the fields as shown in Figure 7.3(b) schematically. It could also process data and react accordingly by stimulating actuators or transferring information via RF communication for instance. The energy required to power these conceptual devices could be harvested from the environment using photovoltaic cells, piezoelectric or thermoelectric units, and stored in micro-batteries embedded in the system. In this context, the integration of novel functionalities and the possibility to bend, fold, or stretch an electronic device open new possibilities [10, 11].

Organic electronics are considered a starting point because organic semiconductors are intrinsically flexible materials. Organic photovoltaic cells [12], flexible displays, and lightning are the major application fields of organic electronics [8]. Currently, many flexible organic components have emerged thanks to printing techniques [13]. These simple and inexpensive methods make it possible to achieve on large surfaces of devices with a single level of interconnections [14, 15]. However, the low mobility of organic semiconductors (of the order of $cm^2{\cdot}V^{-1}{\cdot}s^{-1}$) does not make any use in wave range impossible RF [16–18]. New techniques using the principle of transfer onto a flexible substrate were thus created to give technologies high-performance enhanced mechanical compliance properties. Besides the mechanical flexibility, transfer onto a flexible substrate allows the properties of original devices reported and makes heterogeneous integration possible [19–21].

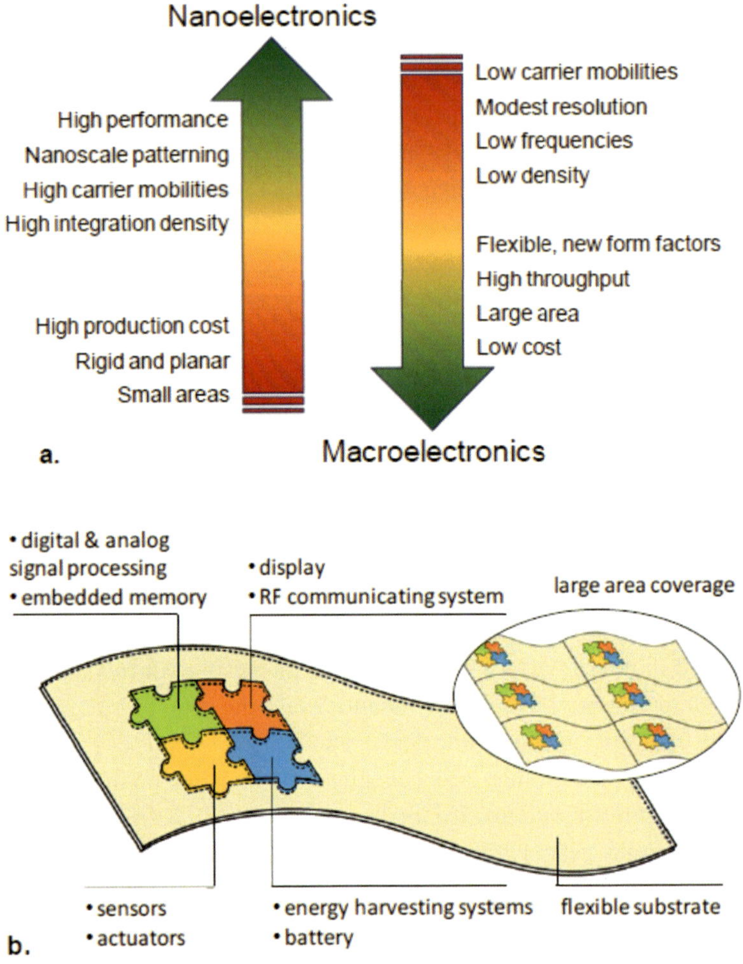

Figure 7.3 (a) Advantages and limitations of current nanoelectronic and macroelectronics fields. (b) Schematic of a fully flexible hybrid system realized via the integration of various sensors for information processing [29].

This chapter falls within this context and aims to present the design and simulation of flexible devices to demonstrate the feasibility of high-frequency flexible building blocks for the development of high-performance foldable electronics. We review the recent developments in the field of high-frequency flexible electronics, the main concepts necessary for understanding the subject, and the possible applications of electronics flexible as well as the associated perspectives and challenges. Starting from the limitations intrinsically associated with current organic electronics that confine it to low-frequency applications [22]. We discuss various solutions suggested to enhance flexible electronics capabilities. Most of them are based on the hybridization of flexible organic substrates with thin inorganic nanostructures featuring high charge carrier mobilities [9, 23]. It will be shown that this approach has enabled the realization of flexible transistors reaching the gigahertz (GHz) range, using either a "bottom-up" [24, 25] or a "top-down" approach [26–28]. The need for communicating and smart foldable devices requiring higher frequency, flexible electronics are highlighted and the hybridization of organic materials with mature conventional CMOS technology is suggested as a promising solution.

Flexible electronics have been one of their variations for twenty years. Intense research work has been carried out to provide mechanical bendability or even stretchability to electronic devices and circuits [23, 30, 31]. This technology can be used in various fields such as biology [32–34], photovoltaics [35–37], or even on-board electronics [31, 38–40]. As a result, sensors, batteries, displays, as well as other elements can be co-integrated in a heterogeneous way to allow the realization of fully flexible intelligent systems [41–45]. As a result, flexible communicating, sensing, actuating, energy harvesting, and storage systems have to be developed to allow the realization of fully flexible smart systems. Low cost and large area, but low-frequency devices and circuits can be processed with current organic semiconductors. They are intrinsically flexible [46], but their frequency performance is currently limited to the megahertz (MHz) range [22]. As higher frequencies are required for numerous applications [4, 47], novel strategies have been developed to combine high frequency and bendability. The integration of chemically synthesized nanotubes [24, 25], or patterned thin films [26–28, 48] on plastic foils enabled the realization of GHz flexible devices. Furthermore, transistors initially fabricated on conventional rigid silicon wafers before thinning and transfer on organic films allow the field of flexible electronics to reach the 100 GHz range [49, 50]. Special attention is currently focused on packaging various functionalities together on flexible films to reach the market of high-performance flexible electronics. However, devices reaching the level of performance required for RF electronics are currently based on the conventional inorganic semiconductor industry. Therefore, there is an urgent need for industrial solutions regarding the handling of ultra-thin wafers. To manufacture flexible electronic devices in conventional semiconductor manufacturing, numerous challenges associated with the handling and processing of thin, flexible wafers have to be studied. Companies such as SÜSS MicroTech [51, 52], Dupont [53, 54], EVGroup [55, 56], IBM [57, 58], and several others are developing industrial solutions for low cost and reliable handling of ultra-thin wafers and conventional processing on these flexible wafers. Numerous applications based on this organic/inorganic hybridization have been reported, leading to promising applications in the biomedical field [59, 60], and/or embedded electronics. Several devices based on the integration of plastic films of poly- or mono-crystalline silicon, compound semiconductors, carbon nanotubes, or graphene demonstrated flexible transistors featuring cutoff frequencies in the GHz range [61–64]. Flexible transistor technologies will, therefore, be our focus in this chapter. Design issues related to geometric, static, and high-frequency figures-of-merit, namely (i) the transistor gate length, (ii) on-state current I_{ON}, off-state leakage current I_{OFF}, and on/off currents ratio, (iii) static transconductance, (iv) current gain, (v) cutoff frequency f_T, and (vi) maximum oscillation frequency f_{max}. To realize components combining high electrical performance with high mechanical bendability, numerous fabrication strategies have been developed during the last decades. Recent progress in the field of high-frequency flexible electronics is shown in Figure 7.4. The main focus has been laid on frequency figures-of-merit, such as the transistor f_T. High-frequency technologies incorporating inorganic materials embedded on plastic films have been developed [24–28, 61–64]. To reach higher frequencies, conventionally processed transistors have been conventionally processed on rigid semiconductor (silicon) wafers prior to transferring on organic films. The field of flexible electronics thus reached the 100 GHz range [49, 50].

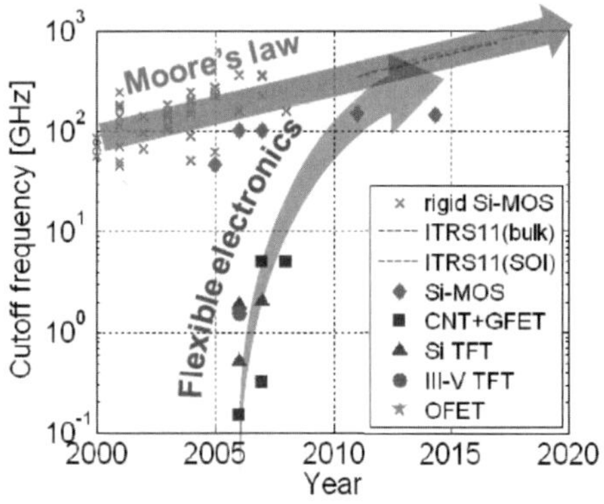

Figure 7.4 Recent progress in the field of high-frequency flexible electronics [29].

7.1 Potential of SOI-CMOS Technology for Flexible Electronics

Several "technology boosters" have been explored by the semiconductor industry to solve scaling limits and also meet the IRDS requirement, such as the use of high-k dielectric gate stacks [7], new channel materials (Ge, III-V) [8, 9], stress engineering [10, 11], or alternative chip structures such as fully depleted silicon-on-insulator (FDSOI) [12–17]. Controlling leakage current [18], improving transport properties inside the channel, and controlling variability (due to process or doping fluctuations) are among the challenges faced by the industry. FDSOI technology is an advantageous candidate structure for continuing the size reduction of the CMOS technology [65, 66]. STMicroelectronics, for instance, has already replaced the traditional bulk MOSFET device with FDSOI devices for their 28 nm node and beyond [67]. Among its remarkable enhancements concerning the bulk MOSFETs, FDSOI observes an improvement in:

- electrostatic control – the possibility of adjusting the threshold voltage by applying a back bias, giving flexibility to circuit designers to combine high performance and low power transistors in the same technology [68–72];

- reduction of parasitic effects and current leakages due to lateral and vertical isolation when compared to junction isolation [73–75];
- decrease of short channel effects since the influence by drain field on the channel is reduced [76];
- thin films and dielectric isolation offering higher transistor packing density and simplified processing [77]; and
- SOI devices yield enhanced switching speed and reduced power consumption. The operating speed is improved as the isolated channel from substrate bias prevents the [68] increase in threshold voltage of stacked SOI transistors [77, 78].

Moreover, the planar FDSOI devices have very similar architecture and fabrication process as their silicon bulk predecessor MOSFET and therefore it simplifies, for instance, the adaptation of the existing device simulation tools. SOI technology is a technology based on wafers composed of a layer of monocrystalline silicon on an insulating film, usually silicon dioxide, the substrate also being in monocrystalline silicon (Figure 7.5(a)) [79]. Of many manufacturing processes exist such as epitaxial growth of silicon on an insulating layer [80, 81], the recrystallization of a thin layer of silicon using a laser [82], the separation of silicon via the implantation of oxygen [83, 84], or the transfer of a substrate of silicon on another via a bonding [85–88]. From these various strategies, three manufacturing methods have been developed for several decades to produce SOI substrates: SIMOX (separation by implantation of oxygen) [83, 84, 87], BESOI (bonded and etch-back SOI) [85, 86], and SmartCut [87, 88]. The first method is based on the formation of a buried oxide layer, while the two others are based on the bonding of silicon substrates. Starting with two bulk silicon wafers A and B, a thermal oxide is first grown of wafer A, followed by the implantation of hydrogen at a chosen depth below the surface of wafer A to create a weakened layer. The two wafers are then bonded together, encapsulating the oxide layer between two silicon layers, wafer A is then separated from wafer B at the weakened layer. Wafer B finally undergoes thermal annealing and polishing, resulting in an SOI wafer, whereas wafer A can be reused.

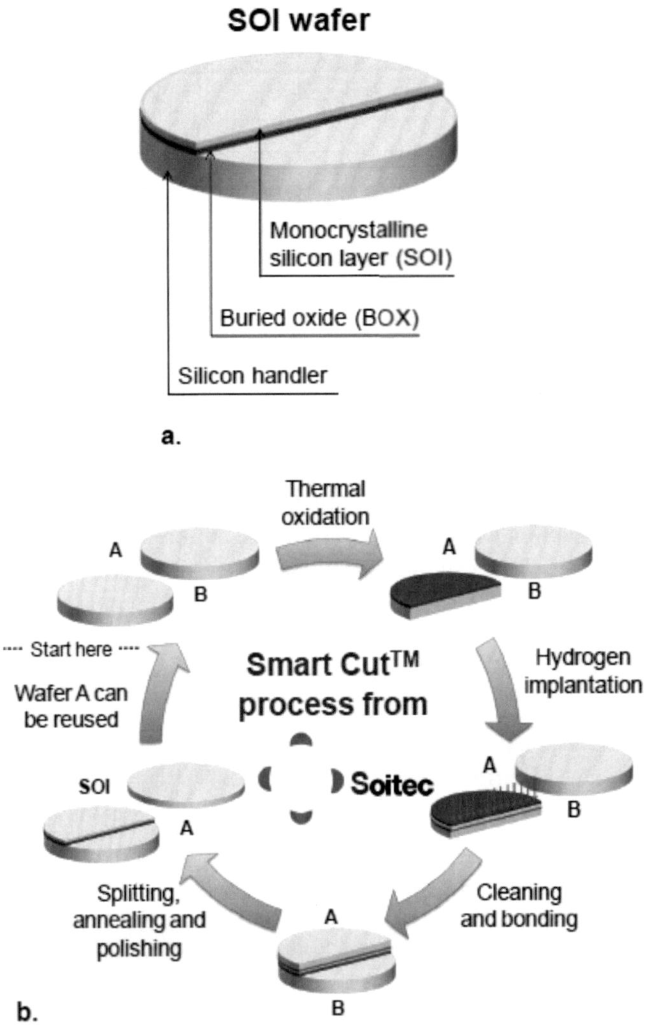

Figure 7.5 (a) Schematic of an SOI wafer and (b) schematic of the SOI fabrication process SmartCut from SOITEC [29].

7.2 SmartCut Process

The SmartCut process from SOITEC (Figure 7.5(b)) makes it possible to produce SOI substrates with excellent thickness uniformity and good crystal quality on the low-cost SOI layer [87, 88]. As shown in

Figure 7.5(b), four steps are required to manufacture SOI substrates [87, 88], namely: (i) ion implantation, (ii) binding of substrates, (iii) separation by heating, and (iv) fine polishing. Two bulk silicon substrates are used. The first is covered with an insulating layer (usually silicon dioxide) then implanted with hydrogen ions. The substrates are then cleaned to remove organic residues, native oxides, and metals that can degrade the bonding quality. The two substrates are then brought into contact and glued by adhesion hydrophilic. Then, a separation step is performed by applying a budget thermal (400–600 °C) on the bonded substrate. During this temperature rise, the ions hydrogen induce the formation of microcavities on the entire surface of the substrate, the separation then takes place at the implanted peak of hydrogen ions. The oxide layer buried and the thin layer of silicon are thus transferred to the second substrate, thus giving the SOI structure. A second thermal budget (around 1,100 °C) allows strengthening the chemical bond between the surface of the second substrate and the dioxide layer of silicon. However, the process generates micro-roughness that must be removed by fine polishing. The first substrate can then be used for a new cycle. Currently, this is the main manufacturing process for SOI substrates [89].

The SOI technology has several advantages over technology on the massive substrate. First of all, the presence of a buried insulation layer increases the resistance to parasitic effects usually observed in bulk technology. Indeed, SOI technology provides lower parasitic capacitances, thus reducing current leakage. In addition, it has complete immunity to the effects of latch-up, improves resistance to radiation, and thanks to the presence of BOX, is less sensitive to temperature variations. In the SOI technology, the BOX reduces parasitic capacitances [80], which has the effect of reducing leakage currents, improving performance, and decreasing energy consumption, as previously indicated. Technologies based on SOI substrates benefit from the fact that the active layer is separated from the silicon support by an insulating layer and that the source and the drain are in contact with the BOX. This vertical insulation protects the active layer from the effects of parasites such as the formation of capacitors between the source, the drain, and the substrate. Therefore, the performance is improved and the energy consumption is reduced. Among other benefits, SOI technology also provides isolation compact and easy lateral between n- and p-type devices. Indeed, in bulk technology, the formation of wells implanted along with the contacts and isolation by Shallow trenches (STI: Shallow Trench Isolation) is required. Trenches deeper than the implanted wells are necessary to ensure reliable insulation.

The reduction of the gate length and therefore of the channel is associated with noise, called short channel effects, resulting in loss of electrostatic control of the gate on the channel. The mainly responsible phenomenon is called drain-induced barrier lowering. For a long transistor, the potential along the channel is almost constant over the entire grid length. The more the length of the channel will decrease, the more the influence of the extensions of the space charge zones of the source and drain regions will increase, thus decreasing the area controlled by the grid. It is the effect of burden-sharing that has the effect of modifying the potential at the center of the channel and thus lowers the source/channel potential barrier. The threshold voltage thus drops in an uncontrolled manner leading to an increased I_{OFF} of the transistor.

7.3 RF CMOS SOI Technology

The SOI technology is a very promising candidate for applications in the field of flexible electronics [89, 90]. Thanks to the presence of a buried oxide layer, high-performance low consumption device energy can be achieved and parasitic effects reduced. In addition, the BOX can be used as a barrier layer in thinning processes. In this section, we will study the 65 nm SOI-CMOS technology. The transistors resulting from this technology are recognized to deliver a level of performance suitable for high frequency, low noise, and low power applications. In addition, the 65 nm CMOS technology provides a good compromise between state-of-the-art performance, large production volume, and a good understanding of its characteristics. Indeed, this mature technology has already been characterized in recent years. Moreover, the understanding of these characteristics after transfer onto flexible film takes advantage of this experience. The following paragraphs describe the static and dynamic characteristics of rigid devices derived from partially depleted SOI substrate technology.

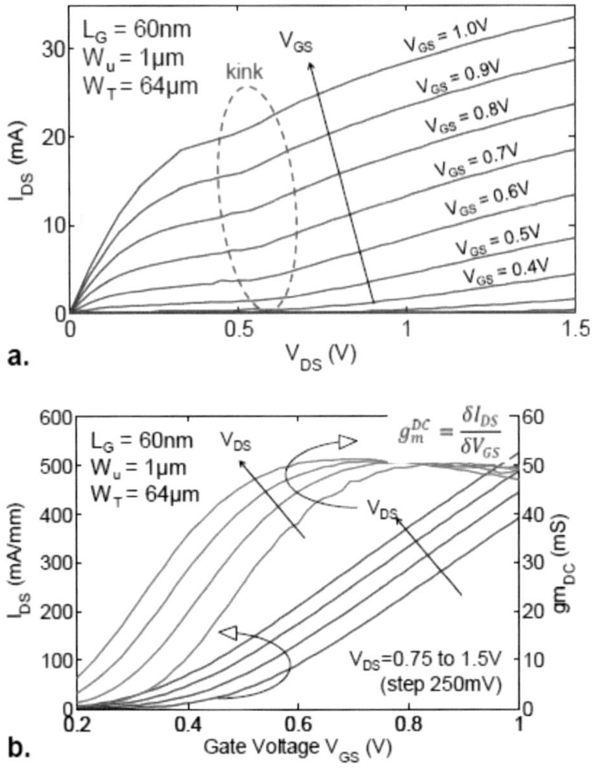

Figure 7.6 (a) Current–voltage I_{DS}–V_{DS} characteristics, highlighting the 'kink effect' in SOI floating body technologies, and (b) drain current I_{DS} and static transconductance g_mDC as a function of gate voltage V_{GS}.

Figure 7.6 shows that between the linear and saturated regime, the slope of the current of drain suddenly increases, it is the kink effect. This phenomenon is present in the SOI floating substrate technology for high drain voltage due to the generation by impact. Indeed, the important electric field at the drain level is the origin of a shock ionization mechanism, thus creating electron/hole pairs. The electrons are evacuated by the drain, while the holes are stored in the substrate of the device (the BOX prevents their passage through the substrate, as in the case of the bulk technology). Due to this, the potential of the body increases, the effective value of the threshold voltage decreases, which has the consequence of increasing the drain current [91].

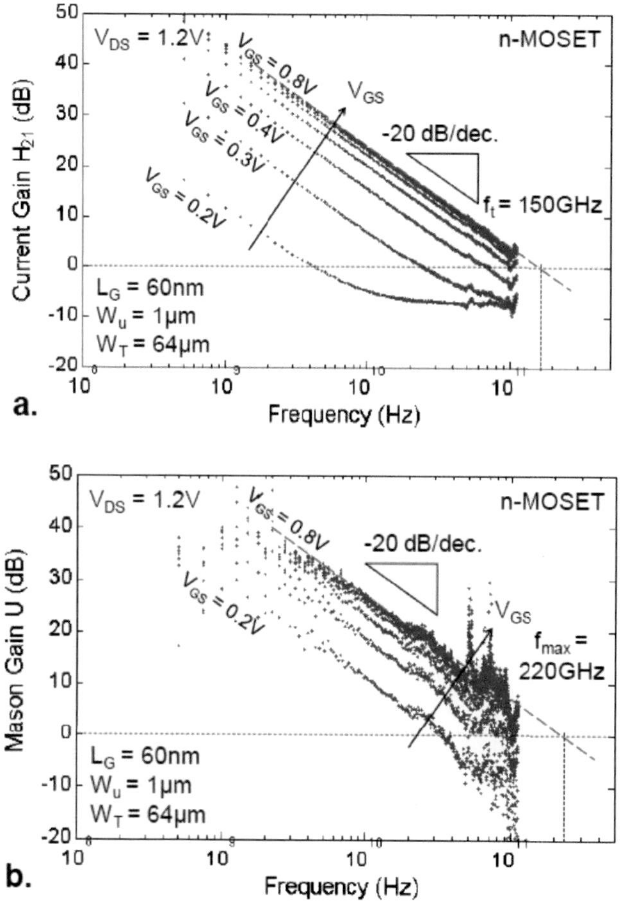

Figure 7.7 (a) H_{21} and (b) U of an SOI n-MOSFET as a function of frequency. V_{GS} varying from 200 mV to 0.8 V.

The current gain H_{21} and unilateral power gain, or Mason gain U can be extracted (Figure 7.7(a)) using the following definitions:

$$|H_{21}| = \frac{2S_{21}}{(1-S_{11})(1+S_{22})+S_{12}S_{21}} \tag{7.1}$$

$$U = \frac{\left|\dfrac{S_{21}}{S_{12}} - 1\right|^2}{2\left[k\left|\dfrac{S_{21}}{S_{12}}\right| - R\left(\dfrac{S_{21}}{S_{12}}\right)\right]} \tag{7.2}$$

$$k = \frac{1 - |S_{11}|^2 - |S_{22}|^2 + |S_{11}S_{22} - S_{12}S_{21}|^2}{2|S_{12}||S_{21}|} \quad (7.3)$$

From these equations, two figures of merits can be extracted (Figure 7.7). The f_T is the frequency where the gain H_{21} is unity (or 0 dB). In the same way, the maximum f_{max} is the frequency where the U is 0 dB. In addition, the slopes of these two gains must be equal to −20 dB/dec, ensuring thus correct extraction of dynamic performance. In the case of an n-transistor MOSFET polarized at V_{DS} = 1.2 V and V_{GS} = 0.8 V, the frequencies obtained are f_T = 150 GHz and f_{max} = 220 GHz.

7.4 SOI CMOS Technology for Flexible Electronics

This section aims to describe the techniques used to realize flexible high-performance CMOS circuits. The method developed is based on a transfer process, allowing the thinning of a thick CMOS circuit 800 μm until a thin film a few microns thick as well as its transfer is obtained on the flexible substrate. The first part of this section will focus on the techniques of thinning used during this process, namely grinding, mechanical-chemical polishing, and selective attack by xenon difluoride (XeF_2). Having described these different stages, the second part of this chapter will study several transfer methods to transfer the thinned circuits on different types of flexible substrates such as plastic films, glass, or metal foil.

Previous developments have demonstrated the high potentialities of the 65 nm node SOI CMOS technology as a building block for high-performance flexible electronics. The combination of high frequency and low noise performance with low power consumption defines a standard for next-generation flexible electronics. It will also be highlighted in the next chapter that the presence of a buried oxide in SOI technology can be an advantage to facilitate the transfer of this technology on a plastic film. The characteristics presented in this section will subsequently be compared in the next chapter with measurements performed on flexible devices to demonstrate the possibility to realize high-performance bendable electronics from the 65 nm node SOI CMOS technology.

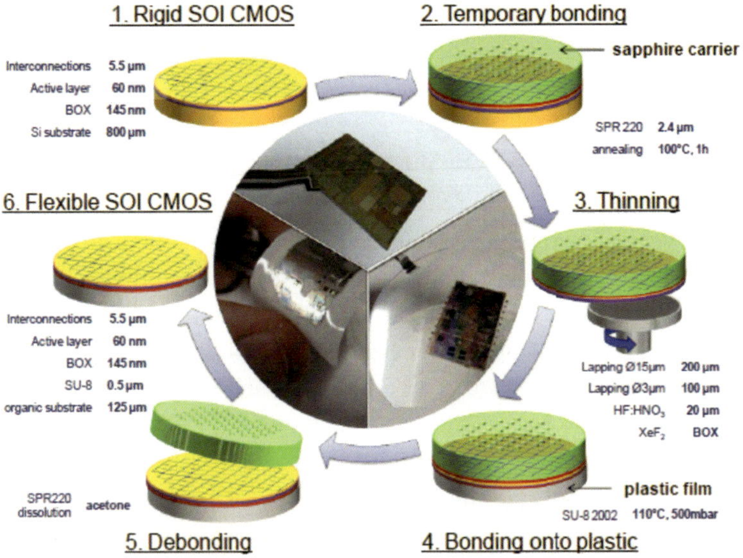

Figure 7.8 Process flow describing the fabrication of flexible RF-CMOS circuits from a conventionally processed SOI CMOS wafer (step 1), on which a temporary carrier is first bonded (step 2), ensuring flatness and rigidity during complete removal of the silicon handler of the SOI wafer (step 3). A flexible film is subsequently bonded on the backside of the thin active die (step 4) before releasing it from the temporary carrier (step 5), resulting in a flexible SOI wafer featuring RF-CMOS circuits [29].

The overview of the process flow of the complete fabrication process is presented in Figure 7.8. In the first step, the starting material corresponds to a processed SOI wafer featuring a 65 nm node RF-CMOS die. This wafer consists of a 780 µm thick high resistivity (HR) silicon handler, a 145 nm thick buried oxide, a 60 nm SOI active layer, and 5.5 µm of backend-of-line (BEOL) interconnection layer stack. After bonding a temporary carrier on the front side of the processed CMOS die to ensure flatness and rigidity (step 2), the initial silicon handler is completely removed by successively using different etching techniques without altering the BOX layer (step 3). A flexible film is then bonded on the backside of the thinned active CMOS die (step 4) and the temporary adhesive layer dissolution leads to the release of the temporary carrier from the active film (step 5). This fabrication process results in a flexible

SOI wafer featuring conventionally processed CMOS dies comprising the interconnection multilayer, the active SOI layer and a BOX layer, the permanent adhesive layer, and the plastic handler.

7.5 SOI RF-MOSFET Characterization

After transfer-bonding thinned SOI CMOS active dies onto organic plastic films, electrical characteristics of these flexible SOI RF-MOSFETs are compared with their rigid counterparts. DC characterizations have been performed considering both rigid and flexible CMOS dies. Figure 7.9 presents static characteristics of n- and p-MOSFETs both on their rigid SOI wafer (blue lines) and after transfer onto an organic film (red dashed lines). Similar behavior is observed for devices on rigid and flexible SOI substrates, leading to the demonstration of flexible transistors with currents and transconductance levels comparable with their rigid counterparts. From the S parameter measurements, cutoff frequencies and maximum oscillation frequencies (H_{21} and U) of the rigid and flexible MOSFETs are plotted as a function of frequency in Figure 7.10. The cutoff frequencies and maximum oscillation frequencies extracted at different gate voltages for n- and p-MOSFETs before (blue lines) and after (red dashed lines) transfer onto a plastic substrate are shown in Figure 7.11. Static figures-of-merit of n- and p-MOSFETs on a rigid substrate and after transfer onto a plastic substrate are shown in Table 7.1.

Table 7.1 Static figures-of-merit of n- and p-MOSFETs on a rigid substrate and after transferred onto a plastic substrate

		I_{ON} μA/μm	I_{OFF} nA/μm	I_{ON}/I_{OFF} ×10⁴	g_mDC μS/μm	V_{TH} mV	DIBL mV	SS mV/dec
n-MOSFET	Rigid	525	12.9	4.1	799	490	130	90
	Flexible	521	13.3	3.9	793	490	155	90
p-MOSFET	Rigid	−272	−2.67	10.2	499	570	90	90
	Flexible	−270	−2.65	10.2	495	570	90	90

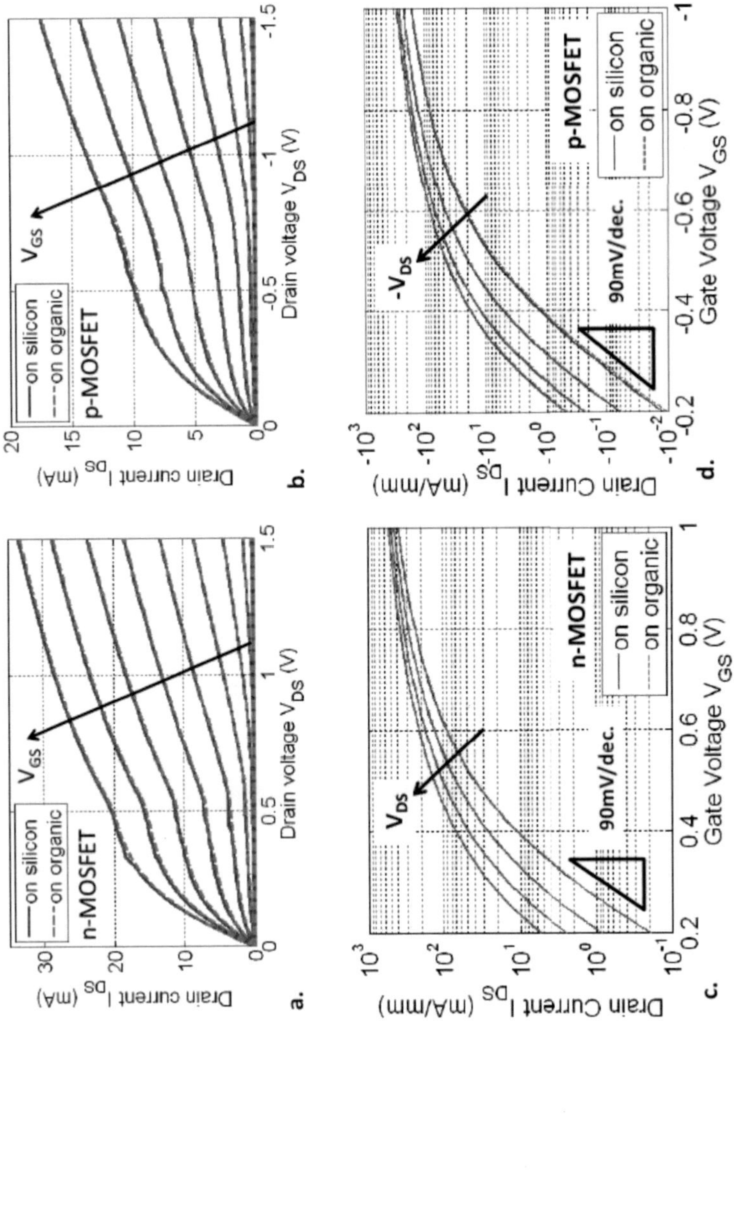

Figure 7.9 Static characteristics of n- and p-MOSFETs on their initial rigid substrate (blue lines) and after transfer onto a plastic film (red dashed lines), demonstrating a high level of on-state drain current and static transconductance.

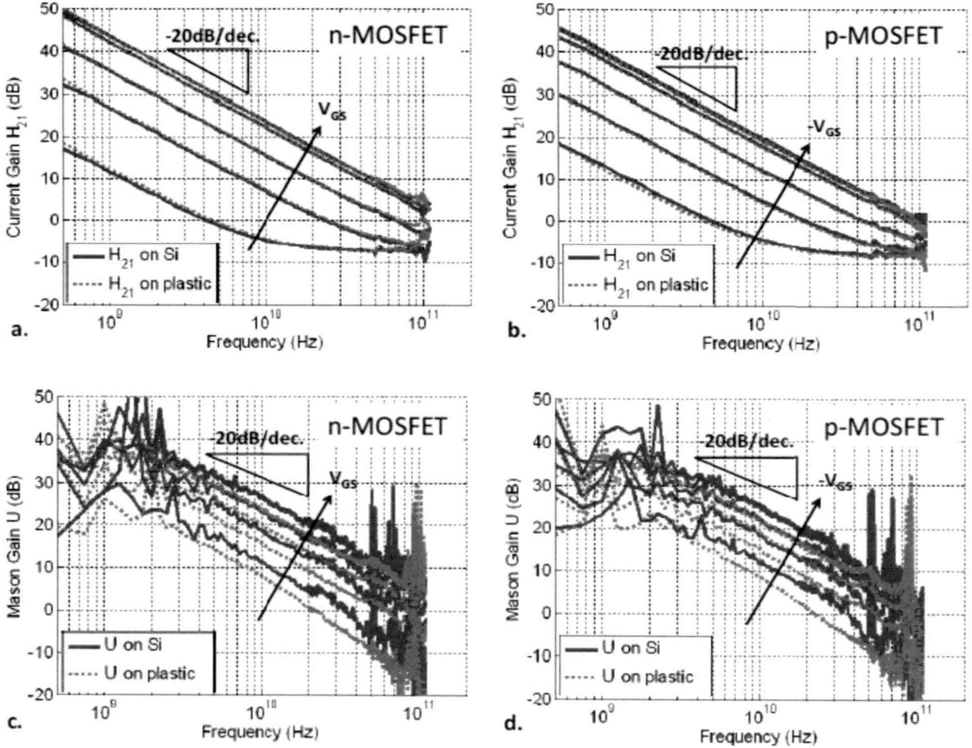

Figure 7.10 H_{21} and U measured on rigid (blue lines) and flexible (red dashed lines) n- and p- MOSFETs, biased at a drain voltage V_{DS} = 1.2 V, a gate voltage V_{GS} = 0.8V, and a resulting drain current I_{DS} = 270 mA/mm, respectively a drain voltage V_{DS} = −1.5V, a gate voltage V_{GS} = −1.0V, and a resulting drain current I_{DS} = −270 mA/mm.

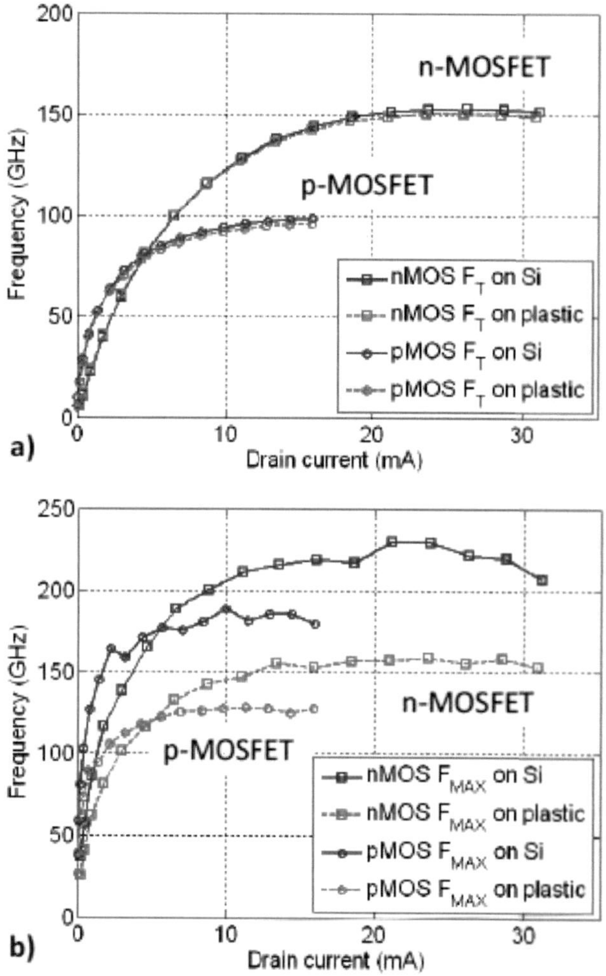

Figure 7.11 Cutoff frequencies and maximum oscillation frequencies extracted at different gate voltages for n- and p-MOSFETs before (blue lines) and after (red dashed lines) transfer onto a plastic substrate.

7.6 Simulation of FDSOI RF MOSFETs

GaAs or silicon bipolar technology is commonly used in the majority of RF applications. Advances in CMOS process technology, on the other hand, have continued to reduce the gate length of MOSFETs, thereby increasing the f_T [92, 93]. Bulk-silicon CMOS is becoming a viable technology option in analog and RF applications for portable wireless communication systems due to the relatively high f_T values. TCAD simulation of SOI CMOS devices for analog and RF applications is presented to obtain optimized devices at low power and low voltage when scaling down the device size. In particular, a technique for designing analog and RF circuits that takes into account body bias effects has been suggested.

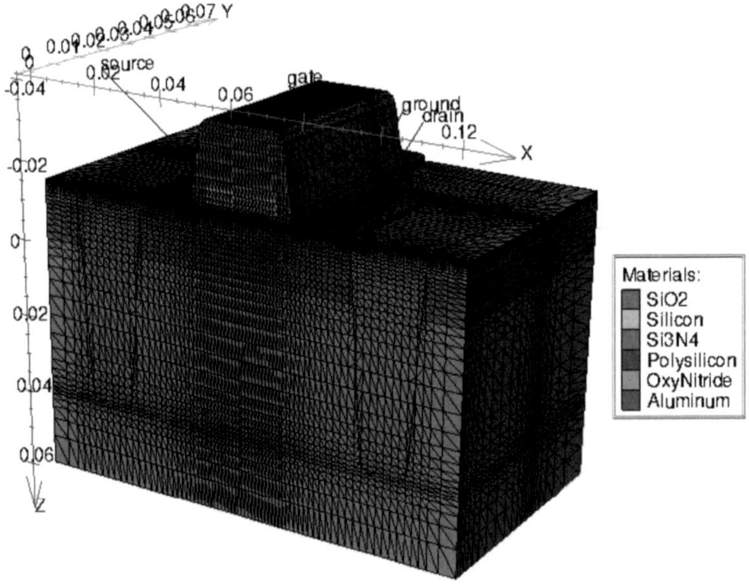

Figure 7.12 VictoryMesh generated ultra-thin body FDSOI n-MOSFET in 3D (20 nm undoped channel).

As discussed above, a potential candidate for aggressive CMOS scaling has been FDSOI, which uses a very thin silicon body on an ultra-thin buried oxide. A careful selection of physical models is needed for the accurate modeling of this type of FDSOI device. In this application example, VictoryDevice [94] simulation tool is used to model and simulate a 20 nm undoped, ultra-thin body, and buried oxide FDSOI bulk-Si n-MOSFET (see Figure 7.12 for structure). The FDSOI architecture, with a gate length in the sub-100 nm range, is susceptible to non-local effects such as velocity overshoot. As such the energy balance model can be used. The energy balance (EB) transport model outperforms the traditional drift-diffusion model, and the non-local effects have been simulated using the EB model. Also for energy balance simulations, the carrier temperature may be used instead of the electric field as the driving force for the field-dependent mobility model. The electron temperature is the driving force for the high-field saturation model.

The inclusion of quantum confinement of electrons becomes essential in the ultra-thin channel of the 20 nm FDSOI MOSFETs. The quantum confinement effects on electron transport can be incorporated in the simulation using the Bohm quantum potential model [95]. The Bohm quantum potential's main quantization path can be set to *x*, *y*, or *z*. The direction of quantization was set in the *z*-direction in the simulation. Due to the low thermal conductivity of the buried oxide, self-heating effects are typical in FDSOI structures. VictoryDevice incorporates lattice heating via self-consistent solutions to the lattice heat flow equation. We compare the I–V characteristics of the 20 nm FDSOI n-MOSFETs simulated using the drift-diffusion model with those simulated using the energy balance transport model alone and in combination with the quantum transport and lattice heating models, the non-local, quantum confinement, and self-heating effects (see Figure 7.13(a and b)).

7.7 Strain Engineering in RF Flexible CMOS

In the previous section, it was demonstrated that the thinning and transfer process does not degrade transistors' properties in terms of static and high-frequency operation in straight flat configuration [96, 97]. However, apart from performance conservation, the question arises: how mechanical deformation can affect the operation of complex flexible electronic systems? In this section, we shall consider the mechanical aspects due to the bending of the FDSOI CMOS devices on plastic substrates. We shall study the impacts of external mechanical stresses on the electrical properties of the devices. It should be noted that in the perspective of commercial applications, the bending can be random and their impact should, as much as possible, be minimized. It may be noted that the cylindrical bending (see Figure 7.14) causing tension in the drain current will be considered. From the application's point of view, stable electrical properties even upon deformation are required.

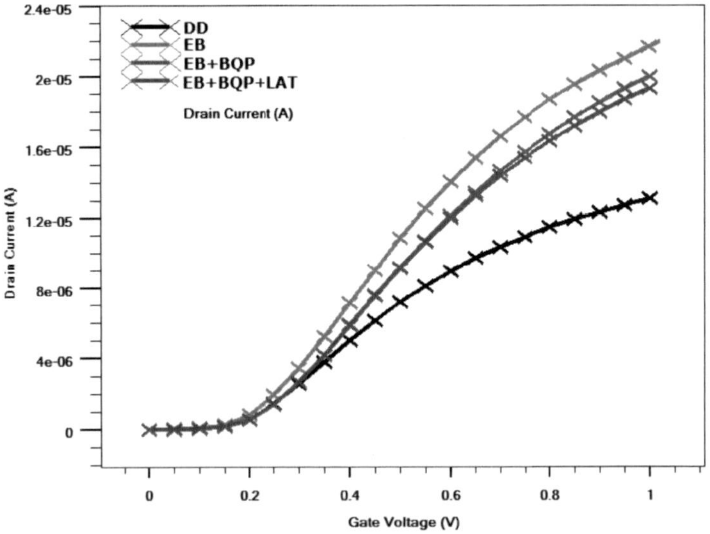

(a)

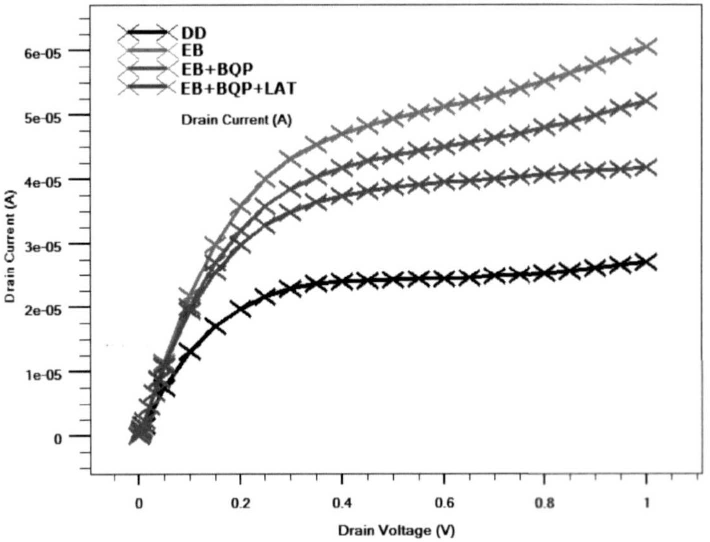

(b)

Figure 7.13 (a) I_{DS}–V_{GS} characteristics of ultra-thin body FDSOI n-MOSFET in 3D. Drain current comparison for drift-diffusion model, energy balance transport model alone, and combination of quantum transport and lattice heating models. (b) I_{DS}–V_{DS} characteristics of ultra-thin body FDSOI n-MOSFET in 3D. Drain current comparison for drift-diffusion model, energy balance transport model alone, and combination of quantum transport and lattice heating models.

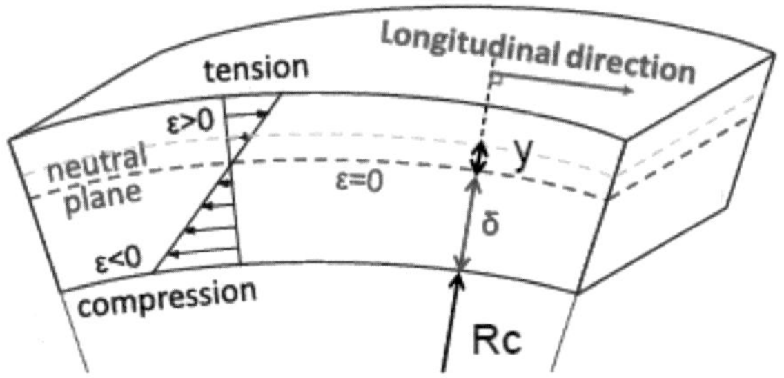

Figure 7.14 Schematic of the neutral plane in a cylindrically bent thin film over a curvature (radius RC). Neutral plane location is highlighted along with the longitudinal strain distribution.

The basic mechanical principles described in this section are mainly based on reference [98]. We shall focus on linear mechanics, introduce the notions of strain and stress tensors, linear elasticity theory, and neutral plane in a multilayer stack (see Figure 7.14), and finally, the piezoresistivity in silicon will be considered. The electrical performance will be simulated, considering linear relationships between electron mobility and figures-of-merit (drain current, cutoff, and maximum oscillation frequencies). Strain and stress components (profiles) in the SOI MOSFETs will be simulated. We briefly summarize the main equations involved in the mechanical models used in stress simulation. The stress generated by the bending of the flexible CMOS devices was simulated based on the theory of finite deformations. The impact of the deformations on the electrical properties is then estimated via the mobility variation due to the piezoresistive coefficients of silicon, assuming that static figures of merit, I_{DS}, and f_T are proportional to the mobility of majority carriers.

Considering an infinitesimal cube of matter, stress is defined as a force by the unit of area and can be applied on each of its faces by its surrounding (see Figure 7.15). The nine-component, second-rank stress tensor, σ consists of three normal stress components, σ_{ij} (i.e., orthogonal to the face plane), and six shear stress components, σ_{ij} (i.e., in this plane). The σ_{ij} component refers to the stress applied along the x_i axis on the face orthogonal to the x_j axis. By definition, positive σ_{ii} corresponds to tensile stress, whereas negative σ_{ii} corresponds to compressive stress. The stress tensor is symmetric, because its antisymmetric part corresponds to torque, not present in a solid at equilibrium [98]. The stress tensor, therefore, only contains six independent components. Relationships between a force $\vec{\mathbf{F}}$ applied on a surface orthogonal to vector $\vec{\mathbf{n}}$, and the stress tensor σ are provided in Eqs. (7.4–7.6).

$$\vec{\mathbf{F}} = \begin{bmatrix} F_1 \\ F_2 \\ F_3 \end{bmatrix} \quad \text{and} \quad \vec{\mathbf{n}} = \begin{bmatrix} n_1 \\ n_2 \\ n_3 \end{bmatrix} \tag{7.4}$$

$$\begin{matrix} F_1 \\ F_2 \\ F_3 \end{matrix} = \begin{bmatrix} \sigma_{11} & \sigma_{12} & \sigma_{13} \\ \sigma_{21} & \sigma_{22} & \sigma_{23} \\ \sigma_{31} & \sigma_{32} & \sigma_{33} \end{bmatrix} \begin{bmatrix} n_1 \\ n_2 \\ n_3 \end{bmatrix} \tag{7.5}$$

i.e.,

$$F_1 = \sigma_{ij} n_j \ (i = 1, 2, 3) \tag{7.6}$$

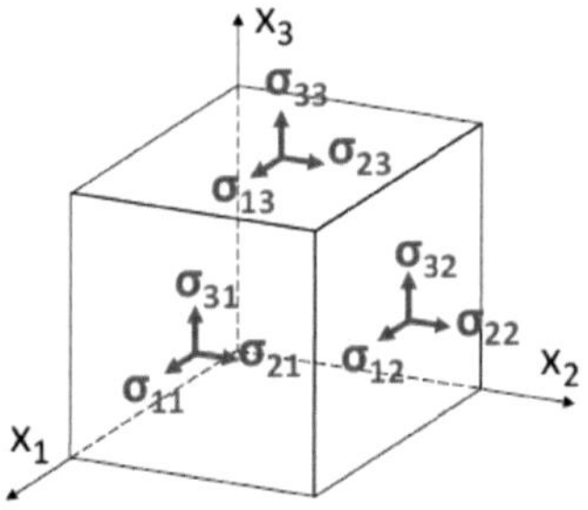

Figure 7.15 Stress tensor components in an infinitesimal cube of matter.

Because strain and stress are interrelated, strain is also a second-rank tensor, defined as dimensionless displacement ratio. If two points A and B separated by an infinitesimal position vector $\overrightarrow{\mathbf{dx}}$, are both subjected to a displacement $\vec{\mathbf{u}}(x)$ resulting in novel positions A' and B', strain can be defined as the ratio of $\mathbf{d}\vec{\mathbf{u}}(x)$ over $\overrightarrow{\mathbf{dx}}$. The strain appears only if both points experience different displacements, i.e., if $\mathbf{d}\vec{\mathbf{u}}(x)$ is non-zero. It should be noticed that this representation assumes that strain is infinitesimal. The general 3D concept is taken into account in strain-displacement equations (Eqs. (7.7–7.9)) that defines the second-rank infinitesimal strain tensor ε. This definition (Eq. (7.4)) excludes rotation and therefore only considers pure material deformation. As a result, the strain tensor ε is a symmetric tensor, as the stress tensor σ.

$$\varepsilon_{ij} = \frac{1}{2}\left(\frac{\partial u_i}{\partial x_j} + \frac{\partial u_j}{\partial x_i}\right) \tag{7.7}$$

$$\begin{bmatrix} du_1 \\ du_2 \\ du_3 \end{bmatrix} = \begin{bmatrix} \varepsilon_{11} & \varepsilon_{12} & \varepsilon_{13} \\ \varepsilon_{21} & \varepsilon_{22} & \varepsilon_{23} \\ \varepsilon_{31} & \varepsilon_{32} & \varepsilon_{33} \end{bmatrix} \begin{bmatrix} dx_1 \\ dx_2 \\ dx_3 \end{bmatrix} \tag{7.8}$$

i.e.,

$$du_i = e_{ij}dx_j \text{ (i=1, 2, 3)} \tag{7.9}$$

The linear elastic theory assumes that stress is directly proportional to strain, and related by Hooke's law (Eq. (7.10)), which can also be inverted to express strain in terms of stress, as in Eq. (7.11).

$$\sigma_{ij} = C_{ijkl} \in_{kl} \tag{7.10}$$

$$\in_{ij} = S_{ijkl}\sigma_{kl} \tag{7.11}$$

The stiffness C, and compliance S tensors are both fourth-rank tensors as they link together second-rank tensors $\in$ and σ. Because strain and stress tensors $\in$ and σ are symmetric, the 81 components of the two fourth-rank tensors can be reduced to 36 independent coefficients. Therefore, similar to the single-column notation of $\in$ and σ, a 6×6 matrix representation of the stiffness C and compliance S tensors can be used. Due to crystal symmetry reasons, when considering silicon (which is a cubic crystal), only three independent stiffness and compliance coefficients need to be considered. The 6×6 matrix representation of the stiffness C and compliance S tensors can therefore be simplified to Eq. (7.12) and Eq. (7.13) [98]:

$$\begin{bmatrix} \sigma_1 \\ \sigma_2 \\ \sigma_3 \\ \sigma_4 \\ \sigma_5 \\ \sigma_6 \end{bmatrix} = \begin{bmatrix} C_{11} & C_{12} & C_{12} & 0 & 0 & 0 \\ C_{12} & C_{11} & C_{12} & 0 & 0 & 0 \\ C_{12} & C_{12} & C_{11} & 0 & 0 & 0 \\ 0 & 0 & 0 & C_{44} & 0 & 0 \\ 0 & 0 & 0 & 0 & C_{44} & 0 \\ 0 & 0 & 0 & 0 & 0 & C_{44} \end{bmatrix} \begin{bmatrix} \in_1 \\ \in_2 \\ \in_3 \\ \in_4 \\ \in_5 \\ \in_6 \end{bmatrix} \tag{7.12}$$

$$\begin{bmatrix} \in_1 \\ \in_2 \\ \in_3 \\ \in_4 \\ \in_5 \\ \in_6 \end{bmatrix} = \begin{bmatrix} S_{11} & S_{12} & S_{12} & 0 & 0 & 0 \\ S_{12} & S_{11} & S_{12} & 0 & 0 & 0 \\ S_{12} & S_{12} & S_{11} & 0 & 0 & 0 \\ 0 & 0 & 0 & S_{44} & 0 & 0 \\ 0 & 0 & 0 & 0 & S_{44} & 0 \\ 0 & 0 & 0 & 0 & 0 & S_{44} \end{bmatrix} \begin{bmatrix} \sigma_1 \\ \sigma_2 \\ \sigma_3 \\ \sigma_4 \\ \sigma_5 \\ \sigma_6 \end{bmatrix} \tag{7.13}$$

When considering homogeneous, isotropic materials, Young's modulus E (expressed in Pa), and Poisson ratio v (dimensionless) are often considered to characterize mechanical properties. Orientation-dependent E and v coefficient can also be defined for anisotropic materials (as silicon) by analogy. They are related to the compliance constants. For homogeneous, isotropic materials following the linear elastic theory, Hooke's law can be equivalently expressed using Young's modulus E and Poisson ratio v, or the Lamé coefficients λ and μ, defined in Eq. (7.14) and Eq. (7.15). Likewise, the parameters of piezoresistivity of the silicon are strongly dependent on the crystallographic orientation. These parameter values are defined below, linking the constraints in silicon with the variation in the mobility of the charge carriers.

$$\lambda = \frac{vE}{(1+v)(1-2v)} \tag{7.14}$$

$$\mu = \frac{E}{2(1+v)} \tag{7.15}$$

7.8 Piezoresistivity

Piezoresistivity theory describes stress-induced resistivity changes [98]. In strained silicon, the resistivity can be expressed as a second-rank tensor ρ and related to the stress tensor σ by the piezoresistivity fourth-rank tensor π (assuming infinitesimal strain), in Eq. (7.16). As the resistivity tensor ρ is symmetric (and also the stress tensor σ), it can be expressed in a single-column form, and the piezoresistivity tensor π can be reduced in a 6×6 matrix form, leading to the reduced equation (Eq. (7.17)).

$$\frac{\partial \rho_{ij}}{\rho_0} = \pi_{ijkl}\sigma_{kl} \tag{7.16}$$

$$\frac{\partial \rho_i}{\rho_0} = \pi_{ij}\sigma_j = -\frac{\partial \mu_i}{\mu_0} \tag{7.17}$$

Numerical values of the piezoresistivity coefficients π_{ij} associated with an electron inversion layer of silicon are given by Eqs. (7.18–7.20). These values are in the [100], [010], [001] coordinate system. Transistors feature a channel direction along with the <110> crystal orientation. As a result, the piezoresistivity coefficients referred to as longitudinal π_L, transverse π_T, and out-of-plane $\pi_\perp$ are expressed in this coordinate system in equations (Eqs. (7.18–7.20)).

$$\pi_{11} = -84.0 \times 10^{-11}\,\text{Pa}^{-1} \tag{7.18}$$

$$\pi_{12} = -34.0 \times 10^{-11}\,\text{Pa}^{-1} \tag{7.19}$$

$$\pi_{44} = -17.0 \times 10^{-11}\,\text{Pa}^{-1} \tag{7.20}$$

$$\pi_L = \frac{1}{2}(\pi_{11} + \pi_{12} + \pi_{44}) \tag{7.21}$$

$$\pi_T = \frac{1}{2}(\pi_{11} + \pi_{12} - \pi_{44}) \tag{7.22}$$

$$\pi_\perp = \pi_{12} \tag{7.23}$$

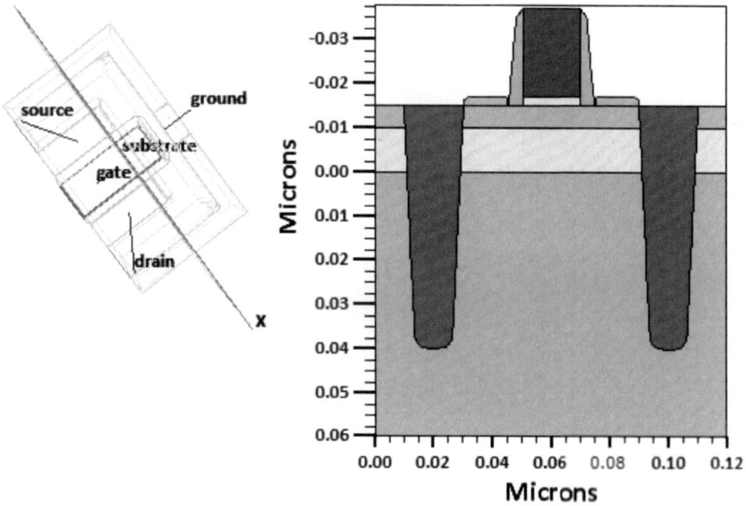

Figure 7.16 2D cut plane has been taken on the XZ plane for FDSOI n-MOSFET in 3D.

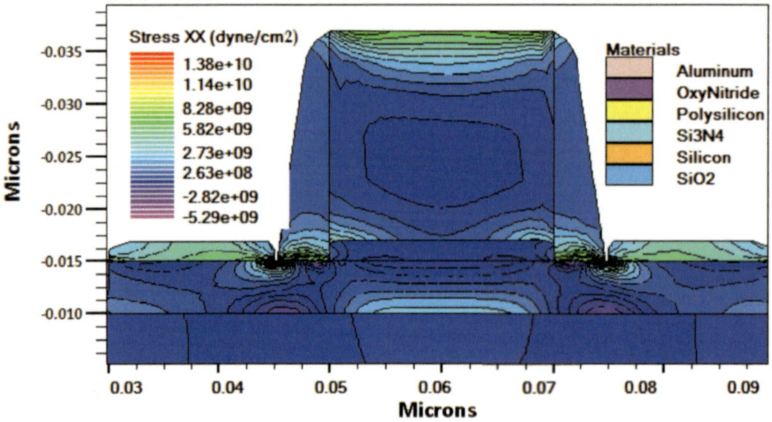

Figure 7.17 Contour plot: XX-stress. 2D profiles shown are taken on XZ-plane at $y = 0$.

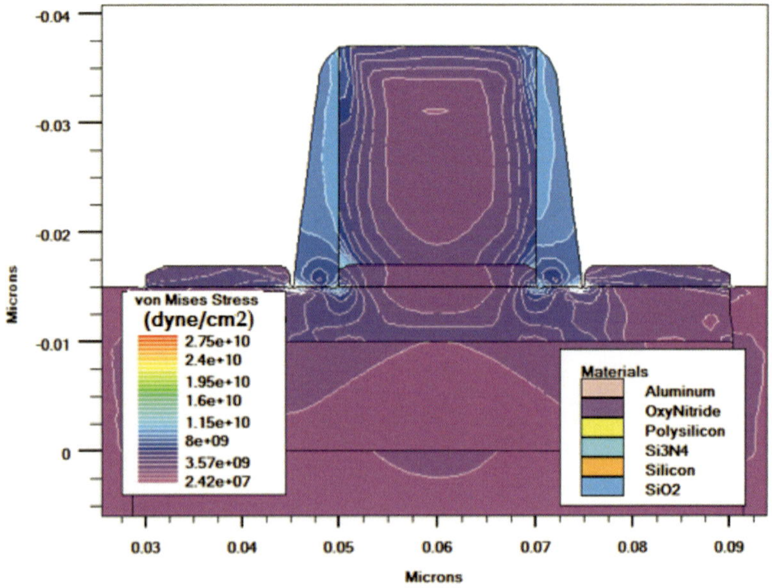

Figure 7.18 von Mises stress contour plot. 2D contour shown is taken on XZ-plane at $y = 0$.

7.9 Performance of Deformed RF CMOS Transistors

A commonly used technique for the analyses of flexible electronic devices involves cylindrical bending. The geometrical description of cylindrical bending is shown in Figure 7.14. Cylindrical bending provides high-performance stability [29, 99]. Using neutral plane engineering, it is possible to accommodate aggressive bending configurations while maintaining a minimal strain level in the active channel of transistors. The thickness of the flexible substrate is selected to reduce the strain level developed in the silicon active layer under bending conditions. Electrical properties of cylindrically bent FDSOI CMOS devices will be compared with electrical measurements performed on CMOS devices before bending.

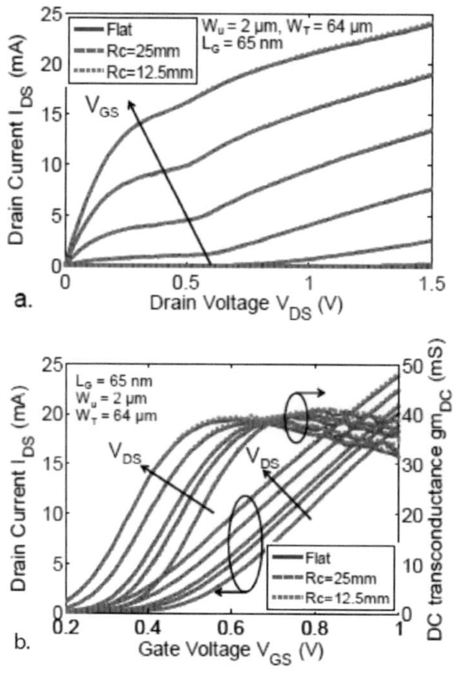

Figure 7.19 DC characteristics of flexible n-MOSFETs under cylindrical bending: (a) I_{DS}–V_{DS} characteristics, and (b) I_{DS}–V_{GS} and g_mDC, measured flatwise (blue lines), under flexure on semi-cylinders with a 25 mm curvature radius (green dashed lines), and a 12.5 mm curvature radius (red dotted lines).

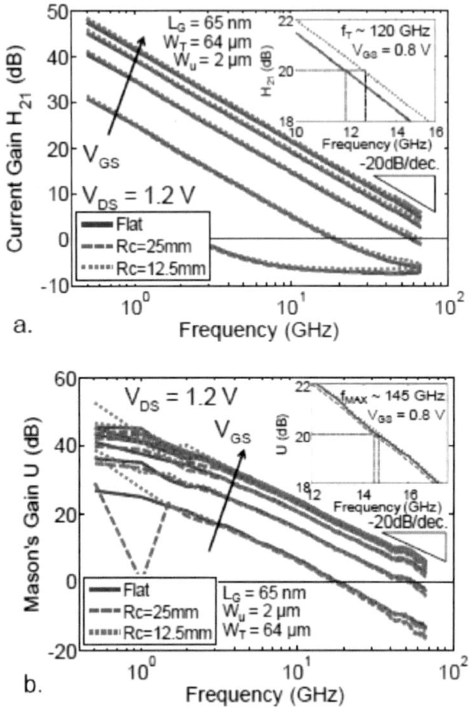

Figure 7.20 HF characteristics of flexible n-MOSFETs under cylindrical bending: (a) H_{21} and (b) U, measured for the flat device, and under bending. In all cases, a good agreement is observed between the measurements and the theoretical −20 dB/dec slope.

Static characteristics presented in Figure 7.19 (a) and Figure 7.19 (b) show I_{DS}–V_{DS}, I_{DS}–V_{GS}, and g_mDC characteristics measured on a flexible transistor (featuring a gate length of 65 nm, a unitary gate finger width of 2 µm and a total width of 64 µm) both on flat and bent configurations. The little discrepancy between DC characteristics measured in flat and bent configurations can be noticed from these two plots. Measurement data in the flat configuration are represented by solid blue lines, whereas characteristics of bent MOSFETs on cylinders with a 25 mm and 12.5 mm curvature radius are drawn in green dashed and red dotted lines, respectively. Figure 7.20(a) and Figure 7.20(b) show almost identical unity H_{21} and U on flat and bent configurations. An f_T ~ 120 GHz and f_{max} ~ 145 GHz are observed for flat conditions, and after cylindrical bending with a 25 mm and a 12.5 mm curvature radii. It is important to note that the plots follow the theoretical slope of −20 dB/dec showing the reliability of the extracted frequencies f_T and $f_{\max}$. These observations on the static and microwave characteristics validate that the neutral plan engineering analysis [100] results in an acceptable level of the variations of electrical properties.

7.10 Summary

As discussed in this chapter, both the silicon-based micro- and nano-electronic industries and the organic and printed electronics community are looking forward to bendable, or foldable devices and circuits. The increasing interest in higher performance (in terms of frequency and power consumption) and more complex devices are now feasible. Several fabrication strategies to enable the realization of devices and circuits combining high electrical performance and high mechanical bendability are demonstrated. It may be concluded that the field of flexible electronics could benefit from the high potential of existing rigid technologies. The transfer-bonding leads to the realization of flexible devices with little degradation. It is demonstrated that industrial SOI MOSFETs thinned to the micrometer-scale and transferred onto a flexible plastic substrate may be used to fabricate high-performance devices with high stability under bending. Slight variations of electrical properties (< 5 %) even after bending, pave the way to promising applications for foldable high-frequency systems taking the advantages of new heterogeneous integration toward a novel solution for high-performance, mechanically flexible, and performance stable integrated electronics.

References

1. R. R. Schaller, "Moore's law: Past, present, and future," *IEEE Spectr.*, vol. 34, no. 6, pp. 52–55, 57, Jun. 1997. doi: 10.1109/6.591665.
2. G. E. Moore, "Cramming more components onto integrated circuits," *Electronics*, vol. 38, no. 8, p. 114, 1965. doi: 10.1109/N-SSC.2006.4785860.
3. G. E. Moore, "No exponential is forever: But 'Forever' can be delayed! [semiconductor industry]," in *2003 IEEE International Solid-State*

Circuits Conference, 2003. Digest of Technical Papers. ISSCC., 2003, pp. 20–23, vol.1. doi: 10.1109/ISSCC.2003.1234194.

4. IEEE, "International Roadmap for Devices and Systems," 2018.
5. ITRS Executive Summary - 2015 Edition, "International Technology Roadmap for Semiconductors 2015," 2015.
6. IEEE, "International Roadmap for Devices and Systems (IRDS) 2020 Edition: Executive Summary."
7. B. Vigna, "More than Moore: Micro-machined products enable new applications and open new markets," in *IEEE InternationalElectron Devices Meeting, 2005. IEDM Technical Digest.*, 2005, p. 8. doi: 10.1109/IEDM.2005.1609252.
8. S. Kirchmeyer, "The OE-A roadmap for organic and printed electronics: Creating a guidepost to complex interlinked technologies, applications and markets," *Transl. Mater. Res.*, vol. 3, no. 1, p. 010301, Mar. 2016. doi: 10.1088/2053-1613/3/1/010301.
9. R. H. Reuss *et al.*, "Macroelectronics: Perspectives on technology and applications," *Proc. IEEE*, vol. 93, no. 7, pp. 1239–1256, 2005. doi: 10.1109/JPROC.2005.851237.
10. C. J. Bettinger, "Recent advances in materials and flexible electronics for peripheral nerve interfaces," *Bioelectron. Med.*, vol. 4, no. 1, Dec. 2018. doi: 10.1186/s42234-018-0007-6.
11. J. H. Koo, J. Seo, and T. Lee, "Nanomaterials on flexible substrates to explore innovative functions: From energy harvesting to bio-integrated electronics," *Thin Solid Films*, vol. 524. pp. 1–19, Dec. 01, 2012. doi: 10.1016/j.tsf.2012.10.001.
12. L. Bian, E. Zhu, J. Tang, W. Tang, and F. Zhang, "Recent progress in the design of narrow bandgap conjugated polymers for high-efficiency organic solar cells," *Progress in Polymer Science*, vol. 37, no. 9. Pergamon, pp. 1292–1331, Sep. 01, 2012. doi: 10.1016/j.progpolymsci.2012.03.001.
13. M. Hambsch *et al.*, "Uniformity of fully gravure printed organic field-effect transistors," *Mater. Sci. Eng. B Solid-State Mater. Adv. Technol.*, vol. 170, no. 1–3, pp. 93–98, Jun. 2010. doi: 10.1016/j.mseb.2010.02.035.
14. E. Halonen *et al.*, "Evaluation of printed electronics manufacturing line with sensor platform application," *2009 Eur. Microelectron. Packag. Conf. Rimini*, pp. 1–8, 2009.
15. Organic Electronics Association, OE-A, "Roadmap for Organic and Printed Electronics," vol. 5th Edition, 2013.

16. Y. Y. Noh, N. Zhao, M. Caironi, and H. Sirringhaus, "Downscaling of self-aligned, all-printed polymer thin-film transistors," *Nat. Nanotechnol.*, vol. 2, no. 12, pp. 784–789, Nov. 2007. doi: 10.1038/nnano.2007.365.

17. J. M. Park *et al.*, "Designing thermally stable conjugated polymers with balanced ambipolar field-effect mobilities by incorporating cyanovinylene linker unit," *Macromolecules*, vol. 49, no. 8, pp. 2985–2992, May 2016. doi: 10.1021/acs.macromol.5b02761.

18. Y. Yuan *et al.*, "Ultra-high mobility transparent organic thin film transistors grown by an off-centre spin-coating method," *Nat. Commun.*, vol. 5, no. 1, pp. 1–9, Jan. 2014. doi: 10.1038/ncomms4005.

19. A. Lecavelier des Etangs-Levallois *et al.*, "Invariance of DC and RF characteristics of mechanically flexible CMOS technology on plastic," Springer, Cham, 2014, pp. 81–103.

20. Y. Zhai, L. Mathew, R. Rao, D. Xu, and S. K. Banerjee, "High-performance flexible thin-film transistors exfoliated from bulk wafer," *Nano Lett.*, vol. 12, no. 11, pp. 5609–5615, Nov. 2012. doi: 10.1021/nl302735f.

21. K. Zhang, J. H. Seo, W. Zhou, and Z. Ma, "Fast flexible electronics using transferrable silicon nanomembranes," vol. 45, no. 14. IOP Publishing, 2012, p. 143001.

22. M. Kitamura and Y. Arakawa, "High current-gain cutoff frequencies above 10MHz in n-Channel C 60 and p-Channel pentacene thin-film transistors," in *Japanese Journal of Applied Physics*, Jan. 2011, vol. 50, no. 1 PART 3, p. 01BC01. doi: 10.1143/JJAP.50.01BC01.

23. J. N. Burghartz, W. Appel, C. Harendt, H. Rempp, H. Richter, and M. Zimmermann, "Ultra-thin chips and related applications, a new paradigm in silicon technology," in *2009 Proceedings of the European Solid State Device Research Conference*, 2009, pp. 29–36. doi: 10.1109/ESSDERC.2009.5331441.

24. N. Chimot, V. Derycke, M. F. Goffman, J. P. Bourgoin, H. Happy, and G. Dambrine, "Gigahertz frequency flexible carbon nanotube transistors," *Appl. Phys. Lett.*, vol. 91, no. 15, p. 153111, Oct. 2007. doi: 10.1063/1.2798583.

25. J. Vaillancourt *et al.*, "All ink-jet-printed carbon nanotube thin-film transistor on a polyimide substrate with an ultrahigh operating frequency of over 5 GHz," *Appl. Phys. Lett.*, vol. 93, no. 24, p. 243301, Dec. 2008. doi: 10.1063/1.3043682.

26. L. Sun *et al.*, "12-GHz thin-film transistors on transferrable silicon nanomembranes for high-performance flexible electronics," *Small*, vol. 6, no. 22, pp. 2553–2557, 2010. doi: 10.1002/smll.201000522.

27. H. C. Yuan, G. K. Celler, and Z. Ma, "7.8-GHz flexible thin-film transistors on a low-temperature plastic substrate," *J. Appl. Phys.*, vol. 102, no. 3, p. 034501, Aug. 2007. doi: 10.1063/1.2761782.

28. Y. Sun *et al.*, "Gigahertz operation in flexible transistors on plastic substrates," *Appl. Phys. Lett.*, vol. 88, no. 18, p. 034501, May 2006. doi: 10.1063/1.2198832.

29. A. Lecavelier Etangs-Levallois, "Report de technologie SOI-CMOS sur substrat flexible: une approche convergente vers les hautes fréquences et la stabilité des performances sous déformation mécanique," PhD Thesis, Univ. Lille, 2013.

30. H. Richter *et al.*, "Technology and design aspects of ultra-thin silicon chips for bendable electronics," in *2009 IEEE International Conference on Integrated Circuit Design and Technology, ICICDT 2009*, 2009, pp. 149–154. doi: 10.1109/ICICDT.2009.5166284.

31. V. Lakafosis, A. Rida, R. Vyas, L. Yang, S. Nikolaou, and M. M. Tentzeris, "Progress towards the first wireless sensor networks consisting of inkjet-printed, paper-based RFID-enabled sensor tags," *Proc. IEEE*, vol. 98, no. 9, pp. 1601–1609, 2010. doi: 10.1109/JPROC.2010.2049622.

32. Y. Chuo, D. Hohertz, C. Landrock, B. Omrane, K. L. Kavanagh, and B. Kaminska, "Large-area low-cost flexible plastic nanohole arrays for integrated bio-chemical sensing," *IEEE Sens. J.*, vol. 13, no. 10, pp. 3982–3990, 2013. doi: 10.1109/JSEN.2013.2262997.

33. G. T. Hwang *et al.*, "In vivo silicon-based flexible radio frequency integrated circuits monolithically encapsulated with biocompatible liquid crystal polymers," *ACS Nano*, vol. 7, no. 5, pp. 4545–4553, May 2013. doi: 10.1021/nn401246y.

34. D. H. Kim, R. Ghaffari, N. Lu, and J. A. Rogers, "Flexible and stretchable electronics for biointegrated devices," vol. 14. Annu Rev Biomed Eng, 2012, pp. 113–128.

35. S. H. Eom *et al.*, "High efficiency polymer solar cells via sequential inkjet-printing of PEDOT:PSS and P3HT:PCBM inks with additives," *Org. Electron.*, vol. 11, no. 9, pp. 1516–1522, Sep. 2010. doi: 10.1016/j.orgel.2010.06.007.

36. A. Lange, W. Schindler, M. Wegener, K. Fostiropoulos, and S. Janietz, "Inkjet printed solar cell active layers prepared from chlorine-free solvent systems," *Sol. Energy Mater. Sol. Cells*, vol. 109, pp. 104–110, Feb. 2013. doi: 10.1016/j.solmat.2012.10.011.

37. J. Lee *et al.*, "Stretchable semiconductor technologies with high areal coverages and strain-limiting behavior: Demonstration in high-

efficiency dual-junction GaInP/GaAs photovoltaics," *Small*, vol. 8, no. 12, pp. 1851–1856, Jun. 2012. doi: 10.1002/smll.201102437.

38. J. Zhang, G. Y. Tian, A. M. J. Marindra, A. I. Sunny, and A. B. Zhao, "A review of passive RFID tag antenna-based sensors and systems for structural health monitoring applications," *Sensors (Switzerland)*, vol. 17, no. 2. MDPI AG, p. 265, Feb. 2017. doi: 10.3390/s17020265.

39. A. Rida, L. Yang, R. Vyas, and M. M. Tentzeris, "Conductive inkjet-printed antennas on flexible low-cost paper-based substrates for RFID and WSN applications," *IEEE Antennas and Propagation Magazine*, vol. 51, no. 3. pp. 13–23, 2009. doi: 10.1109/MAP.2009.5251188.

40. R. Vyas *et al.*, "Paper-based RFID-enabled wireless platforms for sensing applications," in *IEEE Transactions on Microwave Theory and Techniques*, Jan. 2009, vol. 57, no. 1, pp. 1370–1382. doi: 10.1109/TMTT.2009.2017317.

41. A. Van Der Horst, D. Van Der Voort, B. Mimoun, M. Rutten, F. van de Vosse, and R. Dekker, "A novel flexible thermoelectric sensor for intravascular flow assessment," *IEEE Sens. J.*, vol. 13, no. 10, pp. 3883–3891, 2013. doi: 10.1109/JSEN.2013.2252461.

42. J. Liang, L. Li, X. Niu, Z. Yu, and Q. Pei, "Elastomeric polymer light-emitting devices and displays," *Nat. Photonics*, vol. 7, no. 10, pp. 817–824, Oct. 2013. doi: 10.1038/nphoton.2013.242.

43. B. Mimoun, V. Henneken, A. Van Der Horst, and R. Dekker, "Flex-to-rigid (F2R): A generic platform for the fabrication and assembly of flexible sensors for minimally invasive instruments," *IEEE Sens. J.*, vol. 13, no. 10, pp. 3873–3882, 2013. doi: 10.1109/JSEN.2013.2252613.

44. S. Choi, H. Lee, R. Ghaffari, T. Hyeon, and D.-H. H. Kim, "Recent advances in flexible and stretchable bio-electronic devices integrated with nanomaterials," *Adv. Mater.*, vol. 28, no. 22, pp. 4203–4218, Jun. 2016. doi: 10.1002/adma.201504150.

45. L. Xie, G. Yang, M. Mäntysalo, L. L. Xu, F. Jonsson, and L. R. Zheng, "Heterogeneous integration of bio-sensing system-on-chip and printed electronics," *IEEE J. Emerg. Sel. Top. Circuits Syst.*, vol. 2, no. 4, pp. 672–682, 2012. doi: 10.1109/JETCAS.2012.2223554.

46. T. Sekitani *et al.*, "Flexible organic transistors and circuits with extreme bending stability," *Nat. Mater.*, vol. 9, pp. 1015–1022, 2010. doi: 10.1038/nmat2896.

47. IEEE, "International Roadmap for Devices and Systems, 2020 Update, Beyond CMOS," 2020.

48. C. Wang *et al.*, "Self-aligned, extremely high frequency III-V metal-oxide semiconductor field-effect transistors on rigid and flexible substrates," *Nano Lett.*, vol. 12, no. 8, pp. 4140–4145, 2012. doi: 10.1021/nl301699k.

49. H. L. Kao *et al.*, "Low noise RF MOSFETs on flexible plastic substrates," *IEEE Elecron Dev. Lett.*, vol. 26, no. 7, pp. 489–491, 2005. doi: 10.1109/LED.2005.851238.

50. H. L. Kao *et al.*, "DC-RF performance improvement for strained 0.13 μm MOSFETs mounted on a flexible plastic substrate," in *EEE MTT-S International Microwave Symposium Digest*, 2006, pp. 2043–2046. doi: 10.1109/MWSYM.2006.249856.

51. D. Tonnies *et al.*, "Introduction of a unified equipment platform for UV initiated processes in conjunction with the application of electrostatic carriers as thin wafer handling solution," *European Microelectronics and Packaging Conference, 2009*. pp. 1–6, 2009, Accessed: Dec. 05, 2020.

52. J. Hermanowski, "Thin wafer handling: Study of temporary wafer bonding materials and processes," 2009. doi: 10.1109/3DIC.2009.5306550.

53. T. Itabashi, "Integrated materials enabling TSV/3D-TSV," in *Proceedings - Electronic Components and Technology Conference*, 2009, pp. 1759–1763. doi: 10.1109/ECTC.2009.5074254.

54. T. Itabashi and M. P. Zussman, "High temperature resistant bonding solutions enabling thin wafer processing (characterization of polyimide base temporary bonding adhesive for thinned wafer handling)," in *Proceedings - Electronic Components and Technology Conference*, 2010, pp. 1877–1880. doi: 10.1109/ECTC.2010.5490702.

55. J. Charbonnier *et al.*, "Integration of a temporary carrier in a TSV process flow," in *Proceedings - Electronic Components and Technology Conference*, 2009, pp. 865–871. doi: 10.1109/ECTC.2009.5074114.

56. S. Pargfrieder, P. Kettner, M. Privett, and J. Ting, "Temporary bonding and debonding enabling TSV formation and 3D integration for ultra-thin wafers," Jan. 2009, pp. 1301–1305. doi: 10.1109/eptc.2008.4763610.

57. B. Dang *et al.*, "CMOS compatible thin wafer processing using temporary mechanical wafer, adhesive and laser release of thin chips/wafers for 3D integration," in *Proceedings - Electronic Components and Technology Conference*, 2010, pp. 1393–1398. doi: 10.1109/ECTC.2010.5490820.

58. K. Tamura, K. Nakada, N. Taneichi, P. Andry, J. Knickerbocker, and C. Rosenthal, "Novel adhesive development for CMOS-compatible thin wafer handling," in *Proceedings - Electronic Components and Technology Conference*, 2010, pp. 1239–1244. doi: 10.1109/ECTC.2010.5490859.

59. D.-H. Kim *et al.*, "Materials for multifunctional balloon catheters with capabilities in cardiac electrophysiological mapping and ablation therapy," *Nat. Mater.*, vol. 10, pp. 316–323, 2011. doi: 10.1038/nmat2971 ID.

60. D.-H. Kim *et al.*, "Dissolvable films of silk fibroin for ultrathin conformal bio-integrated electronics," *Nat. Mater.*, vol. 9, pp. 511–517, 2010. doi: 10.1038/nmat2745.

61. J. Vaillancourt, X. Lu, X. Han, and D. C. Janzen, "High-speed thin-film transistor on flexible substrate fabricated at room temperature," *Electron. Lett.*, vol. 42, no. 23, pp. 1365–1366, 2006. doi: 10.1049/el:20062295.

62. X. Han, D. C. Janzen, J. Vaillancourt, and X. Lu, "Printable high-speed thin-film transistor on flexible substrate using carbon nanotube solution," *Micro Nano Lett.*, vol. 2, no. 4, pp. 96–98, 2007. doi: 10.1049/mnl:20070033.

63. J.-H. H. Ahn *et al.*, "High-speed mechanically flexible single-crystal silicon thin-film transistors on plastic substrates," *IEEE Electron Device Lett.*, vol. 27, no. 6, pp. 460–462, Jun. 2006. doi: 10.1109/LED.2006.874764.

64. H. C. Yuan and Z. Ma, "Microwave thin-film transistors using Si nanomembranes on flexible polymer substrate," *Appl. Phys. Lett.*, vol. 89, no. 21, p. 212105, Nov. 2006. doi: 10.1063/1.2397038.

65. I. Ferain, C. A. Colinge, and J. P. Colinge, "Multigate transistors as the future of classical metal-oxide-semiconductor field-effect transistors," *Nature*, vol. 479, no. 7373, Nature Publishing Group, pp. 310–316, Nov. 17, 2011. doi: 10.1038/nature10676.

66. T. Skotnicki, *Competitive SOC with UTBB SOI*, 2011, pp. 1–61.

67. N. Planes *et al.*, "28 nm FDSOI technology platform for high-speed low-voltage digital applications," in *Digest of Technical Papers - Symposium on VLSI Technology*, 2012, pp. 133–134. doi: 10.1109/VLSIT.2012.6242497.

68. C. Gallon *et al.*, "Ultra-thin fully depleted SOI devices with thin BOX, ground plane and strained liner booster," in *Proceedings - IEEE International SOI Conference*, 2006, pp. 17–18. doi: 10.1109/SOI.2006.284410.

69. C. Fenouillet-Beranger *et al.*, "Impact of a 10nm ultra-thin BOX (UTBOX) and ground plane on FDSOI devices for 32nm node and below," in *ESSDERC 2009 - Proceedings of the 39th European Solid-State Device Research Conference*, 2009, pp. 89–92. doi: 10.1109/ESSDERC.2009.5331588.

70. C. Fenouillet-Beranger *et al.*, "FDSOI devices with thin BOX and ground plane integration for 32 nm node and below," *Solid. State. Electron.*, vol. 53, no. 7, pp. 730–734, 2009. doi: 10.1016/j.sse.2009.02.009.

71. A. Asenov, "Random dopant induced threshold voltage lowering and fluctuations in sub-0.1 /spl mu/m MOSFET's: A 3D 'atomistic' simulation study," *IEEE Trans. Electron Devices*, vol. 45, no. 12, pp. 2505–2513, 1998. doi: 10.1109/16.735728.

72. T. A. Karatsori *et al.*, "Analytical compact model for lightly doped nanoscale ultrathin-body and box SOI MOSFETs with back-gate control," *IEEE Trans. Electron Devices*, vol. 62, no. 10, pp. 3117–3124, 2015.

73. D. Flandre *et al.*, "Fully-depleted SOI CMOS technology for lowvoltage low-power mixed digital/analog/microwave circuits," *Analog Integr. Circuits Signal Process.*, vol. 21, pp. 213–228, 1999. doi: 10.1023/A:1008321919587.

74. T. Ernst and S. Cristoloveanu, "Buried oxide fringing capacitance: A new physical model and its implication on SOI device scaling and architecture," in *1999 IEEE International SOI Conference, SOI 1999 - Proceedings*, 1999, pp. 38–39. doi: 10.1109/SOI.1999.819847.

75. T. Ernst, C. Tinella, C. Raynaud, and S. Cristoloveanu, "Fringing fields in sub-0.1 μm fully depleted SOI MOSFETs: Optimization of the device architecture," *Solid. State. Electron.*, vol. 46, no. 3, pp. 373–378, Mar. 2002. doi: 10.1016/S0038-1101(01)00111-3.

76. V. Barral *et al.*, "Strained FDSOI CMOS technology scalability down to 2.5nm film thickness and 18nm gate length with a TiN/HfO_2 gate stack," in *Technical Digest - International Electron Devices Meeting, IEDM*, 2007, pp. 61–64. doi: 10.1109/IEDM.2007.4418863.

77. Y. Kim *et al.*, "Challenges for nanoscale MOSFETs and emerging nanoelectronics," *Trans. Electr. Electron. Mater.*, vol. 11, no. 3, pp. 93–105, 2010. doi: 10.4313/TEEM.2010.11.3.093.

78. A. Soussou, "Modeling and characterization of electrical effects of Ge integration in Metal/High-k/SiGe MOS structures," PhD Thesis, Université Grenoble Alpes, 2014.

79. J. P. Colinge and J.-P. Colinge, *Silicon-on-Insulator Technology: Materials to VLSI*, Springer Berlin Heidelberg, 1997.

80. G. K. Celler and S. Cristoloveanu, "Frontiers of silicon-on-insulator," *J. Appl. Phys.*, vol. 93, no. 9, pp. 4955–4978, May 2003. doi: 10.1063/1.1558223.

81. L. Jastrzebski, J. F. Corboy, and R. Soydan, "Issues and problems involved in selective epitaxial growth of silicon for SOI fabrication," *J. Electrochem. Soc.*, vol. 136, no. 11, pp. 3506–3513, Nov. 1989. doi: 10.1149/1.2096495.

82. J. P. Colinge *et al.*, "Fabrication of thin silicon-on-insulator flims using laser recrystallisation," *Electron. Lett.*, vol. 21, no. 23, pp. 1102–1103, 1985. doi: 10.1049/el:19850782.

83. K. Izumi, M. Doken, and H. Ariyoshi, "C.m.o.s. devices fabricated on buried sio2 layers formed by oxygen implantation into silicon," *Electron. Lett.*, vol. 14, no. 18, pp. 593–594, Aug. 1978. doi: 10.1049/el:19780397.

84. P. L. F. Hemment *et al.*, "Novel dielectric/silicon planar structures formed by ion beam synthesis," *Nucl. Inst. Methods Phys. Res. B*, vol. 21, no. 1–4, pp. 129–133, Jan. 1987. doi: 10.1016/0168-583X(87)90811-1.

85. M. Kimura, K. Egami, M. Kanamori, and T. Hamaguchi, "Epitaxial film transfer technique for producing single crystal Si film on an insulating substrate," *Appl. Phys. Lett.*, vol. 43, no. 3, pp. 263–265, Aug. 1983. doi: 10.1063/1.94320.

86. W. P. Maszara, "Silicon-on-insulator by wafer bonding: A review," *J. Electrochem. Soc.*, vol. 138, no. 1, pp. 341–347, Jan. 1991. doi: 10.1149/1.2085575.

87. M. Bruel, "Silicon on insulator material technology," *Electron. Letts.*, vol. 31, no. 14, pp. 1201–1202, Jul. 1995. doi: 10.1049/el:19950805.

88. M. Bruel, B. Aspar, and A. J. Auberton-Hervé, "Smart-cut: A new silicon on insulator material technology based on hydrogen implantation and wafer bonding," *Japanese J. Appl. Physics, Part 1 Regul. Pap. Short Notes Rev. Pap.*, vol. 36, no. 3 SUPPL. B, pp. 1636–1641, Mar. 1997. doi: 10.1143/jjap.36.1636.

89. G. K. Celler, "SOI technology driving the 21st century ubiquitous electronics," *ECS Trans.*, vol. 19, no. 4, pp. 3–14, Dec. 2019. doi: 10.1149/1.3117387.

90. A. Lecavelier Des Etangs-Levallois *et al.*, "Radio-frequency and low noise characteristics of SOI technology on plastic for flexible electronics," *Solid. State. Electron.*, vol. 90, pp. 73–78, Dec. 2013. doi: 10.1016/j.sse.2013.02.049.

91. N. Waldhoff, C. Andrei, D. Gloria, S. Lepilliet, F. Danneville, and G. Dambrine, "Improved characterization methology for MOSFETs up to 220 GHz," in *IEEE Transactions on Microwave Theory and*

Techniques, Jan. 2009, vol. 57, no. 1, pp. 1237–1243. doi: 10.1109/TMTT.2009.2017359.

92. S. P. Voinigescu, S. W. Tarasewicz, T. MacElwee, and J. Ilowski, "Assessment of the state-of-the-art 0.5 μm bulk CMOS technology for RF applications," in *Technical Digest - International Electron Devices Meeting*, 1995, pp. 721–724. doi: 10.1109/iedm.1995.499320.

93. Q. Huang, F. Piazza, P. Orsatti, and T. Ohguro, "The impact of scaling down to deep submicron on CMOS RF circuits," *IEEE J. Solid-State Circuits*, vol. 33, no. 7, pp. 1023–1036, Jul. 1998. doi: 10.1109/4.701249.

94. Silvaco, International, VictoryDevice User Manual, 2018.

95. G. Baccarani, E. Gnani, A. Gnudi, S. Reggiani, and M. Rudan, "Theoretical foundations of the quantum drift-diffusion and density-gradient models," *Solid. State. Electron.*, vol. 52, no. 4, pp. 526–532, 2008. doi: https://doi.org/10.1016/j.sse.2007.10.051.

96. A. L. Des Etangs-Levallois *et al.*, "150-GHz RF SOI-CMOS technology in ultrathin regime on organic substrate," *IEEE Electron Device Lett.*, vol. 32, no. 11, pp. 1510–1512, Nov. 2011. doi: 10.1109/LED.2011.2166241.

97. Y. Tagro *et al.*, "High frequency noise potentialities of reported CMOS 65 nm SOI technology on flexible substrate," in *2012 IEEE 12th Topical Meeting on Silicon Monolithic Integrated Circuits in RF Systems, SiRF 2012 - Digest of Papers*, 2012, pp. 89–92. doi: 10.1109/SiRF.2012.6160147.

98. Y. Sun, S. E. Thompson, and T. Nishida, *Strain effect in semiconductors: Theory and device applications*, Springer, US, 2010.

99. R. Chen, J. Chow, C. Taylor, J. Meth, and S. Sitaraman, "Adaptive curvature flexure test to assess flexible electronic systems," in *2018 IEEE 68th Electronic Components and Technology Conference (ECTC)*, 2018, pp. 236–242. doi: 10.1109/ECTC.2018.00044.

100. S.-Y. Y. Kim, J. H. Bong, C. Kim, W. S. Hwang, T.-S. S. Kim, and B. J. Cho, "Mechanical stability analysis via neutral mechanical plane for high-performance flexible Si nanomembrane FDSOI device," *Adv. Mater. Interfaces*, vol. 4, no. 21, p. 1700618, Nov. 2017. doi: 10.1002/admi.201700618.

Chapter 8

Simulation at Atomic Scale

Arup Ratan Saha
Synopsys India, Hyderabad 500032, Andhra Pradesh, India
saha.arup2020@gmail.com

Electronic design automation (EDA) has been the key enabler of the fabless semiconductor industry. However, EDA and intellectual property (IP) are not so much important as providing big profit for leading semiconductor manufacturers. Technology CAD (TCAD) fulfills a big gap between the design group and process engineers when fabrication complexities arise in the semiconductor industry involving the risk of higher cost and manufacturability. The history of the EDA industry may be summarized in different phases.

From the earliest days of integrated circuits until the mid-1970s, chips were designed manually, with no automation of any part of the design, layout, verification, or mask preparation. As the chips become complex, it became clear that some method of automation would be required. Once the industry had a design flow that was bifurcated into front-end and back-end, chips continued to grow in complexity. Both factors led to increased specialization of all aspects of the application-specific integrated circuits (ASIC) flow and opened

Fabless Semiconductor Manufacturing: In the Era of Internet of Things
Edited by Chinmay K. Maiti

ISBN 978-981-4968-27-0 (Hardcover), 978-1-003-31497-4 (eBook)
www.jennystanford.com

the door for a new crop of EDA companies. By the late 1980s, a lot of semiconductor designs were done using the ASIC methodology, with the system companies doing the front-end and the semiconductor companies doing the back-end physical design.

The EDA industry was driven by two factors. Firstly, it became unfeasible for every semiconductor company to develop all at their place and route tools internally, so companies discontinued their internal development in favor of buying EDA tools from external EDA companies. It became possible for system companies to do their back-end design in addition to the front-end to complete the physical layout of their designs. The EDA industry was in the transition to synthesis-based design, which only became mainstream in the mid-1990s. The graphical schematic-based design was replaced, except for analog and some other specialized areas, by language-based synthesis with astounding leaps in design productivity.

The final era of full-service EDA companies belongs to a strategy called grow by accumulating fundamental startup companies. Moreover, companies now focus on design-technology-co-optimization and advance R&D on atomic-level simulation. As EDA company-built portfolios of software to cover the entire design flow with best-in-class point tools from several companies. That is why the fundamental simulation (ab-initio) is now also a part of the EDA industry to expand business on the newly growing market for atomic-scale devices.

8.1 Semiconductors at Atomic Level

8.1.1 Single and Multi-Electron System

The one-electron system is the simplest bound system that occurs in nature and it is the simplest case to study atomic behavior quantum mechanically. Using eigenvalues and functions, obtained by solving the Schrodinger equation of one electron atom, a few basic properties of the atom can be found out: (a) the probability density-function (DF) that gives us a detailed picture of the structure of the atom, (b) the orbital angular momentum of the atom, (c) the electron spin and other relativity effects, and (d) the rate at which the atom makes the transition from its excited states to its ground state.

The one-electron system consists of positively charged nuclei and a negatively charged electron moving under the influence of Coulomb attraction. The 3D character of the atom causes difficulties due to its complicated nature in the mathematical modeling domain. However, the procedure is a straightforward extension of the simple 1D system. The fact that one-electron atom contains two particles causes no difficulty in case one considers only the reduced mass technique. In classical mechanics, the motion of the reduced mass electron about the stationary nucleus in the model atom duplicates the motion of the electron relative to the nucleus in the actual atom.

8.1.2 Molecules

Two and more atoms can combine to form a stable molecule and the details of interatomic forces necessary to bind atoms into the molecule, the nature of energy level, and spectra of the molecule are important before one understands molecular level statistics. A very large number of atoms may join to make a solid, just like a few do it to form a molecule. The phenomena of molecular binding are very relevant to the properties of solid. A well-established and rapidly expanding field of science is molecular astronomy, which involves the measurement of molecular spectra originating from interstellar or intergalactic matter. Measurement of molecular spectra is the primary source of information about important properties of nuclei as well as a molecule.

A molecule is a stable arrangement of a group of nuclei and electrons. The exact arrangement can be described by electromagnetic force and quantum mechanical laws. Another picture of a molecule as a stable structure formed by the association of two or more atoms. In this case, the atoms retain their identity, whereas in the first case it is not. When a molecule is formed by using two atoms, the inner shell electron of each atom may be tightly bound to the nucleus or barely distributed. The outermost loosely bound electrons are known as valance electrons, are influenced by particles. The wave function is significantly modified when the atoms are brought together. The interatomic force is the electromagnetic interaction; hence the valance electron plays the important role in molecular binding. There are two main types of molecular binding, i.e., the ionic bond and the covalent bond. NaCl is a molecule held together by an ionic bond, whereas binding in an H_2 molecule is a covalent bond having the energy of the Coulombic interaction of ions.

8.2 Elementary Particles and Nuclear Models

The study of all elementary particles is important to understand the border between the interaction fields of nuclear physics and elementary particles. The history of quantum physics revealed that the macroscopic structure of matter, i.e., atoms are composed of nuclei and electrons and considered to be elementary particles. All matters appeared to be composed of a variable composition of a small number of elementary particles, the neutrino, the proton, and the electron. Later it was found that there are also muons and π-mesons. Present-day researchers noticed several discoveries of particles related to other mesons, neutrons, and protons.

To understand the nuclear model, the user should gather some information about nuclear forces acting between nucleons. Similarly, to model nuclei and corresponding forces, the user must consider the following facts: (a) nucleon forces are strong, (b) these forces are applicable in short-range, (c) nucleon forces are attractive, and (d) nucleon forces are charge independent.

8.3 Models for Electron Dynamics

Metallic and semiconductor nano-objects operate in different regimes, as the electron density is larger than several orders of magnitudes of the former. Consequently, the typical time, space, and energy scales can be different. However, if one considers the effective electron mass and dielectric constant, the relevant dimensionless parameters turn out to be rather similar. It is, therefore, not surprising that the electron dynamics of both types of nanostructures can be described using similar models. A global overview of the various relevant models for atomistic simulation is currently used in the industry [1], as shown in Figure 8.1.

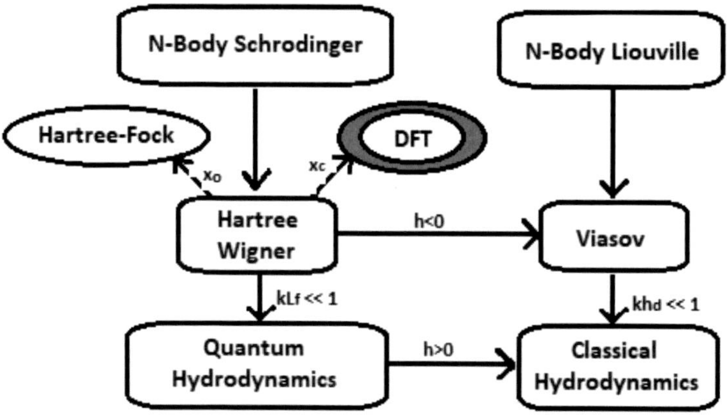

Figure 8.1 Overview of N-body interaction and model cards.

8.4 Ab-Initio Calculation: DFT and SIESTA Models

Computer simulation code SIESTA [2] was developed as an approach to compute the electronic properties and perform atomistic simulations of complex materials from the first-principles study. Very large systems, with an unprecedented number of atoms, can be studied while keeping the computational cost at a reasonable level [3]. This has made it a widely used tool for the study of materials. It has been applied to a large variety of systems, including surfaces, adsorbates, nanotubes, nanoclusters, biological molecules, amorphous semiconductors, ferroelectric films, and low-dimensional metals.

The contribution of first-principle calculations to several fields in physics, chemistry, and material science, and the recent progress in biology are more important than ever. On one side a steady increase in computational power, on the other hand, continuous progress in methodology in terms of accuracy and efficiency can improve programming algorithms and approximations. As larger and more complex systems are falling within the range of applications, some fundamental barriers are always present. One of such barriers is the scaling of computational efforts with system size scaling. In density function theory (DFT) [4], quantum mechanical formulation of the electronic structure of atomic system scale as the cube of the number of atoms or electrons present in the system. This assumption makes it very difficult to reach a system size larger than a few hundred atoms, hence it becomes a barrier for the study of problems in nanoscale materials.

The key for achieving linear scaling is the explicit use of locality, which is the property of a region in the system with perturbation and is adequately far from the system. The well-known computer algorithm, the SIESTA method, is based on DFT and can be used in both local density approximation (LDA) and generalized gradient approximation (GGA) functions [5]. The function specifically includes spin polarization, collinear, and non-collinear effects. The DFT equations are solved using the self-consistent field method, whereas the one particle Schrodinger equation is solved the energy and density matrix for the ground state. This is performed by cube-scaling approximation (diagonalization) for the system with multiple atoms or with a linear scaling algorithm. Once the density matrix has been obtained, the self-consistent field procedure continues with the calculation of the new Hamiltonian matrix. The matrix elements of the different terms of that Hamiltonian are calculated in one of the two different ways. The terms that involved integral over two atoms are only computed as a function of the distance between centers of the system. The corresponding results are stored in a table to be interpolated later with very little use of time and memory of the computational resource. The other terms are calculated with the help of a uniform grid of the points in real space.

8.4.1 Case Study 1: Carbon Nanostructure Fullerenes (C_{60}) and Nanotubes

The discovery of C_{60} and other fullerenes and carbon nanotubes are the most incredible findings in the last decade. One of the first applications of the initial stage of SIESTA was on the prediction and solving of the shape of large hollow carbon fullerenes. The structure of nested fullerenes had been observed to be nearly perfect spherical while irradiated with a high-energy electron beam. The computer algorithm, SIESTA, is used to calculate the lowest energy structure of iso-hedral fullerenes up to C_{60}. After the simulation user can identify that the structure is polyhedral instead of spherical and this polyhedral character becomes pronounced as the cluster size increases.

8.4.2 Case Study 2: Properties of Nanotube

The structural, elastic, and vibrational properties of single-wall nanotubes with different radii and chirality can be analyzed using SIESTA. The elastic properties show a similar trend with a small deviation from ideal behavior. The strain energy and other fundamental properties calculated using the ab-initio method follow very closely the behavior predicted by the elastic theory. The calculated Young modulus does not show any appreciable difference with that of graphene and does not depend on the chirality or radius.

8.4.3 Case Study 3: The Electronic States in Finite Nanotube

The electronic properties of nanotubes exhibit a wide range of very interesting phenomena. The structure and topology of nanotubes, armchair (n, n) are metallic, while zigzag (n, 0) tubes can be a semiconductor. However, some phenomena are purely related to the nanoscale size of this material, such as the quantum confinement effect. A conducting armchair tube, cut into a finite length, should behave as a finite 1D box in which the conducting electrons are confined. This 1D quantum confinement should then yield electron states, which are standing waves in contrast to the running waves of an infinite system. Atomic simulation using SIESTA helps confirm the interpretation of the experimental results, in terms of the standing wave.

8.5 Time-Dependent Density Functional Theory

Calculating the electronic and magnetic properties of materials is not a simple task even in the ground state. Materials are composed of atoms held together by chemical bonds insured by the valence electrons. Involving so many particles gives rise to a complex many-body problem. One of the early proposed simplifications to deal with such complexity is the Born–Oppenheimer approximation (BOA). The BOA consists of dividing the total solid-state problem into two parts: the motion of the electrons in a stationary lattice and that of the ions in a uniform space charge of electrons. The total Hamiltonian *H*, which represents the total energy of a realistic system can be written as:

$$H = H_e + H_i + H_{e-i}, \tag{8.1}$$

where, H_e, H_i, and H_{e-i} are the electrons, the ions, and the electrons–ions interaction parts of the Hamiltonian, respectively. The solution of this Hamiltonian would predict any physical or chemical property of the material. This problem cannot be solved as so many approximations should be made to hope for some reasonable description of the properties of materials. The solution of this (*N*) many-body problem will be achieved by solving the following Schrodinger equation:

$$\left[T_e + T_n + V_e(r) + V_n(R) + V_e N(r,R)\right] * \Psi(x,R) = E\Psi(x,R), \tag{8.2}$$

where, *(r, R)* represents the electron positions and spin variables, and Ψ (*x*, *R*) is the eigenfunctions of *H* with eigenvalue *E*. In the above equation, the separation of the degrees of freedom of the electrons and that of the nucleus can be expressed with the following assumption:

$$\Psi(x,R) = \varphi(x,R).\chi(R), \tag{8.3}$$

where, $\chi(R)$ is a nuclear wave function and $\phi(x, R)$ is an electronic wave function that depends parametrically on the nuclear positions (*R*). This BOA is justified by the fact that nuclei are several thousand times heavier than electrons because a proton or a neutron is about 1,836 times more massive than an electron. The electrons follow the nuclear motion adiabatically, i.e., they are dragged along with the nuclei with almost no relaxation time. This is a reasonable approximation because the non-adiabatic effects that do not allow the electrons to follow in this instantaneous manner are usually small. Users can focus on the motion of the electrons, as described in Eq. (8.3), based on the Hartree–Fock (HF) approximation. Consider an electron gas that is embedded in a homogeneous, positively charged medium or a rigid lattice of positively charged ions; this problem is very difficult to solve because of the complexity of the electron–electron interaction. In the absence of this interaction, the many-body problem would decouple into a one-body problem, which describes the motion of an electron in an effective potential (the one-electron approximation).

8.6 Electron–Phonon Collision

Phonons can scatter through several mechanisms as they travel through the material. These scattering mechanisms are phonon–phonon scattering, phonon–impurity scattering, phonon–electron scattering, and phonon–boundary scattering. Each scattering mechanism can be characterized by a relaxation rate $1/\tau$, which is the inverse of the corresponding relaxation time. In crystalline materials, U-process is a scattering process that results in a wave vector and falls outside the first Brillouin zone. The periodic materials have a Brillouin zone, and any point outside the first Brillouin zone can also be expressed as a point inside the zone. The wave vector is then mathematically transformed to a point inside the first Brillouin zone. This transformation allows for scattering processes that would otherwise violate the conservation of momentum. For phonon–phonon scattering, effects by normal processes (processes that conserve the phonon wave vector, termed as N processes) are ignored in favor of Umklapp processes (U processes). Since normal processes vary linearly with $\omega$and Umklapp processes vary with ω^2, Umklapp scattering dominates only at high frequency. Phonon–electron scattering can also contribute when the material is heavily doped. The corresponding relaxation time is given as a function of parameters like n_e is conduction electrons concentration, ε is deformation potential, ρ is mass density, and m^* is effective electron mass [6]. It is usually assumed that contribution to thermal conductivity by phonon–electron scattering is negligible. All scattering processes can be combined using Matthiessen's rule. Then the combined relaxation time τ_c can be written as:

$$\frac{1}{\tau_C} = \frac{1}{\tau_U} + \frac{1}{\tau_M} + \frac{1}{\tau_B} + \frac{1}{\tau_{ph-e}}. \tag{8.4}$$

8.7 Particle Dynamics for Metallic and Semiconductor Nanostructure

Presently, the modeling and numerical simulation of the nonlinear electron dynamics and transport in metallic and semiconductor nanostructures, such as metallic nanoparticles, thin films, quantum wells, and quantum dots, represents a considerable challenge for both fundamental and applied condensed matter physics. Experimentally, it is now possible to monitor the femtosecond dynamics of an electron gas confined in metallic nanostructures such as thin films, nanotubes [6], metal clusters, and nanoparticles. Therefore, meaningful comparisons between experimental measurements and numerical simulations based on microscopic theories are becoming possible.

The same type of collective electron motion is also observed in semiconductor nanostructures, such as quantum wells and quantum dots. Although the spatial and temporal scales differ by several orders of magnitudes due to the large difference in the electron density, the relevant dimensionless parameters take similar values in both cases [7]. For instance, the effective Wigner–Seitz radius is of order unity for both metallic and semiconductor nano-objects. Therefore, the electron dynamics can be investigated using similar models and both types of nano-objects are likely to share several similar dynamical properties. A quantum hydrodynamic model based on velocity moments of the corresponding Wigner distribution function has been proposed in reference [7]. This approach should lead to considerable gains in computing time in comparison with simulations based on conventional methods, such as density functional theory. A promising approach to the development of a quantum computer relies on small semiconductor devices, such as quantum dots and quantum wells [7]. To implement basic qubit operations, most proposed schemes make use of the electron spin states, so that a thorough understanding of the spin dynamics is a major prerequisite.

8.8 Ab-Initio to Circuit Simulation: Roadmap of First-Principle Calculation

From the first principle or fundamental proposition, the user can extract several bulk properties of new type materials like mobility, bandgap, vacancy, defects, deformation potential, and effective mass. To complement the existing EDA simulator beyond the 7 nm technology node, nano-device, and process simulators with unique functionalities, based on atomic-scale modeling will be essential soon. The treatment of electrons and holes as semiclassical particles with an effective mass was usually enough. For most devices, the drift-diffusion equation provides a simple but adequate description of carrier transport. Device dimensions have now shrunk to the nanoscale. The properties of materials can be engineered by intentional strain and side effects due to quantum confinement. Devices contain a countable number of dopants and are sensitive to structure at the atomic scale. To describe carrier transport in nanoscale devices, engineers must learn how to think about charge carriers as quantum mechanical entities at the atomic scale, rather than as semiclassical particles in continuum fields [8, 9]. The non-equilibrium Green's function (NEGF) approach can be used to model devices from ballistic to diffusive limits [10].

Ab-initio theoretical methods are increasingly being used to study the properties of defects in covalent materials. The dynamical properties like defect-diffusion, and defect reaction, as well as electronic properties, can be studied using this method. Most electrical and optical properties of semiconductors such as crystalline silicon are determined by the type of concentration of defect present in the grown crystal. The HF method in the molecular cluster and density function-based molecular dynamics (MD) simulation can be explored in those specific cases. More defects are introduced during the processing of the device. The growth of the surface layer during oxidation, silicidation, or epitaxial growth; injects self-interstitial into the bulk. Ion implantation introduces damage, etching process introduces hydrogen, whereas chemical-mechanical polishing often introduces Cu-like ions. In general, defects of impurities disrupt the periodicity of the crystal, introduce local strain, and result in a new energy level in the energy gap. Depending on the Fermi level and the position of defect level, some defects may exist in several charge states. Moreover, defects can be mobile hence a range of reactions may occur between them. Such a reaction creates new, more complicated defects with different electrical and optical properties than the original ones. As the size of the active area of the device shrinks, the trace of unwanted defects can affect their properties and it is becoming critical to understand how they diffuse, interact with the crystal, with each other, and with other defects. To understand such behavior and predict future state, a combination of theoretical or simulation studies is important. For this purpose, scientists use the "ab-initio" or "first-principle" method to understand the fundamental properties of advanced materials [11].

According to the first-principle theory, the "ab-initio" expression has never been precisely defined. The fancy terminology usually means that no parameter adjusted to experimental data has been used. Whereas "semi-empirical" methods fit parameters such as potentials, ionization energy, or bond length to measured values. While those methods are excellent at predicting quantities, their predictive power is unknown for many other situations. However, it is important to remember that all the methods contain one kind of parameter. These include basic set size and plane-wave type, the cut-off radii for ab-initio pseudopotentials, and the amount of electron correlation including the size of cluster or periodic supercell. However, none of those parameters are fitted to experimental data. Whereas a much better modeling technique and algorithm have been developed recently, much experience has been gained and the computers themselves have tremendously increased operational speed, power, and memory size to perform high-quality simulation. The predictive power is such that it helps them to refine models and a few predictions are nearly quantitative, for example, equilibrium configuration, binding energy, diffusion paths, and vibrational frequencies for various charge states of isolated defects.

The application of ab-initio simulation is grouped into three major segments:

(a) The two commonly used ab-initio techniques are based on DF and HF theories. The DF theory replaces many electrons with a single particle via an effective potential either within the LDA or using the GGA. The method gives excellent total energies and total densities without a one-electron wave function. But the HF theory can solve many-electron problems upfront and the calculations are much more time-consuming than the DF method [12].

(b) MD simulations are used to study the diffusion of defects, local vibrational modes, as well as re-defect distribution and reactions. In this calculation, the temperature is introduced via the kinetic energy of nuclei in classical mode. The electronic problem is solved quantum mechanically and the ionic motion is calculated using classical mechanics [13].

(c) The geometries are optimized with gradient technique and symmetry assumption is performed using the approximate ab-initio method through HF level simulation. In this application, clever basis sets transformation is used to reduce the number of four centers integral with some loss of accuracy. The method reproduces the results of ab-initio HF calculations at a fraction of computational expense without introducing semi-empirical parameters. To perform geometry optimization for large clusters and low symmetry, using ab-initio, would be computationally expensive [14].

8.9 Numerical Techniques (Algorithms) for Atomistic Simulation

(a) **MD algorithm**: In this, atoms are in constant motion, but this logic is computationally extensive. In atomic arrangements with lower potential energy, atoms move faster, but total energy tends to grow slowly with time and total energy (potential + kinetic) should be conserved.

(b) **Monte Carlo (MC) algorithm**: It is a type of resource-restricted algorithm that returns solutions based on probability. MC methods generally follow multiple steps: First, determine the statistical properties of possible inputs. Second, generate many sets of possible inputs that follow the statistical properties and perform a goal-based calculation. Finally, it analyzes results, errors, etc. via statistical methods [15].

The kinematics of Newtonian physics is the mathematical formalism and infrastructure that can be used to describe the motion of the particle. Newtonian mechanics is only valid in the limit that velocity is low. Once the particle/atom moves fast enough, special relativity takes over. A very massive object with enormous masses distorts space-time, which is the domain of general relativity. Once the user moves into the atomic dimension, the matter is comprised of waves, which become apparent at length scales approaching the diameter of an atom. In that scenario, consider quantum mechanics from that point and consider small quantities of energy, such as single photons of light.

8.10 Semiclassical Ballistic Transport: Atomistic Simulation Methodology

Atomistic simulation is commonly used to explain thermal transport across interfaces. Users must define the thermal conductance of a grain boundary and interface using NEMD or Green's function-based phonon transmission to understand the macroscopic mechanism in nanoscale devices. Bulk conductivity and interface conductance, like phonon scattering at reservoir boundaries, reduce conductivity. Hence, the atomic simulation is necessary to understand correct bulk conductivity values from MD simulations [16, 17].

Before the start of such simulation, the user must set up the system as a device configuration and use Green's functions technique to calculate the propagation of an incoming phonon. Similarly, one calculates the phonon transmission spectrum, $\tau(\omega)$, using the NEGF technique. The semi-infinite electrodes of the system are at equilibrium, but the results with different phonon occupations lead to a non-equilibrium phonon distribution. NEGF phonon transport includes ballistic transport, quantum effects, elastic scattering due to disorder at interface and boundaries but it excludes, and inelastic phonon–phonon scattering for interfacial conductance, specifically in low temperatures related applications [18, 19].

8.10.1 Semiclassical Transport

Engineers have commonly treated the electrons as semiclassical particles that move through a device under the influence of an electric field and random scattering potentials. In momentum space, the equation of motion looks like Newton's law for a classical particle.

$$\frac{dhk}{dt} = \Delta_r E_C(r,t), \tag{8.5}$$

where, k is the crystal momentum and E_C is the bottom of the conduction band energy. In position space, the equation of motion is:

$$\frac{dr}{dt} = \frac{1}{h} * \Delta_{rk} E_C(k), \tag{8.6}$$

where, E_C describes the band structure of the semiconductor in momentum space k. Device simulation for carrier transport in nanodevices is usually performed with two options: (a) the field-dependent solution of the Poisson equation using mobilities and diffusion coefficients (drift-diffusion approach), and (b) the local kinetic energy dependent solution of self-consistent Poisson equation using mobilities and diffusion coefficients (energy transport or hydrodynamic approaches). Drift-diffusion and energy transport approaches for treating carrier transport in semiconductor devices assume that carriers can be treated as semiclassical particles and there is a lot of scattering. Both assumptions are losing validity as the devices shrink to nano-regime [20].

To compute the electron density, current, average velocity, etc., within the device, simulation engineers have two options: (a) The carriers as semiclassical particles and solve the Boltzmann equation to obtain the carrier distribution function. (b) To treat carrier quantum mechanically. In the first approximation, the local density-of-states are like that of a bulk semiconductor but shifted up or down by the local electrostatic potential. To find how the k-states within the ballistic device are occupied, one must solve the Boltzmann transport equation (BTE). Under equilibrium conditions, the proper function of total energy is the Fermi function as:

$$f\{E\} = \frac{1}{\left\{1 + \exp\left(\frac{E - E_f}{k_B T}\right)\right\}}, \tag{8.7}$$

where, the Fermi level E_f and temperature T are constant in equilibrium. The ballistic device is as far from equilibrium and the relevant steady-state Boltzmann equation is the same equation as in equilibrium. The solution is again a function of the carrier's total kinetic energy where ballistic transport can be viewed as a special kind of equilibrium. Each k-state is in equilibrium with the lead from which it was populated. Using this reasoning, one can compute the distribution function at any location within the device as well as at any moment of it (e.g., carrier density, carrier velocity, etc.) [21]. Quantum mechanically the electron is a wave and the wave function is obtained by solving Schrodinger's equation.

$$\left[-\left(h^2/2m\right)\Delta^2 + V(r)\right]\Psi(r) = E.\Psi(r) \tag{8.8}$$

In quantum mechanical modeling, consider electron waves incident from the left and right nodes. The electron wave function in device region D can be thought to arise from the following:

(a) Waves incident from the left contact (L) of form e^{ikx}, $\Psi_D{}^{(L)}$
(b) Waves incident from the right terminal (R) of the form e^{-ikx}, $\Psi_D{}^{(R)}$
(c) Localized and quasi-localized states within the device are filled up by scattering due to electron–phonon and electron–electron interaction, represent as $\Psi_D^{(LOC)}$

The quantum mechanical density of states at energy E due to wave's incident from the left ($LDOS_L$) and right ($LDOS_R$) are:

$$\text{LDOS}_L(x,E) = 2\sum |\Psi_D^{(L)}(x)|^2 \tag{8.9}$$

$$\text{LDOS}_R((x,E)) = 2\sum |\Psi_D^{(R)}(x)|^2 . \tag{8.10}$$

The electron density can be formulated as an integral summation of this wave equation, as:

$$n(x) = \int \text{LDOS}_L(x,E).f_L(E).dE + \int \text{LDOS}_R(x,E).f_{LR}(E).dE. \tag{8.11}$$

Quantum transport with scattering

Accurate modeling of nanodevices should have the ability to capture:

(a) Interference effects
(b) Quantum mechanical tunneling
(c) Discrete energy levels due to confinement in 2D and 3D device geometries
(d) Scattering mechanisms (electron–phonon and electron–electron)

The first three effects can be modeled by solving Schrodinger's equation while in the semiclassical device modeling, the Boltzmann equation accounts for the energy and momentum relaxation due to scattering mechanisms. The wave function of an electron loses phase coherence due to scattering that has an internal degree of freedom. Once the device dimensions become comparable to the scattering lengths, phonons and other phase-breaking mechanisms become important to describe quantum transport. In quantum mechanical device modeling, the NEGF approach is necessary to account for energy, momentum, and quantum mechanical phase relaxation. Compared to the available 3D modeling possible in the drift-diffusion framework, the computational modeling of quantum transport is not mature enough. Most work on quantum transport is based on solving the quantum transport equations in reduced dimensions using detailed device physics. The NEGF approach is based on the rigorous many-body quantum theory. It is designed to describe carrier transport with scattering [22]. In practice, the NEGF approach provides a good description for many devices, when scattering is strong and potential variations are slow. There is no major roadblock to including scattering mechanisms, determining the self-energy, and computation of the Green's functions are required to make quantum transport simulation more practical.

8.10.2 Why Do We Need Atomistic Simulation?

To understand the microscopic mechanisms

Atomistic simulations, which are the most commonly used methods in the nano-mechanics area, are valuable computational methods for the investigation of magnetic, electrical, chemical, and mechanical properties of carbon nanostructures since these modeling approaches can precisely trace atomic position and capture microscale physical mechanisms like buckling. There has been already much research on carbon nanostructures using atomistic simulations. The thermal stability, the elastic and plastic properties of CNTs, and twisting effect, as well as buckling on CNTs bundles using MD simulation are important. The second-generation of reactive empirical bond-order potential and Lennard–Jones potential are also useful to describe the atomic interaction.

Despite the precise trace of the atomic displacement of CNTs, these atomistic simulations are highly expensive in terms of computational resources. Therefore, their applications are extremely limited to a simple system with a small number of atoms and do not meet the demands of engineering applications. For example, Liew et al. [23] spent 36 h studying the buckling behavior of a (10, 10) SWCNT containing 2,000 atoms using MD simulation, whereas the investigation of a four-walled CNT containing 15,097 atoms required four months. The understanding of the effects of hypervelocity impacts (using atomistic simulation) emerge from atomic interactions is important. High-quality semiconductor crystals can be produced by solid-phase epitaxial growth at the amorphous-crystal interface. Despite extensive experimental studies, the microscopic mechanisms that lead to crystallization are not yet known. MD simulations of a-Si(001) amorphous-crystal interface, using an accurate empirical interatomic potential, give activation energy and a shape for the activated state that are in reasonable agreement with experimental measurements.

To understand and optimization of heat conduction in nanoscale devices

Thermal conductance G (inverse of thermal resistance) quantifies the ability of the entire system to transport thermal energy from the hot side to the colder part of the device. Theoretically, it is defined as [24]:

$$G = -\frac{Q}{\{T_{\text{hot}} - T_{\text{cold}}\}}. \tag{8.12}$$

Now thermal conductivity is a normalized bulk property, which is calculated as k:

$$k = -\frac{Q}{A * \Delta T} = \frac{G * L}{A}. \tag{8.13}$$

Thermal transport can be simulated via different techniques: (a) Green-Kubo equilibrium molecular dynamics, (b) relaxation time approximation of the BTE, (c) non-equilibrium molecular dynamics (NEMD), and (d) NEGF-based phonon transmission. The first two options are applicable for simulating thermal transport in the bulk region, whereas the last two are suitable for interfaced modeling.

8.11 Atomic-Scale Modeling in Advanced Semiconductor Design and Manufacturing

The trend of reducing device sizes results in a very small number of impurity atoms that determine the threshold voltage of a transistor. It is hard to believe that a limit will be reached soon where small discretized distribution can no longer be accurately modeled with a continuum process simulation approach. An MC-based diffusion, implantation, oxidation, epi-growth, silicidation type of simulation with an atomistic interface will provide an alternative to the continuum strategy and that will be the accurate approach in the future. MD and subsurface lattice KMC simulation techniques are important for atomic-level process simulation like diffusion, deposition, epitaxial growth, etc.

Simulations of the Si device process flow provide valuable design guidelines for new devices during the design phase. These models are being improved by including better portrayals of the physical mechanisms and more accurate values of the rates and configuration parameters. Atomistic simulation techniques can be divided into three categories, depending on the level of detail included in the model. Models including quantum calculations of electron distributions, and the first-principles methods are the most comprehensive [25]. They involve high computing costs but have the advantages that energies of specific atomic configuration can be calculated directly, without any fitting parameters. Although the methods include some uncontrolled approximations and are limited to small numbers of atoms, direct information on saddle points, binding energies for dopant, and point defect interactions are available in such calculations. Fitting simulations to experimental data may also provide a unique set of atomic interaction energies, provided that the experiments are carefully designed [26].

Classical MD simulations employ empirical force laws, which are adjusted to fit a combination of experimental data and first-principles calculations [27]. These methods are much more efficient than those discussed above since the force law calculation replaces the complex calculations of the electron distribution. Using these methods, it is possible to simulate systems containing tens of thousands of atoms for nanoseconds or more. The simulation of an implantation process is accomplished by alternating the implantation of an atom with the MC diffusion process, and the time of the diffusion is determined by the dose rate of the implant. The high concentration of point defects in the very early stages has been reduced dramatically due to the recombination of Frenkel pairs. At doses of more than $1e13/cm^2$, the simulation results are consistent with experimental evidence while using the "plus one" model, which suggests that rapid recombination will eliminate the Frenkel pairs at an early stage and only the interstitial corresponding to the implanted ion will contribute significantly in transient enhance diffusion (TED). However, at lower doses, the Frenkel pairs can make an important contribution to TED [26, 27]. The MC simulations of TED also evaluate the number of times an interstitial visits a site in the target lattice for completion of annealing [28]. The continuum approach of modeling dopant diffusion during process simulation is used to solve a system of partial differential equations (PDEs) that describe the transport of the dopants and the dose conservation as a function of projection range (Rp). This approach has been useful in the past, but the sub-micron device manufacturers might face the difficulty of using it in the future. The shrinking thermal budget significantly reduces diffusion and, therefore, reduces the need to accurately model dopant diffusion. Whereas, dopant activation phenomena, including the formation of a variety of dopant defect pairs and extended defects of different configurations, often do not reach thermodynamic equilibrium. Hence, atomic-scale modeling becomes the first choice for researchers and device design engineers.

8.12 Atomistic Kinetic MC Simulation

For atomic-level process simulation, multiple blueprints are required to characterize the electrical, mechanical, chemical, and physical properties of new materials, as well are used to extract model parameters. In the MC approach, the user needs to introduce the probabilities for the additional reactions. These probabilities are calculated based on the binding energies obtained from experiments, MD, or ab-initio study. Conventional KMC simulation can operate in two different modes during process simulation:

(a) The atomistic model handles data spontaneously during the simulation and is expected to be the most accurate.

(b) The non-atomistic mode allows KMC simulation to be used during defect re-distribution. The optimization part of a simulation may allow the user to take advantage of the efficiency of the continuum solver during equilibrium, where data is transferred back and forth to the continuum solver.

A predictive atomistic process simulator must include a crystalline-amorphization model. However, accounting for each particle and position during the amorphization is not convenient due to the requirements of massive computer resources during process modeling. Notwithstanding this, amorphization involves the destruction of the lattice structure and without a lattice base, the KMC method is opened to a quasi-atomistic approach. In a real situation, the recrystallization of thin layers in a fin-like structure may be displayed as an arrow tip-like shape, which gradually slows down the solid-phase-epitaxial-growth and forms polycrystalline silicon in the regions not yet recrystallized [25]. To simulate the doped deposition process in an atomistic way, dopants are passed from lattice-KMC to KMC, to allow for bulk diffusion. Point defects injected during interstitial movement are introduced from the interface during the lattice-KMC process, and it continues to diffuse until there is any recombination with a vacancy site. Such diffusion events can be a few orders of magnitude faster than lattice-KMC events. The recrystallization process is usually implemented as a special event performed by the amorphous defects in KMC simulation. At a given temperature, every amorphous defect can recombine all its internal damage and recrystallize. The conventional KMC model assigns a recrystallization rate for each amorphous defect, to perform recrystallization simulations [29, 30]. An MC-based diffusion simulation provides a valuable alternative to the continuum approach. Computational resources required for MC diffusion simulation are decreasing with device dimension because they are proportional to the number of dopants and defects in the device.

8.13 Atomistic Simulation for Semiconductor Devices

To perform carrier transport simulation at the atomic level, the user should have a proper theoretical background on drift-diffusion to full/sub-band Boltzmann transport models and corresponding interaction with effective mass, k.p models, as well as DFT and NEGF models [31–33]. Flexible, predictive, and quick simulations are possible with the self-consistent framework of NEGF and state-of-the-art band structure calculations [34, 35]. The simulation of a nanoscale FET incorporating atomic precision and inelastic carrier interactions has been possible with the recent introduction of the Victory Atomistic tool [36] inherited from NEMO5 [37, 38]. Device designers can now simulate futuristic devices in an atomistic CAD environment, which opens new possibilities. Furthermore, important quantum transport quantities can be easily extracted and visualized with advanced visualization tools at the atomic level, allowing for the accurate extraction of transistor parameters [39].

Quantization uses the density gradient model in nanodevices simulation. In sub-14 nm technologies, users cannot neglect quantum correction where gate oxide thicknesses are much smaller than 3 nm. For very short gate lengths, the isotropic density gradient model can overestimate the leakage current in the transport direction (source/drain tunneling). In those cases, the optimized parameters associated with the density gradient model can help suppress the current in the transport direction. For InGaAs nanowire/FinFET devices with a gate width less than 10 nm, people must use the modified local density approximation (MLDA) model to activate size quantization with the multivalley model. In the case of a 2D structure, the depletion at corner regions correlates better with the results of the Schrödinger equation and the layer thickness can be extracted automatically. The quantization parameter of the density gradient model depends on the layer thickness and the fin geometry too. Therefore, it is desirable to calibrate the solutions from the Schrödinger equation in test structures with a similar channel geometry [34].

8.14 BTE Solver

To solve the quantum transport equation for next-generation device simulation on an atomic scale, the first choice will be the BTE solver. The subband-based BTE solver in modern device simulators is based on the physical models to simulate quasi-ballistic transport phenomena of nanodevices. The solver takes into consideration of strong quantum confinement effects and quasi-ballistic transport along the channel of small FinFET and nanowires devices. The solution of the Schrödinger equation on the 2D cross-section of the 3D domain along the channel and the distribution functions of carriers in each subband are delivered from the BTE solver. This information is combined with the subband dispersion and the wave functions from the Schrödinger solver. Similarly, the valley models specify the band model and parameters to use when solving the Schrödinger equation. For example, (a) 2-band k.p models can be used for the description of electrons in E-k space, (b) 6-band k.p valley model gives an accurate description of holes, and (c) 3-band k.p valley model approximates 6-band k.p, where the split off-band is not fully captured due to spin-orbit coupling.

Scattering models are used to model specific, microscopic scattering phenomena such as phonon and surface roughness scattering. The scattering models can be used in mobility calculations for 2D structures, including elastic and non-elastic nonpolar phonon scattering, polar optical phonon scattering, alloy scattering, surface roughness scattering, and Coulomb scattering from bulk impurities, and interface charge, as well as traps. The quantum transport equation solver, known as the NEGF solver, mainly considers quantum-mechanical effects such as confinement, source-to-drain tunneling, and coherence effects like resonant tunneling. The use of the NEGF solver is most appropriate for devices with a well-defined channel, in which geometric quantum confinement plays a strong role. This is typically the case in small FinFETs, nanowire devices, and ultrathin-body FDSOI structures.

8.15 Simulation Strategy beyond the End of Technology Roadmap

Silicon-based transistors are being scaled down continuously to increase density and speed. Multi-gate FinFETs have strengths of high robustness on short channel effects (SCEs) and superior scalability using conventional processes [40]. However, the ultimate minimum feature-sized device operating at room temperature is expected to be 1.5 nm according to Heisenberg's uncertainty principle and Shannon-von expression [41]. The fabricated sub-5 nm gate-all-around FinFET is approaching this fundamental limit. For the ultimately scaled transistor, GAA nanowire/nano-sheet FinFET is known to be the best structure to provide scalability and flexibility in device design. In these scaled technologies, transistors are becoming more prone to process variation due to the shrinking of device footprints. The migration to fully depleted channel and multi-gate structures such as FinFET and FDSOI technology has provided a respectable short channel effect and substantially reduced random variation, which is an essential enabler for RAM scaling. However, looking to 5 nm and beyond technologies, scalability of such technologies is resumed questioning if an incremental improvement of SCEs needs to be maintained. Therefore, gate-all-around nanowire FETs allow continued gate length scaling and increase SRAM read stability.

8.16 Bio-Inspired Device and Systems

As dimensional scaling of CMOS will eventually approach fundamental limits, several new information processing devices and micro-architectures for both existing and new functions are being explored. This is the driving interest in new devices for information processing, new technologies for heterogeneous integration of multiple functions, and new models for the logic and memory system architecture. The semiconductor industry is facing some difficult challenges related to extending integrated circuit technology to new applications and to beyond the end of CMOS technology scaling. One of the first challenges will be the integration of high-speed, high-density, and low-power memory technology onto the CMOS platform. Similarly, extending CMOS scaling with alternative channel materials is also progressing. Another option is to extend ultimately scaled CMOS as a platform technology into new domains of functionalities and bridge the gap between novel devices and unconventional architectures following atomistic computing standards.

Nanowire field-effect transistors are structures in which the conventional planar MOSFET channel is replaced with a semiconducting nanowire. Nanowire gate-all-around geometry is of interest primarily due to its superb gate electrostatics that can allow further gate-length scaling. Despite the promising results that are mostly obtained from academic research laboratories, bottom-up nanowire transistors still face significant challenges before commercialization is feasible. Emerging CMOS devices will likely have new issues at the interfaces between the device level and higher levels of computer design. Many of the emerging device classes that are currently being explored are not intended simply due to replacements for CMOS devices but will require new types of circuit designs, new functional module architectures, and even new software to utilize the capabilities of the new devices. Smart nanostructures are susceptible to various environmental or biological parameters. They offer great potential for numerous biomedical applications such as monitoring, diagnosis, repair, and treatment of human biological systems. In addition to drug delivery, which has been extensively reported and reviewed, increasing interest is being observed in using smart nanostructures to develop various novel techniques of sensing, imaging, tissue engineering, bio-fabrication, nanodevices, and nanorobots for the improvement of healthcare and other applications in the biotechnology sector. Recent developments have been made in the use of biological systems in molecular self-assembly, spatial positioning, micro construction, biocomposite fabrication, nanomachines, and biocomputing.

8.17 Summary

In terms of atomistic simulation, lattice discretization has become a very popular method to approximate physical models, which are too complicated to be solved analytically. However, despite their success, there are certain features of continuum dynamics, which are generally not addressed clearly. Interesting properties of the physical system could come from either of them. For instance, nontrivial topological properties of the configuration space may have deep consequences even for simple dynamics. According to current industry trends, the data traffic explosion for the server to mobile to IoT expansion will further increase in demand. Hence, industry leaders will be considering the usage of atomistic simulation for cost optimization and enhancing profit margins while developing futuristic devices.

Dimensional scaling is now under pressure, "More than Moore" is now required by either atomic-level material engineering or by planned usage of design and system technology co-optimization (DTCO, STCO) techniques. Another technology option at the 5 nm node, is advanced patterning (Smart Layout optimization) technology, extending FinFET with stress engineering in scaled-Si and non-Si devices obtained by using alternative high mobility materials, FEOL, and BEOL parasitic resistance and capacitance mitigation with advanced meteorology. The ballistic ratio keeps improving with scaling and is projected to improve toward 2 nm design rules for FinFET or nano-sheet/wire devices. One important side effect is that fundamental physics plays an increasing role, and there is a lesser need for calibration on the future nodes.

Due to present market trends and requirements, it is evident that this is the right time to explore atomic-level fundamentals of semiconductor-based systems. The importance of ab-initio calculation for advanced material engineering along with atomistic simulation resources, application-driven simulation strategies, and MD/DFT simulation competencies are explored in this chapter. Finally, this chapter also looks at the current trends in atomistic simulation methodology, adopted by several industry leaders for future technology nodes and revolutionary devices.

References

1. M. W. C. Dharmawardana, "A review of studies on strongly-coupled coulomb systems since the rise of DFT and SCCS-1977," *Contrib. to Plasma Phys.*, vol. 55, no. 2–3, pp. 85–101, Feb. 2015. doi: 10.1002/ctpp.201400073.
2. R. Strasser, R. Plasun, and S. Selberherr, "Practical inverse modeling with SIESTA," in *1999 International Conference on Simulation of Semiconductor Processes and Devices, SISPAD'99 (IEEE Cat. No.99TH8387)*, 1999, pp. 91–94. doi: 10.1109/SISPAD.1999.799268.

3. L. N. Zhao *et al.*, "Atomistic modeling of the electrostatic and transport properties of a simplified nanoscale field effect transistor," *J. Comput. Electron.*, vol. 7, no. 4, pp. 500–508, 2008. doi: 10.1007/s10825-008-0260-0.
4. W. Kohn and L. J. Sham, "Self-consistent equations including exchange and correlation effects," *Phys. Rev.*, vol. 140, pp. A1133–A1138, 1965. doi: 10.1103/PhysRev.140.A1133.
5. J. P. Perdew, K. Burke, and M. Ernzerhof, "Generalized gradient approximation made simple," *Phys. Rev. Lett.*, vol. 77, no. 18, pp. 3865–3868, 1996. doi: 10.1103/PhysRevLett.77.3865.
6. J. Zou and A. Balandin, "Phonon heat conduction in a semiconductor nanowire," *J. Appl. Phys.*, vol. 89, no. 5, pp. 2932–2938, Mar. 2001. doi: 10.1063/1.1345515.
7. P. Zoller *et al.*, "Quantum information processing and communication: Strategic report on current status, visions and goals for research in Europe," *Eur. Phys. J. D*, vol. 36, no. 2, pp. 203–228, Nov. 2005. doi: 10.1140/epjd/e2005-00251-1.
8. R. Lake, G. Klimeck, R. C. Bowen, and D. Jovanovic, "Single and multiband modeling of quantum electron transport through layered semiconductor devices," *J. Appl. Phys.*, vol. 81, no. 12, pp. 7845–7869, Jun. 1997. doi: 10.1063/1.365394.
9. G. Mil'nikov, N. Mori, and Y. Kamakura, "Equivalent transport models in atomistic quantum wires," *Phys. Rev. B - Condens. Matter Mater. Phys.*, vol. 85, no. 3, p. 035317, Jan. 2012. doi: 10.1103/PhysRevB.85.035317.
10. J. R. Barker, "2D/3D NEGF modeling of the impact of random dopants /dopant aggregation in silicon nano-transistors," in *2008 International Conference on Simulation of Semiconductor Processes and Devices*, 2008, pp. 337–340. doi: 10.1109/SISPAD.2008.4648306.
11. L. Yang and A. J. Minnich, "Thermal transport in nanocrystalline Si and SiGe by ab initio based Monte Carlo simulation," *Sci. Rep.*, vol. 7, no. 1, pp. 1–11, Mar. 2017. doi: 10.1038/srep44254.
12. "Applications of ab-initio molecular dynamics simulations in chemistry and polymer science | SpringerLink." https://link.springer.com/chapter/10.1007/978-3-642-58360-5_25 (accessed May 20, 2021).
13. G. ~B. Martin, B. Kirtman, and F. ~J. Spera, "Ab initio MD simulations of geomaterials with \raisebox{-0.5ex}\textasciitilde1000 atoms," in *AGU Fall Meeting Abstracts*, Dec. 2009, vol. 2009, pp. MR31A-1637.
14. L. Messina, N. Castin, C. Domain, and P. Olsson, "Introducing ab initio based neural networks for transition-rate prediction in kinetic Monte

Carlo simulations," *Phys. Rev. B*, vol. 95, no. 6, p. 064112, Feb. 2017. doi: 10.1103/PhysRevB.95.064112.

15. L. D. Hernández, S. Moral, and A. Salmerón, "A Monte Carlo algorithm for probabilistic propagation in belief networks based on importance sampling and stratified simulation techniques," *Int. J. Approx. Reason.*, vol. 18, no. 1–2, pp. 53–91, Jan. 1998. doi: 10.1016/S0888-613X(97)10004-4.
16. P. C. Howell, "Comparison of molecular dynamics methods and interatomic potentials for calculating the thermal conductivity of silicon," *J. Chem. Phys.*, vol. 137, no. 22, p. 224111, Dec. 2012. doi: 10.1063/1.4767516.
17. D. P. Sellan, E. S. Landry, J. E. Turney, A. J. H. McGaughey, and C. H. Amon, "Size effects in molecular dynamics thermal conductivity predictions," *Phys. Rev. B - Condens. Matter Mater. Phys.*, vol. 81, no. 21, p. 214305, Jun. 2010. doi: 10.1103/PhysRevB.81.214305.
18. J. S. Wang, N. Zeng, J. Wang, and C. K. Gan, "Nonequilibrium Green's function method for thermal transport in junctions," *Phys. Rev. E - Stat. Nonlinear, Soft Matter Phys.*, vol. 75, no. 6, p. 061128, Jun. 2007. doi: 10.1103/PhysRevE.75.061128.
19. T. Markussen, A. P. Jauho, and M. Brandbyge, "Electron and phonon transport in silicon nanowires: Atomistic approach to thermoelectric properties," *Phys. Rev. B - Condens. Matter Mater. Phys.*, vol. 79, no. 3, p. 035415, Jan. 2009. doi: 10.1103/PhysRevB.79.035415.
20. S. Jin and K. A. Novak, "A semiclassical transport model for two-dimensional thin quantum barriers," *J. Comput. Phys.*, vol. 226, no. 2, pp. 1623–1644, Oct. 2007. doi: 10.1016/j.jcp.2007.06.006.
21. D. K. Ferry and R. O. Grondin, "Semiclassical carrier transport models," in *Physics of Submicron Devices*, Springer, US, 1991, pp. 137–171.
22. G. D. Mahan, *Many-Particle Physics*, Springer, US, 2000.
23. "Mechanical behaviors of carbon nanotubes | ScienceDirect." https://www.sciencedirect.com/book/9780323431378/mechanical-behaviors-of-carbon-nanotubes (accessed Mar. 14, 2021).
24. "Thermal conductance: An overview | ScienceDirect Topics." https://www.sciencedirect.com/topics/computer-science/thermal-conductance (accessed May 21, 2021).
25. J. Zhu, T. D. dela Rubia, L. Yang, C. Mailhiot, and G. H. Gilmer, "Ab initio pseudopotential calculations of B diffusion and pairing in Si," *Phys. Rev. B - Condens. Matter Mater. Phys.*, vol. 54, no. 7, pp. 4741–4747, Aug. 1996. doi: 10.1103/PhysRevB.54.4741.

26. T. Diaz De La Rubia and G. H. Gilmer, "Structural transformations and defect production in ion implanted silicon: A molecular dynamics simulation study," *Phys. Rev. Lett.*, vol. 74, no. 13, pp. 2507–2510, Mar. 1995. doi: 10.1103/PhysRevLett.74.2507.

27. L. Pelaz *et al.*, "B diffusion and clustering in ion implanted Si: The role of B cluster precursors," *Appl. Phys. Lett.*, vol. 70, no. 17, pp. 2285–2287, Apr. 1997. doi: 10.1063/1.118839.

28. L. Pelaz, G. H. Gilmer, V. C. Venezia, H. J. Gossmann, M. Jaraiz, and J. Barbolla, "Modeling of the effects of dose, dose rate, and implant temperature on transient enhanced diffusion," *Appl. Phys. Lett.*, vol. 74, no. 14, pp. 2017–2019, Apr. 1999. doi: 10.1063/1.123742.

29. N. Strecker, V. Moroz, and M. Jaraiz, "Introducing Monte Carlo diffusion simulation into TCAD tools," in *Proc. Intl. Conf. Modeling and Simulation of Microsystems*, 2002, pp. 462–465.

30. I. Martin-Bragado and V. Moroz, "Modeling of {311} facets using a lattice kinetic Monte Carlo three-dimensional model for selective epitaxial growth of silicon," *Appl. Phys. Lett.*, vol. 98, no. 15, p. 153111, Apr. 2011. doi: 10.1063/1.3580771.

31. A. Razavieh, P. Zeitzoff, and E. J. Nowak, "Challenges and limitations of CMOS scaling for FinFET and beyond architectures," *IEEE Trans. Nanotechnol.*, vol. 18, pp. 999–1004, 2019. doi: 10.1109/TNANO.2019.2942456.

32. J. Z. Huang, H. Ilatikhameneh, M. Povolotskyi, and G. Klimeck, "Robust mode space approach for atomistic modeling of realistically large nanowire transistors," *J. Appl. Phys.*, vol. 123, no. 4, p. 044303, Jan. 2018. doi: 10.1063/1.5010238.

33. S. Steiger, M. Povolotskyi, H. H. Park, T. Kubis, and G. Klimeck, "NEMO5: A parallel multiscale nanoelectronics modeling tool," *IEEE Trans. Nanotechnol.*, vol. 10, no. 6, pp. 1464–1474, Nov. 2011. doi: 10.1109/TNANO.2011.2166164.

34. P. Blaise *et al.*, "Nanoscale FET: How to make atomistic simulation versatile, predictive, and fast at 5nm node and below," in *2020 International Conference on Simulation of Semiconductor Processes and Devices (SISPAD)*, 2020, pp. 249–252. doi: 10.23919/SISPAD49475.2020.9241651.

35. D. A. Lemus, J. Charles, and T. Kubis, "Mode-space-compatible inelastic scattering in atomistic nonequilibrium Green's function implementations," *J. Comput. Electron.*, 2020. doi: 10.1007/s10825-020-01549-8.

36. Silvaco, Victory Atomistic Tool, 2020.
37. K. Wang *et al.*, "NEMO5: Predicting MoS2 heterojunctions," in *2016 International Conference on Simulation of Semiconductor Processes and Devices (SISPAD)*, 2016, pp. 221–224. doi: 10.1109/SISPAD.2016.7605187.
38. J. Charles *et al.*, "Incoherent transport in NEMO5: Realistic and efficient scattering on phonons," *J. Comput. Electron.*, vol. 15, no. 4, pp. 1123–1129, Dec. 2016. doi: 10.1007/s10825-016-0845-y.
39. "Technology Computer Aided Design (TCAD) | Synopsys." https://www.synopsys.com/silicon/tcad.html (accessed May 21, 2021).
40. F.-L. Yang *et al.*, "5nm-gate nanowire FinFET," in *Digest of Technical Papers, Symposium on VLSI Technology, 2004.*, 2004, pp. 196–197. doi: 10.1109/VLSIT.2004.1345476.
41. V. V. Zhirnov, R. K. Cavin, J. A. Hutchby, and G. I. Bourianoff, "Limits to binary logic switch scaling: A gedanken model," in *Proceedings of the IEEE*, 2003, vol. 91, no. 11, pp. 1934–1939. doi: 10.1109/JPROC.2003.818324.

Index